Radio Wave Propagation

Radio Wave Propagation

Lucien Boithias
C.N.E.T.

Translated by

David Beeson

Foreword by

L-J Libois

McGraw-Hill Book Company
New York St. Louis San Francisco
Montreal Toronto

Original French language edition
(Propagation des ondes radioélectriques:
Dans l'environnement terrestre)

Revised and updated 1987

English edition first published 1987
by North Oxford Academic Publishers Ltd,
a subsidiary of Kogan Page Ltd, 120 Pentonville Road,
London N1 9JN

Published in the U.S.A. by
McGraw-Hill Inc.

ISBN 0-07-006433-4

Printed and bound in Great Britain

Contents

Foreword

In 1887 Heinrich Hertz, professor of physics at the University of Karlsruhe, wrote to his famous compatriot Helmholtz, who was working on light and electricity:

'... I have just taken a new step forward which, in my view, establishes an incontrovertible link between light and electricity. I cannot wait to discuss it with you.

In the first place I learnt by a stroke of luck that waves several metres long are not the only ones we can produce, but that we can use far shorter waves which are vastly easier to handle. My tests worked out very well.'

This letter may be regarded as the birth of 'Hertzian waves' or radio waves. It is interesting to note that Hertz, who was engaged in a series of experiments using parabolic mirrors, was already thinking in terms of using very short waves; it would however take more than half a century for microwave techniques to be applied to communications using radio links.

The work carried out by Hertz provided spectacular confirmation of the theory proposed in 1864 by the brilliant mathematician James Clerk Maxwell that using purely theoretical arguments light and electromagnetic waves were identical phenomena. This famous theory, summarized in four remarkably simple equations called Maxwell's equations, was regarded at the time Maxwell first propounded it as no less revolutionary than Einstein's theory of relativity would be 40 years later.

However, if Hertz could lay claim to being the father of radio waves called for that reason 'Hertzian waves' in some countries, it never occurred to him to use them as a means of communication. At that time it was rare for pure scientists to turn their attention to applied science.

'This discovery', wrote Maurice Ponte, 'would be applied by others. Scientists who were also to some extent engineers, engineers who were masters of science, would some years later create wireless telegraphy.'

When mention is made of wireless telegraphy as an invention we think in particular of Edouard Branly in France and Alexandre Stepanovitch Popov in Russia. But it is to Guglielmo Marconi that credit is due for taking wireless telegraphy out of the laboratory at the end of the last century. Marconi was a very active technician of immense talent. The experiments he carried out in 1896 were incontestably the first in the area of radio communication.

Maxwell, a theoretician of genius, Hertz, a talented experimentalist, and Marconi, a man of firm action, were the founders of radiocommunication.

Since the beginning of the twentieth century, progress in our knowledge of phenomena associated with propagation and the use of radio waves has been extraordinary. In December 1901 the first radiotelegraphic link was established between Great Britain and Newfoundland by Marconi. In 1902 Oliver Heaviside, in England, and Kennelly, in the United States, proposed that waves could travel round the Earth by reflecting off the ionized layers of the upper atmosphere; this hypothesis, taken up and defended in France by André Blondel and Henri Poincaré, would be confirmed some years later. The discovery of the 'ionosphere' formed the foundation for the development of intercontinental telecommunications for half a century. The Second World War, with its marked effect on technical progress, was responsible for new radiocommunication methods: radio relay links in the first place, using techniques developed directly from radar, and then, more recently, satellite telecommunications.

Theoretical and experimental studies of propagation phenomena have been taking on increasing importance for the construction of telecommunication systems using radio relay links. Telecommunication engineers have also contributed to the description of these phenomena and their applications. Particular mention of the work carried out in France by Pierre David, whose study on wave propagation published in 1955 forms the basis for training radio and telecommunications engineers in this field, must be made. A new edition of this book, updated with the help of Jean Voge, appeared in 1966.[1] No other general work on propagation has been published in France since that date.

The greatest French contribution to this field since the Second World War has, however, been that made by research workers at the CNET and, in particular, by Lucien Boithias, a scientist of national and international repute.

Lucien Boithias, whose remarkable talents as a researcher I came to value highly at the CNET, began his work on propagation in the troposphere immediately after the Second World War. At that time, radiocommunication and broadcasting used almost exclusively radio waves of frequency less than 30MHz. As we have already pointed out, the ionosphere was already playing a crucial part in intercontinental communications, and was being studied by many theoretical and experimental scientists. On the other hand, systematic study of propagation in the troposphere, although it had begun in 1945 in the United States and in the early 1950s in France, had advanced far less. For this reason, Lucien Boithias was asked to set up the 'troposphere department' at the CNET to meet predicted trends in the development of line of sight radio relay networks. Then, in the mid 1950s the discovery of 'tropospheric scatter' propagation phenomenon led to renewed interest in non-line of sight communications. This is a field in the study of propagation to which the group of engineers at the CNET, including most notably Jean Voge, Lucien Boithias and François du Castel, made a major original contribution.[2]

However, propagation by reflection from the ionosphere and studies associated with it were still crucial for intercontinental communications. It was only in

---

[1]David, P; Voge, J *Propagation des ondes* Editions Eyrolles, Paris, 1966.

[2]du Castel, F *Propagation troposphérique et faisceaux hertziens trans-horizon* Collection Technique et Scientifique du CNET, Editions Chiron, Paris, 1961.

1962, when the first satellite telecommunications link between Andover and Pleumeur Bodou was established on 11 July 1962, that a new era began for long range telecommunications. New fields of research appeared for those working on propagation of radio waves. In particular, the ionosphere was no longer seen simply as a medium which reflects waves but also as one through which they pass.

Moreover, the available frequencies are becoming saturated, and previously undetected forms of interference are beginning to appear. Consequently, telecommunications engineers are having to recommend the use of increasingly high frequencies (greater than 10GHz), both for terrestrial radio links and for satellite communications. The use of such frequencies, which may at first appear attractive, poses special problems, such as absorption by the atmosphere, particularly in rainfall. Further research is needed to analyse these phenomena and obtain a clearer understanding of them.

We might have thought, a few years ago, that we knew everything there was to know about propagation. Now we realize that nothing could be further from the truth. The propagation of waves, and the phenomena associated with it, such as distortion and noise, form a research field whose value for the development of radio telecommunication links is obvious, and for which a state of the art review is extremely useful. It is such a review that Lucien Boithias provides in this book.

As the author points out in the Preface the book is aimed mainly at users. So, for example, only theories leading to practical results of immediate application are mentioned. On the other hand, this book contains a considerable quantity of information for technicians; generally, the data presented have been obtained by semi-empirical methods based on statistical analysis in particular.

As far as the layout of the book is concerned, Lucien Boithias deliberately separates line of sight from non-line of sight propagation. This a very sensible separation from the point of view of the text's educational nature. This book is therefore divided into 11 chapters. In the Preface he gives a brief history of the field and discusses the different aspects of propagation studies.

The first six chapters deal with line of sight propagation, from the traditional starting point of propagation in open space, and the fundamental physical phenomena involved. Note, in particular, the value to users of Chapter 6, devoted to a study of distortion (fades, multi-paths, etc.) which can reduce transmission quality. The author describes prediction and prevention methods which can be applied to alleviate this problem.

Having dealt with line of sight propagation, Lucien Boithias describes in Chapters 7, 8 and 9 the three major mechanisms affecting non-line of sight propagation: knife-edge diffraction, tropospheric scatter and reflection from the ionosphere. Chapters 10 and 11 examine phenomena associated with noise, particularly over great distances, and to distortion of all kinds affecting radio communication. Appendix 1 summarizes the most important probabilistic distributions for the study of propagation.

Lucien Boithias therefore provides a study which is both structured and complete, whereas at the same time being easily accessible. Every chapter contains useful references, as well as the important graphs and tables needed by the user. The work is undoubtedly an essential reference text for those who are interested in understanding and applying propagation phenomena. Let me finally say that Lucien Boithias' research has often been used as the basis for official documents drawn up by international organizations, in particular the Comité Consultatif International des Radiocommunications (CCIR), which is a leader in this field.

The CNET scientific and technical communications series already comprises many essential titles for telecommunications researchers, engineers and technicians. In my view a basic work on propagation, giving a complete, clear, accessible overview of the field was missing. Lucien Boithias's book fills that gap. I take considerable pleasure, as a former director of the CNET, in congratulating not only the author, but also the publishers for taking the initiative in making this valuable text available.

L -J LIBOIS

Conseiller Maître à la Cour des Comptes
Directeur Général Honoraire des Télécommunications

October 1986

Preface

When I began to take an interest in wave propagation thirty years ago France had not yet established so much as a kilometre of a centimetric radio relay wave network. 'Radio' essentially meant broadcasting on long and medium waves or-called short waves (HF) for intercontinental communications. In other words, in most cases frequencies below 30MHz were used.

Beyond this frequency, the electromagnetic spectrum was vast and virtually unused. Metric waves (VHF) were beginning to be used for certain links with few channels, and in France used by the 441 line television transmitter on the Eiffel Tower, operating on 46MHz for picture transmission and 42MHz for sound transmission. Work was beginning on the installation of television transmitters in the 162-216MHz range, and the first in Paris and Lille had been established on the same frequency, as it was obvious for many people that at this frequency propagation took place along a straight line 'as for light waves', and consequently there could be no interference between the two transmitters. This pioneering period, when in order to receive a television picture you generally had to build your own receiver, gave me the opportunity to win a trophy by receiving TV transmissions from Paris 380km away. Reception obviously had certain acrobatic qualities, and not only from the electromagnetic point of view, since the antenna was set up on top of a church spire. Furthermore, the image received on the oscilloscope tube had a green tinge to it, which was somewhat surprising for black and white picture transmission. I have to admit that propagation phenomena were generally kind to me then.

As a result of research into radar during the war, centimetric waves (SHF) were beginning to be better known in certain frequency bands. These bands were generally referred to by code names at the time (*S* band, *X* band), which still survive. The use of radar, particularly over sea, highlighted ducted propagation phenomena. The first known example was the detection of the Arabian coast by radar in Bombay 1500km away.

The start of the 1950s saw the introduction of the first microwave radio link in France, and the first studies into direct line of sight propagation carried out by the CNET, both in the North of France and on the Mediterranean coast. Propagation phenomena were still new concepts for many people, to the extent that a notable figure, to whom it was being explained that a radio link could not be kept free above a forest because the trees were blocking the first Fresnel ellipsoid asked if the trees could not be pruned to the shape of a Fresnel ellipsoid.

In 1956 the Comité Consultatif International des Radiocommunications (CCIR) defined standards of quality for microwave radio relay systems, and these standards were much stricter than those that had been considered sufficient for transmission using dekametric wavelengths (HF). Time percentages were no longer between 50 and 90%, but between 99.9 and 99.99%, while signal: noise ratio requirements shot upwards by several tens of decibels.

Consequently, propagation problems took on a new importance and the necessary levels of predicting accuracy grew considerably. A major decision was taken in 1955. On the one hand, as the French telecommunications agency did not intend any immediate further development of the radio relay network that had already been established, studies into line of sight propagation became less urgent. On the other hand, it had recently been discovered that it was possible to obtain propagation over several hundred kilometres by tropospheric scatter. This discovery immediately stimulated considerable interest both among theoreticians and telecommunication scientists. With hindsight, we can see that theoretical work led to no usable results, and that only simple radiometeorological considerations, based on comparing the results of measurements in various climates, made it possible to find the characteristics of this type of propagation and set up reliable prediction methods.

On this subject, it is worth noting the importance of measurements on propagation made by the CNET in France and Africa. In particular, the plans for radio broadcasts to Africa on metric or dekametric waves drawn up by the UIT in 1963 were entirely based on the propagation measurements by the CNET, since no other results were available than those generated by my group at that time and we had also worked for many years in some ten countries in Africa, under less than laboratory conditions.

At around 1970, there was a sudden return to the earlier preoccupation with line of sight communications, but radio relay equipment had in the meantime changed profoundly. On the one hand transistors were replacing traditional valves, while on the other digital modulation was beginning to replace analog modulation. There followed a rapid growth of radio relay networks throughout the world, leading to a saturation of frequency bands lower than 10GHz, and to renewed interest in studies into propagation at higher frequencies. At such frequencies rain begins to play an important role. Propagation scientists suddenly began to take an interest in rainfall levels, in the shape of water droplets, in their speed of descent, etc.

Digital modulation also had repercussions for the study of propagation because of the bandwidth necessary. On the one hand, the existence of multiple propagation paths could lead to unacceptable distortion in propagation delays and to capacity limitations, while on the other it was necessary, in order to save on the electromagnetic spectrum used, to transmit two orthogonally polarized signals. As propagation properties were likely to limit the possibility to do this, it was necessary to carry out new studies.

At about the same time, the arrival of telecommunication satellites led to new propagation problems, and a distinction could be drawn between terrestrial links, corresponding to horizontal paths and space links corresponding to slant paths for which the propagation trajectory penetrates the troposphere and ionosphere. Certain propagation anomalies were reduced but others appeared, and in any case more precise knowledge of the phenomena involved was required than for terrestrial links, as in general space equipment allows only a very small fading margin. Other satellites, such as those used for geodesic measurements, make it necessary to have precise knowledge of the interference due to the neutral or ionized atmosphere.

As for the use of the ionosphere for telecommunications, the introduction of satellites led to fundamental changes, as radiowaves in the dekametric (HF) band disappeared almost completely from the intercontinental communications where they had been, for some decades, the only technical means available. They are still often used, above all for unimportant links over short distances in developing countries, and also for military purposes. The consequence is that in books dealing with radio wave propagation for telecommunications, sections describing reflection from the ionosphere, previously a major aspect of the field, can now be reduced considerably, while it has become necessary, on the other hand, to deal with the question of propagation through the ionosphere.

Another consequence of the introduction of satellites is the increased importance of problems associated with interference at long distances, even on super high frequencies. Consequently, propagation phenomena likely to create intense fields at long distances had to be studied in considerable detail.

It is therefore clear that propagation studies have by no means reached a fixed format and, although the basic phenomena are still the same (it is very unlikely that Maxwell's equations will ever be revised), applications have developed considerably, with certain fields of study losing in importance while many others have appeared or become far more significant.

Work on propagation has a number of aspects that have to be considered at the outset since the form a discussion of propagation takes will depend on the relative importance that the author attaches to each of these aspects.

One point of view, often qualified as scientific, involves the search for relations between various phenomena. This is an aspect that clearly must not be neglected, even though it can create the illusion that a phenomenon has been 'explained' when in fact only its similarity with a known phenomenon has been shown. It is often also the case that the extent of scientific discussion devoted to a problem depends more on the formal interest of the computations it makes possible, rather than their practical importance. Another danger to be avoided in theoretical studies is the introduction of parameters which are virtually impossible to measure. Clearly, under such conditions theory cannot be contradicted by experiment, but its value is very limited. The history of propagation studies provides examples of such situations.

A more utilitarian viewpoint is also possible, aiming to give practical rules to those who are designing radio telecommunication systems. As the transmission quality of such systems is defined for the different time percentages specified by international organizations such as the CCIR statistical analysis of long duration measurements is obviously crucial for drawing up such practical rules. Statistical studies of this kind are therefore extremely useful, and also constitute a powerful analytical tool, of great use for generating theory. It is also known that statistical studies raise a number of problems, in particular those that concern the quantities to be recorded and the degree of computing to be applied to data. In this computer age it seems obvious that most procedures ought to be done automatically. In fact, however, automated processing requires a level of data collection that is almost impossible to maintain continuously over several years' study, and for which significant probability levels are sometimes of the order of 10^{-2} to 10^{-5}. Manual or semi-automated processing makes it possible to eliminate most recording errors. This compromise is regularly made.

In this book, which is principally intended for users, we do not study theory for its own sake and only mention those theories which lead to practical results. Instead we provide large amounts of directly usable information, often obtained by semi-empirical methods.

I should like to draw a parallel between propagation studies and medicine. In medicine we are concerned with discovering the laws of nature and protecting ourselves against certain harmful consequences of those laws by the appropriate treatment. The medical research worker or doctor is often led to carrying out long expensive analysis, although, to a certain degree at least, such analysis can be avoided or at least reduced by the expertise already available in the field. This is indeed also true in propagation studies and the propagation scientist is consulted about an observed anomaly, and asked if possible, to provide a remedy to the situation. Unfortunately, a suitable treatment is not always available.

The analogy could be taken further. There are, for example, spontaneous cures for propagation deficiencies: we just have to wait a sufficiently long time. This is a general law, since propagation defects have the considerable advantage over hardware defects of being able to 'cure themselves'. However, such defects can sometimes be chronic (e.g. they recur every summer) and this may make servicing necessary, generally in the form of a replacement or modification to an installation, e.g. by adding diversity reception facilities. This 'medical' aspect of propagation has important consequences. As in medicine, we only attain efficiency and value after many years of work. 'Independence', which is viewed favourably in some fields, is harmful here. It is true that independence has certain advantages, for example, it is just as well not to be around when it becomes clear that a communication link will not work because the predictions of the propagation scientist were wrong, but it is nontheless true that even independent workers prefer to consult specialists sometimes.

The way the subjects are distributed throughout this book is a consequence, to a considerable extent, of a deliberate decision to separate line of sight propagation from non-line of sight propagation. Many years' teaching has shown me the advantage of such a distinction. The first six chapters therefore deal with line of sight propagation. Chapter 1 deals with propagation in open space and chapter 2 with the characteristics of the major physical phenomena dealt with in the rest of the book (reflection, refraction, diffraction, scattering). Chapters 3, 4 and 5 examine the effect of the ground and the troposphere. Chapter 6 brings together a number of useful data concerning anomalies met in line of sight links, which are often due to the combined effect of the ground and the atmosphere.

Chapters 7, 8 and 9 are devoted to the three main mechanisms allowing links to be established between two points which are not visible to each other, these mechanisms being diffraction by hills, scatter by the Earth's curvature and the troposphere and reflection by the ionosphere. However, in order not to separate the two sections describing the ionosphere, Chapter 9 also deals with links penetrating the ionosphere.

Chapter 10 gives certain results on propagation mechanisms particularly involved in long distance interference. Not all characteristics of such mechanisms are particularly well known and the number of publications dealing with them is not always proportional to their actual practical importance.

Chapter 11 concerns the problem of electromagnetic disturbance, and mainly interference caused by other transmissions or various forms of radiation. It is in fact traditional to include a chapter of this kind in a discussion of propagation, since propagation plays an important role in determining the level of disturbance, and also the signal: interference ratio at the receiver.

Appendix 1 gives information concerning the main statistical distributions used in propagation studies. This will perhaps avoid readers having to consult specialist works, which as well as being voluminous are often too theoretical for the intended application.

Appendixes 2 and 3 gives both administrative and technical information from Radio Regulations on the various services using electromagnetic waves and for whom, consequently, knowledge of propagation phenomena is essential.

I have attached particular importance to the presentation of figures. Most of them, and in particular all the aligned point nomograms have been entirely calculated or verified. Many of the figures in this work, which also appear in works published by the CCIR, were drawn up by the CNET propagation team and supplied to the CCIR. We thank the CCIR for permission to reproduce these figures.

Choosing the bibliography is always difficult. In the present case, the bibliography is relatively concise as there are few recent books in this field, and articles often handle subjects which are too specialized or give results which are likely to be rapidly out of date. A number of titles are given for each chapter and, moreover, I mention Volumes V and VI of the CCIR, since these volumes, concerned with propagation, are revised every four years and contain many references.

Before concluding this preface I should like to thank my colleagues who inspired some of the ideas in this book, my typist who prepared the manuscript, and the referees who criticized the content.

During the time of writing this book I have been associated with many people with the aim of studying propagation for telecommunication purposes. With this in mind, as I explained above, the main tool used was statistical analysis of long time measurements. This is detailed difficult work requiring constant attention, which is difficult to leave to entirely automated devices given the reliability levels of hardware and to very low time percentages required in statistics. I cannot overstate my gratitude to the people who have generated reliable data and carried out the statistical analysis of those data. I should like to make particular mention of my colleague, J Battesti, who has as a result of our collaboration for so many years and publication of so many articles played a certain role in the preparation of this book. And above all, my thanks are due to him for two extremely precious qualities, total intellectual honesty and considerable patience.

My thanks are due to those readers who sent me their comments and criticisms, and I hope that this translation of the new French edition, which includes new formulas and ammendments to bring the text in line with new developments and recent prediction methods presented to the CCIR, will be as well received as the original French edition.

Lucien Boithias
Telecommunication Chief
CNET
Issy les Moulineaux
France
October 1986

Chapter 1

Wave propagation in free space

1.1 Propagation and waves

The term wave propagation can be divided into two components, propagation and wave, which deserve close analysis. Propagation can be defined in general terms as a transfer of energy without a transfer of matter. The transfer of energy between two points in a certain medium is therefore the result of changes, with time, in the spatial distribution of a non-static field in that medium. This concept, which is valid at the macroscopic level, applies to a number of physical phenomena in the electromagnetic, heat, acoustic and other fields.

The term wave can only be defined by its relationship with the terms field and propagation. Wave propagation is a particular form of propagation, characteristic of certain fields and such that a given element of the field (e.g. a given phase value, an extremum, etc.) which is at a particular point at a certain moment will at a subsequent moment have undergone a displacement with little or none deformation. This property makes it possible to define a propagation velocity. Its value depends on the field parameter chosen (we can, for example, obtain the phase velocity, the group velocity, etc.) and on the characteristics of the medium. If the variations with time of the quantity which characterizes the fields are periodical, the distance travelled in any one period is known as the wavelength.

From the mathematical point of view, this set of properties can be expressed by the fact that the linear second order partial differential equation which represents such a field is a hyperbolic equation. For such a type of equation there exists two groups of real surfaces on which at least two solutions of the equation join up. One of these solutions may be the null solution (absence of the phenomenon). Such a surface is a wave front. Since this surface depends on time, the wave front moves with a definite velocity.

The simple harmonic equation, the telegraphist equation and Maxwell's equations are all hyperbolic, and the associated field shows wave propagation. On the other hand, heat conduction along a metal rod obeys a parabolic law; there is no wave front or propagation velocity associated with this phenomenon, and therefore wave propagation is not involved.

Therefore it can be said that the term wave is no more than a label used for the category of fields for which a propagation velocity may be defined. In fact, the equation of a field and that of the corresponding wave cannot be distinguished.

As well as these terms the term radiation is introduced. It is defined according to the Radio Regulations as the energy flux produced in the form of waves by a source, or the energy itself. Studying radiation therefore means studying its sources, whereas the study of a wave can be carried out independently of its source, and does not assume that the wave can actually be produced by a real source (plane waves, cylindrical waves, etc.). The relationship between sources and waves is nothing more than a study of antennas. This is not the purpose of this book.

1.2 Ducted propagation and radiation propagation in free space

Approaching the question from the point of view of practical applications there are two major types of electromagnetic waves propagation:

1. propagation ducted along a transmission line;
2. propagation of waves radiated from an antenna.

Under the Radio Regulations that part of the electromagnetic wave spectrum below 3000GHz (wavelengths greater than 0.1mm) concerns radio waves. This definition is merely conventional, since the prefix radio (from the Latin to radiate, *radior radiari*) does not imply any restricted frequency range.

Ducting requires material structures and, consequently, it involves attenuation. If we assume that all the elements exhibit linear characteristics and that the line is homogeneous, the percentage energy lost over a section of given length is independent of the energy entering that section. This implies that power along the line decreases exponentially with distance.

The same is not true for radiation propagation, since it can exist without a material component. Consider, therefore, the ideal case of an empty unlimited space containing a transmitting antenna and its associated transmitter only. This is termed open space. Assume too that the dimensions of these elements are negligible with respect to the transmission distances studied. The energy emitted at any given time will, at any time later, be distributed over the surface of a sphere of radius d. The power flux density decreases with respect to $1/d^2$. From a certain point onwards this relation gives a far smaller reduction than does the exponential function for ducted propagation. For example, no transmission line could allow communication to be set up from the Earth to the Moon, which only requires a relatively low power for radio communication.

This difference in behaviour is derived from the fact that, in ducted propagation the decrease in the received level is due to the progressive transformation of electromagnetic energy into heat or another form of energy, whereas in radiation propagation the decrease is the result of the energy being distributed over an increasingly large surface. The $1/d^2$ relation is not defined at the origin, since for radiated waves, the transmitting antenna corresponds to a singular point in the space.

Propagation in free space plays a fundamental role in providing a frame of reference against which to judge all other cases. In particular, the actual received energy level is compared with the level which would have been received in free space.

1.3. Choice of reference antenna: summary of antenna characteristics

The propagation of a wave between two points can only be studied and measured using an antenna at each of these two points. However, the properties of the antennas used must not influence the results. This means that propagation must be defined as occurring between two reference antennas which are also used to define the characteristics of the antennas actually used. A clear distinction has to be made between reference antennas used for these calculations, which it is not necessary therefore to build, and comparison antennas used for measurements on antennas, which must be constructed and whose performance levels must be measured.

1.3.1. The most common reference antennas

Consider two types of radio links. Each type determines a particular type of reference antenna.

1. Links where the antenna's height above ground is considerably greater than the wavelength. This is normally the case for the metre range and shorter waves. Reference antennas are then the Hertzian dipole, the half-wave dipole and the isotropic antenna.
   a) Hertzian dipole: a current element whose length is considerably shorter than that of the wavelength on which it operates. It can be shown that the radiation pattern of such a dipole is a torus whose axis is the dipole and whose cercle de gorge has a zero radius. If this antenna is used as a reference antenna, the direction considered is in the equatorial plane.
   b) Half-wave dipole: a linear radiating element whose length is half the wavelength on which it is operating. Its radiation pattern is very similar to the one above, but the semi-meridian of the torus is not quite circular.
   c) Isotropic antenna: an antenna which, by definition, radiates equally in all directions. Its radiation pattern is therefore a sphere centred on the antenna. It can be shown that such an antenna cannot be built, as isotropic radiation is incompatible with the transverse character of electromagnetic waves. However, this antenna is increasingly used as a reference antenna because of its simplicity.
2. Links between antennas whose height above the ground is very much less than the wavelength. This is generally the case for low frequencies. Only vertical polarization is viable and in this case the reference antennas are:
   a) Short vertical antenna: above the ground which is assumed to be infinitely conducting. This is half of an Hertzian dipole.

b) Quarter wave vertical antenna: above the ground which is assumed to be infinitely conducting. This is half of a half-wave dipole. These two antennas have the same radiation pattern above ground as the symmetrical antennas from which they are derived, but the fact that the power supply to them is radiated in only half the space must be taken into account.

1.3.2. Gain and effective area of an antenna

If a transmission antenna is used instead of a reference antenna, the field in general has a different value. The power P of the transmitter becomes the effective power P', and:

$$G = \frac{P'}{P} \tag{1.1}$$

where G is the infinite gain of the transmitting antenna in the direction considered. In cases where the reference antenna is an isotropic antenna the gain is called the absolute gain or isotropic gain of the antenna. For example, it can be shown that in free space the Hertzian dipole and the half-wave dipole have absolute gains of 1.5 and 1.64 (1.75 and 2.15dB), respectively. Consequently, if one of these dipoles is taken as the reference antenna, the corresponding relative gain of any antenna is measured as a number that is less than the value for the absolute gain of the same antenna by 1.75dB or 2.15dB, respectively.

A receiving antenna, placed at a point of a plane wave where the surface density of the power is S, receives a certain power P proportional to S. The coefficient of proportionality, which has the dimensions of an area, is called the effective area of the antenna. This area is not necessarily simply related to the geometrical dimensions of the antenna. Nevertheless, in most cases dealing with waves in the decimetre range (roughly UHF) or shorter, where the antenna presents a large opening perpendicular to the direction of propagation, the effective area is generally proportional to the area of this opening, the coefficient of proportionality being between 0.5 and 0.7.

Gain is therefore defined for a transmitting antenna and effective area for a reception antenna. Since the same antenna can be used either for transmission or for reception, there is necessarily a relation between these two quantities. Applying the principle of reciprocity, it can be shown that the gain G of an antenna in a particular direction is proportional to the effective area A in the same direction. The coefficient of proportionality is indepedent of the antenna and it can be shown that for a wavelength λ in free space:

$$\frac{G}{A} = \frac{4\pi}{\lambda^2} . \tag{1.2}$$

For an isotropic antenna gain equals 1, and consequently the effective area A_0 of an isotropic antenna equals:

$$A_0 = \frac{\lambda^2}{4\pi}. \tag{1.3}$$

Under non-free space conditions these equations no longer apply. In particular, if a vertical antenna, on an infinitely conducting surface, is compared with a dipole antenna obtained by adding a portion symmetric to it with respect to the surface, with respect to the dipole it is found that:

the gain is twice as great;
the effective area is halved;
the radiation resistance is halved;
the power necessary to create a given field is halved;
the *G/A* ratio is four times greater.

1.4 Electromagnetic fields and attenuation of propagation in free space

A transmitting antenna supplied with power *P* radiates an electromagnetic field *E* in the entire space, and a receiving antenna in this field will receive a certain power *p*. There are therefore two different ways of handling a propagation problem:

1. Calculate the electromagnetic field created at a given point by an antenna fed from a transmitter.
2. Calculate the ratio between the transmitted and received power. This ratio, called the transmission loss, is usually expressed in decibels.

The choice of method depends on the problem being studied. If a transmitter must serve a large number of receivers whose positions are not necessarily known (radio broadcasting), the first method is to be preferred. On the other hand, in a point to point communication (radio relay link) the link forms a whole in which the transmitting and receiving antennas play symmetrical roles; it is then natural to use the second method.

1.4.1. Field in free space created by an isotropic antenna

A transmitter of power *P* supplying an isotropic antenna creates at distance *d* a spherical wave transporting a power flux density *S* such that:

$$S = \frac{P}{4\pi d^2}. \tag{1.4}$$

This power flux density is equal to the amplitude of the Poynting vector.

$$\mathbf{S} = \mathbf{E} \times \mathbf{H}. \tag{1.5}$$

At great distances from the antenna the vectors **E**, **H**, and **S** form a direct tri-rectangular trihedron and the scalar relation is obtained:

$$\frac{E}{H} = \sqrt{\frac{\mu_0}{\varepsilon_0}} = 120\ \pi \text{ ohms}. \tag{1.6}$$

The modulus of **S** therefore takes the value:

$$S = E^2 \sqrt{\frac{\varepsilon_0}{\mu_0}} = \frac{E^2}{120\ \pi}. \tag{1.7}$$

from which the electric field can be deduced:

$$E = \frac{\sqrt{30\ P}}{d}. \tag{1.8}$$

which is often expressed (Fig. 1.1):

$$E_{(\mathrm{mV/m})} = 173 \frac{\sqrt{P_{(\mathrm{kW})}}}{d_{(\mathrm{km})}}. \tag{1.9}$$

The product Ed, which is a voltage and is generally expressed in volts, is known as the cymomotive force (c.m.f.).
If instead of the isotropic antenna, the Hertzian dipole or the half-wave dipole is taken as the reference antenna, the constant 173 in Equation (1.9) must be replaced by 212 or 222, respectively. In other words, for 1kW of radiated power, the cymomotive force (c.m.f.) is:

173 volts for an isotropic antenna;
212 volts for a Hertzian dipole;
222 volts for a half-wave dipole antenna.

Equations (1.8) and (1.9) apply to waves with rectilinear polarization. If polarization is elliptical and if E_x and E_y are the components of the electric field along two orthogonal axes, the first elements in these equations must be replaced by $\sqrt{E^2{}_x + E^2{}_y}$. From this E_x and E_y can be deduced only if the ellipticity ratio is known. For example, for circular polarization E must be replaced by $E\sqrt{2}$.

1.4.2. Propagation in the free space between two isotropic antennas: free space basic transmission loss

This is the term used for the ratio between the power P_t transmitted by an isotropic antenna and the power P_r received by another isotropic antenna at a distance d away and in free space. Now, for any two antennas:

$$\frac{P_r}{P_t} = \frac{GA}{4\ \pi d^2}, \tag{1.10}$$

where G is the gain for the transmitting antenna and A is the effective area of the receiving antenna. For isotropic antennas from Equation (1.3):

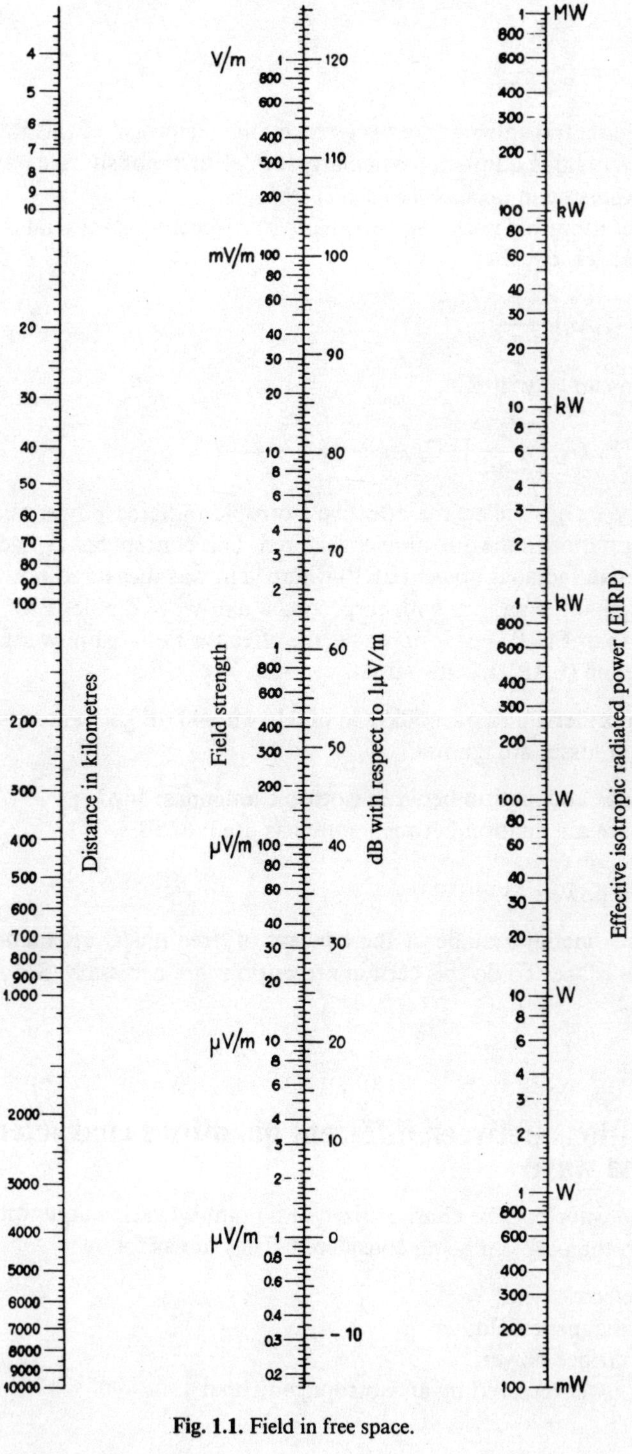

Fig. 1.1. Field in free space.

$$\left(\frac{P_r}{P_t}\right)_{\text{iso}} = \left(\frac{\lambda}{4\,\pi d}\right)^2 . \tag{1.11}$$

Free space basic transmission loss is given by the reciprocal of this ratio, which is generally a very large number, particularly for decimetre or shorter wavelengths; it is usually expressed in decibels (Fig. 1.2).

If transmitting and receiving antennas with respective gains G_t and G_r are used, the power ratio is equal to:

$$\frac{P_r'}{P_t} = G_t\, G_r \left(\frac{\lambda}{4\,\pi d}\right)^2 . \tag{1.12}$$

This relation can be written:

$$P_r = (P_t\, G_t) \left(\frac{\lambda}{4\,\pi d}\right)^2 G_r\,. \tag{1.13}$$

The quantity P_tG_t is called the effective isotropic radiated power (EIRP) of the transmitting station in the direction considered. This concept has replaced the older idea of effective radiated power (ERP) where gain was measured not with respect to an isotropic antenna, but with respect to a half-wave dipole. ERP is therefore 2.15dB less than EIRP. For long waves the effective radiated power from a short vertical antenna (ERPV), is also used.

Example: Consider a link over 50km at 6GHz with 40 dB gain antennas and a 2W transmitter; if losses are ignored:

free space attenuation between isotropic antennas: 142dB;
free space attenuation between antennas used: 62dB;
EIRP: 43dbW;
received power: –59dBW.

N.B. Use is sometimes made of the concept of 'free space' attenuation above a conducting surface. To do this certain precautions are necessary. See in particular Section 1.3.2.

1.5. Relations between different quantities characterizing a plane wave

A plane wave may be characterized at a point by various quantities, selected according to the problem being considered. They are as follows:

E, the electric field;
H, the magnetic field;
S, the surface power;
p, the power received by an isotropic antenna;

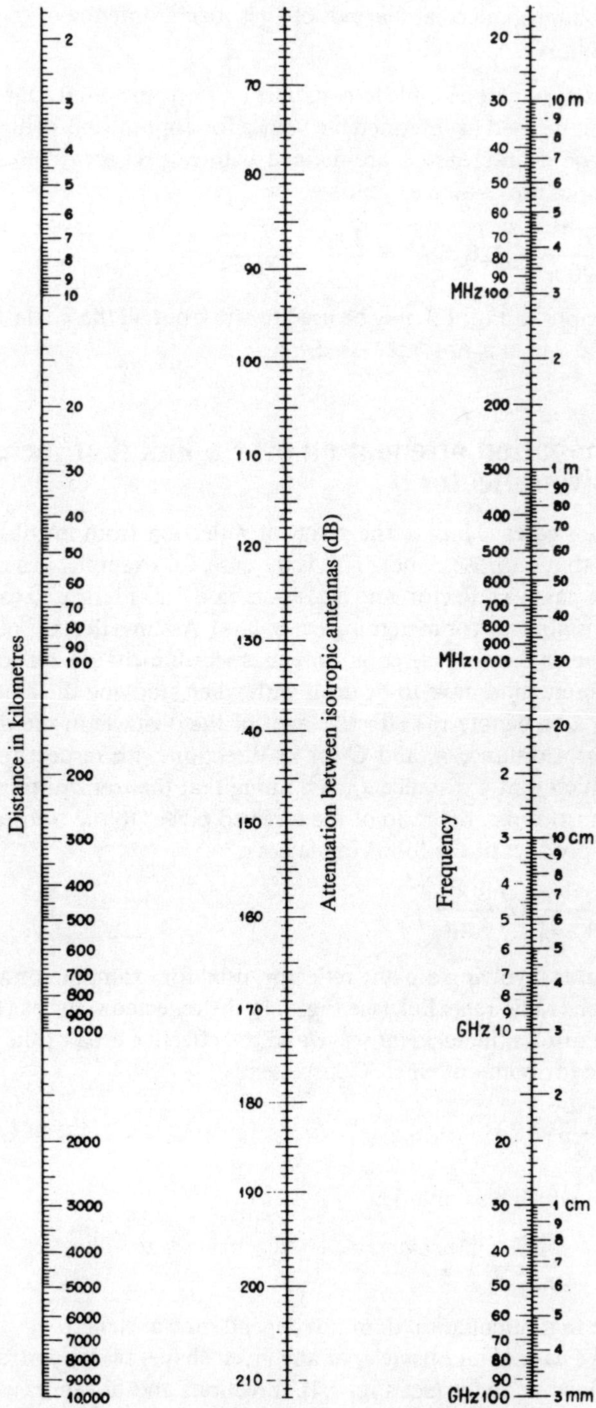

Fig. 1.2. Free space attenuation between isotropic antennas.

v, the voltage induced at the ends of an isotropic antenna of given impedance, e.g. 50 ohms.

Since it is practically impossible to construct an isotropic point source, the last two quantities will be used as intermediate values for applications concerned with any antenna whose characteristics are defined with respect to an isotropic antenna. These quantities are related as follows:

$$S = \frac{E^2}{120\,\pi} = 120\,\pi H^2 = \frac{4\,\pi p}{\lambda^2} = \frac{4\,\pi v^2}{50\,\lambda^2}. \tag{1.14}$$

The nomographs in Fig. 1.3 may be used to work out all these relations except the magnetic field which is not often used.

1.6. Propagation attenuation over a link that includes a passive reflector

In certain cases a link is the result of reflection from an obstacle, whether intended for that purpose or not. This is the case, for example, in a radio relay link using a plane passive reflector, and is also the case for radars used to detect specific shapes (so it is not true for meterological radars). Assume that the obstacle is in the far field of the antennas. Periscopic antennas which include a reflector in the near field are different, and have to be dealt with when studying the antenna itself.

We use Σ to denote the effective area of the obstacle in the direction of the transmitter at a distance d_1 and G for its direction with respect to the receiver's direction, which is at a distance d_2. Assuming that the transmitting and receiving antennas are isotropic, the ratio of the received power to the transmitted power is equal to the product of the following factors.

$$\frac{P_r}{P_t} = \frac{\Sigma}{4\,\pi d_1^2}\, G\left(\frac{\lambda}{4\,\pi d_2}\right)^2. \tag{1.15}$$

1. In cases involving a plane reflector, used for example, for a telecommunications radio relay link (see Fig. 1.4), the reflected wave has the same phase structure as the incident wave and the effective area of the reflector is the same for both antennas. Consequently:

$$G = \frac{4\,\pi\Sigma}{\lambda^2}, \tag{1.16}$$

   from which we conlude:

$$\frac{P_r}{P_t} = \left(\frac{\Sigma}{4\,\pi d_1\, d_2}\right)^2; \tag{1.17}$$

   note that attenuation does not depend on wavelength.
2. In the case of an obstacle, of any given shape, placed in front of standard monostatic radar (see Fig. 1.4), directions and distances are the same for

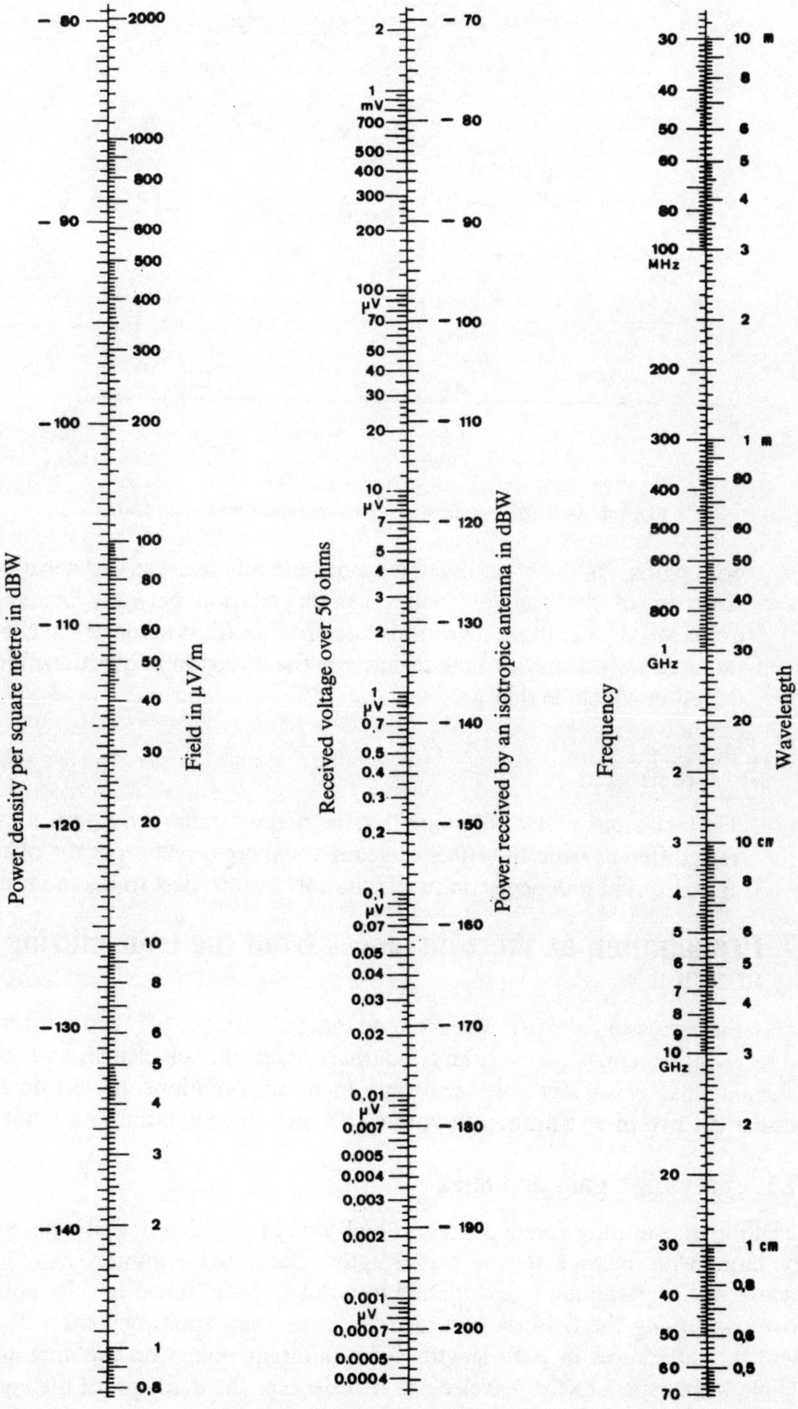

Fig. 1.3. Characteristics of a plane wave.

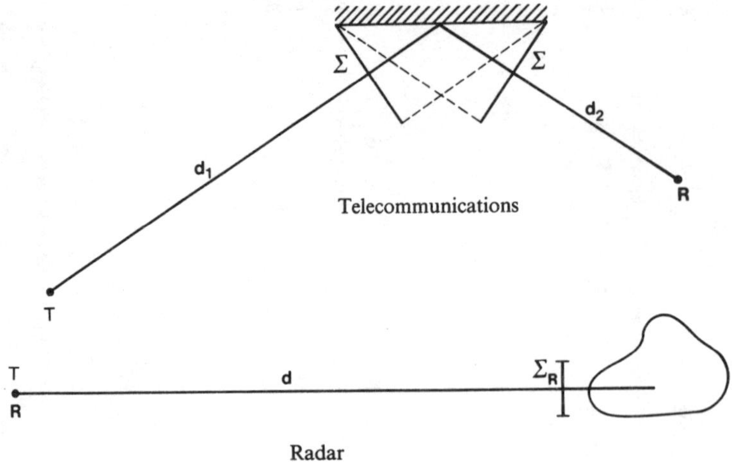

Fig. 1.4. A comparison between telecommunications and radar.

both paths. On the other hand, it is not generally possible to determine the structure of the scattered wave, i.e. the relation between G and Σ is unknown. Under these conditions, the product $G\Sigma$ is taken as the effective radar cross-section Σ_R of the obstacle in the direction in question. We can therefore conclude that:

$$\frac{P_r}{P_t} = \frac{\Sigma_R \lambda^2}{(4\pi)^3 d^4}. \tag{1.18}$$

The inclusion of factor G means that the effective radar cross-section can be very different from the effective geometrical cross-section of the obstacle, and also that it depends on its orientation with respect to the radar beam.

1.7 Propagation at short distances from the transmitting antenna

In the discussion above it was assumed that the distance between transmitting and receiving antennas was very large compared with the wavelength and the size of the antenna. This is not necessarily true under all conditions. In Section 1.7 we consider the two most common situations in which these conditions are not met.

1.7.1. Very high gain antennas

If the transmitting antenna has very high gain in one direction, it is necessarily very large with respect to the wavelength. The most common case is the antenna with a parabolic reflector used in point to point links. Results obtained above concerning the field or transmission losses only apply beyond a distance where the differences in path lengths from different points on the antenna are neglible with respect to the wavelength. If D denotes the diameter of the antenna and λ denotes the wavelength, distance d must satisfy the following condition:

$$d > \frac{2\,D^2}{\lambda}, \quad \text{or} \quad \frac{d}{\lambda} > 2\left(\frac{D}{\lambda}\right)^2 \tag{1.19}$$

Or a condition which can be written in a still more general form as follows:

$$\frac{d}{\lambda} > 0.4\,G\,, \tag{1.20}$$

where G is the antenna gain expressed as a power ratio (not in decibels) in the direction in question.

The minimum distance obtained in this way may be very great in certain cases. For example, for an antenna used in satellite telecommunications, with a diameter of 10m and operating at 14GHz, the minimum distance along the axis of the beam is 9.3km. At a closer distance from the antenna the field along the axis of the beam no longer follows a $1/d$ relation, but oscillates around a nearly constant value.

1.7.2. Very low gain antennas

If the antenna is very small with respect to the wavelength, its radiation is indistinguishable for practical purposes from that of an electric or magnetic dipole with maximum gain of 1.5, and the actual size of the antenna is no longer a contributing factor. This case is met in particular when studying unwanted radiation from certain equipment.

At distances less than the wavelength, radiation no longer has a plane wave structure, i.e. the ratio between the electric field and the magnetic field is no longer given by the equation:

$$\frac{E}{H} = \sqrt{\frac{\mu_0}{\varepsilon_0}} = Z_0 = 120\,\pi \text{ ohms}\,. \tag{1.21}$$

In particular, if the wave is treated as though it was radiated by an electric dipole, the ratio between the electric field and the magnetic field at a short distance d with respect to the wavelength λ, in the equatorial plane of the dipole diameter is given by:

$$\frac{E}{H} = \frac{Z_0}{2\,\pi}\,\frac{\lambda}{d}. \tag{1.22}$$

In this case, a measuring receiver with an antenna sensitive to the electric field (electric dipole) will give a higher reading than if it is equipped with an antenna sensitive to the magnetic field (frame). The opposite is true if the wave is radiated by an element similar to a magnetic dipole, i.e. a ring or a coil, for then at short distances:

$$\frac{E}{H} = 2\,\pi Z_0\,\frac{d}{\lambda}. \tag{1.23}$$

Chapter 2

Fundamental propagation phenomena: summary of physical concepts

2.1. Introduction

For the purposes of studying propagation in free space as described in Chapter 1 it is assumed that the propagation medium is a vacuum housing only the transmitting and receiving antennas of negligible size. Real conditions generally differ significantly from this ideal case. On the one hand, the characteristics of the propagation medium are different from those of a vacuum and can vary from point to point, while on the other hand the propagation medium may contain bodies with very varied electrical characteristics, producing discontinuity surfaces. Here antenna sizes are still considered to be negligible with respect to the distances in question.

From the theoretical point of view, the study of a propagation problem involves solving Maxwell's equations, which can be theoretically determined as soon as we know:

1. the position and the nature of the sources;
2. the electromagnetic characteristics of the propagation medium or media (e.g. the refractive index);
3. the boundary conditions on surfaces delimiting or separating media;
4. initial conditions (these generally vanish as we are usually only concerned with solutions which are sine functions of time).

In practice, this method can only be used at the price of introducing major simplifications, which still leave the problem too complex to handle. It is therefore usually reduced still further into individual models for which far simpler specific methods are already available. Clearly, this leads to problems of reconciling different solutions with each other, and these difficulties have to be overcome for each individual case. Introducing these problems also means considering 'propagation phenomena' or 'propagation mechanisms', such as line of sight propagation, diffraction, etc., which also means interpreting Maxwell's equations, but in theory no such interpretation should be necessary.

2.2. The line of sight propagation model

The simplest model is drawn from optics. The concept of a wave is replaced by that of a trajectory or a 'ray'. Therefore, phenomena such as reflection and refraction are involved. The concept of a wavelength no longer applies and Maxwell's equations are replaced by far simpler relations, such as the equality of angles of incidence and reflection or Snell's law of refraction.

The crucial problem is therefore to determine cases suited to this purely geometrical method. In optics the concept of line of sight is used, and that concept has to be re-analysed to make it applicable to radio waves. We start with the following comment: an electromagnetic wave is characterized not only by electric and magnetic parameters, but also by a geometric parameter, its wavelength. Optics ignores this parameter, which reduces to the wavelength being treated as though it was infinitesimal with respect to the size of all objects lying in the path of the waves. Given the shortness of their wavelengths, this condition is easy to apply to light waves. It is, however, far more difficult to apply to radio waves, where the wavelengths of the shortest waves commonly used are some 10^4 times larger than for light waves, and this factor can be increased to 10^{10} if the whole radio spectrum is covered.

2.3. Definition and criterion for a line of sight link

Great care is therefore needed if the concept of line of sight is applied to radio waves, and in fact it can only be applied negatively by defining a 'non-line of sight' link. In optics such a link would be associated with a zero field strength. In fact, however, the field strength is not zero. To explain this fact, the phenomenon known as diffraction is introduced. In this way, a line of sight link can be defined as a link for which diffraction is negligible, and therefore one for which only reflection, refraction or scattering phenomena apply.

For this definition to be usable a simple criterion to determine whether or not diffraction can be neglected must be established.

Consider the situation in Fig. 2.1. Signals are transmitted from a transmitter T and received at R; P is a plane perpendicular to the straight line TR at point A. Using Maxwell's equations to calculate the field strength at R we only need to know the field strength created by T in all points of plane P. Moreover, if polarization, is ignored, all points on the circle within plane P about a centre at A make the same contribution to the field at R, because they are in phase with each other. Therefore, plane P can be divided into concentric rings and the contribution to the field at R made by each ring, taking into account the phase variation from one to the other, can be calculated. Phase differences are calculated with respect to the origin defined by ray B following the shortest path, TAR. The path through B is longer than the straight line path by a quantity ΔL which can easily be calculated. All points with the same value for ΔL make the same contribution to the field strength at R. If plane P is displaced parallel to itself and ΔL has the same value, point B will generate an ellipsoid of revolution with E and R as foci. Therefore, the space is divided into a

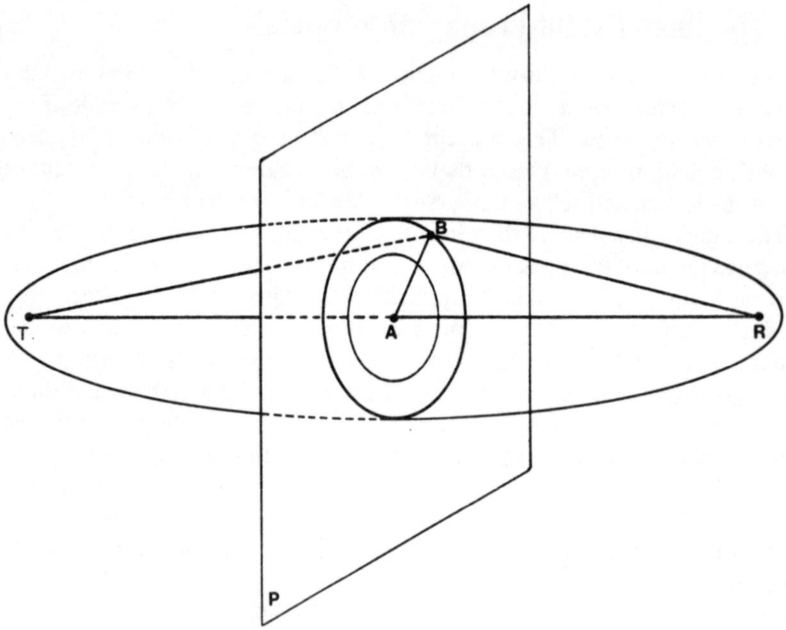

Fig. 2.1. Determination of Fresnel ellipsoids.

series of homofocal ellipsoids.

Furthermore, it is clear, and can be shown by the stationary phase theorem, that most of the energy at R has travelled along TR. As long as:

$$\Delta L = TBR - TAR \tag{2.1A}$$

is less than half the wavelength, the points within a circle of radius AB, summed at R, have the same sign. If the difference is greater than half the wavelength, some of the points will supply a negative contribution, and the field at R will decrease. It increases once the difference is greater than the wavelength, and so on.

$$\text{Let} \quad \Delta L = n\frac{\lambda}{2}. \tag{2.1B}$$

The integer values of n define a group of ellipsoids called Fresnel ellipsoids, and the first of them ($n = 1$) plays a special role, in that it defines the region of space through which most of the energy travels, and therefore the region which needs to be free of obstacles if propagation is not to be 'disturbed'. This makes it possible to define conditions for line of sight propagation. Closer study shows that only about 60% of the first ellipsoid needs to be free of obstacles for the received level to be equal to that in free space (see Chapter 7).

From Fig. 2.1 it is immediately clear that:

$$\Delta L = \sqrt{a^2 + \rho^2} + \sqrt{b^2 + \rho^2} - a - b\,. \tag{2.2}$$

Since ρ is generally small with respect to a and b, a limited expansion can be used:

$$\Delta L = \frac{\rho^2}{2}\left(\frac{1}{a} + \frac{1}{b}\right). \tag{2.3}$$

For a given path length $d = a + b$, ΔL is maximum when $a = b$, i.e. at the mid-point of the path and:

$$(\Delta L)_{\max} = \frac{2\,\rho^2}{d}. \tag{2.4}$$

The first ellipsoid is obtained when the value of $\Delta L$is equal to half the wavelength. The equatorial radius of the ellipsoid is therefore equal to:

$$\rho_e = \frac{1}{2}\sqrt{\lambda d}\,. \tag{2.5}$$

To give an idea of the orders of magnitude involved consider a 50km link operated at a wavelength λ. The following values for ρ_e are given in Table 2.1.

λ	ρ_e	
0.5μm	0.08m	(light waves)
5.0cm	25.00m	(centimetre wave range)
5.0m	250.00m	(metre wave range)
500.0m	2500.00m	(100 metre wave range)

Table 2.1. Relationship between λ and ρ_e.

Table 2.1 shows that the concept of a 'ray' has practical significance for light waves because even on a 50km link all the energy is concentrated within a cylinder a few centimetres in diameter. On the other hand, for the centimetre wave range the diameter of the cylinder is some tens of metres, making it necessary to build antenna towers (fig. 2.2), and for the metre wave range or higher ranges it is virtually impossible to keep the first Fresnel ellipsoid free. In other words, for the latter wavelengths 'line of sight' conditions are virtually impossible to create over a link at the Earth's surface, and diffraction is always involved (see Chapter 7).

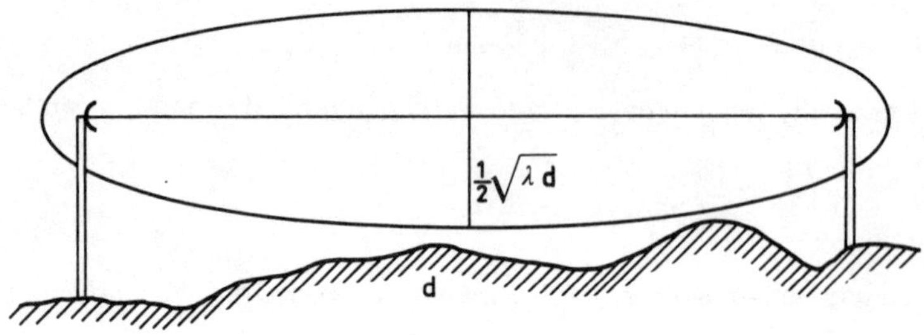

Fig. 2.2. Definition of a line of sight link at the Earth's surface.

2.4. Reflection and refraction

A major role in line of sight propagation is played by reflection and refraction. We assume that the reader is familiar with these phenomena and only a few brief remarks are given here.

Reflection and refraction take place when two different types of medium are separated by a surface which is very large and has very small irregularities with respect to the wavelength of the radiation considered. In particular, the radius of curvature at any point on the surface must be very great with respect to the wavelength. It is for this reason that these phenomena are studied in the first place for a flat and infinitely large separation surface. Surface curvature is then taken into account and the concepts of convergence, divergence and focussing are introduced. Finally, we study perturbations in these phenomena due to the fact that the surface is not extremely large, or irregularities extremely small, with respect to the wavelength. In the same way, with the aim of simplifying analysis, these phenomena are defined in terms of plane waves. It is well known that such a wave is a mathematical solution of Maxwell's equations, but is not a physical solution, as it would transport infinite power. Its value stems from its simplicity, as any wave may be broken down into a summation of plane waves, and finally because, if far enough from its source, any wave has the structure of a plane wave locally.

In all cases, the formulas which express the laws of refraction and diffraction are derived from the following continuity properties which are themselves consequences of Maxwell's equations. At the separation surface between the two media the following four properties apply:

1. the tangential components of the electric and magnetic fields are continuous;
2. the normal components of electric and magnetic induction are continuous.

When considering wave propagation one medium is very likely to be the atmosphere, which can generally be treated as though it were a vacuum for the purposes of reflection or refraction (this is not true for other phenomena). The types of medium in contact with the atmosphere (ground, sea) are characterized by relative permittivity ε (to simplify the notation ε is used to denote permittivity instead of ε_r which is the standard notation) and conductivity σ, but their magnetic permeability can generally be regarded as equal to that of the vacuum. Therefore, it is assumed throughout that relative magnetic permeability is equal to 1. Moreover, it is worth considering first of all the case where σ is zero (dielectric medium), since this simplifies the situation.

2.4.1. Reflection and refraction in dielectric media

If media are purely dielectric, there is no attenuation and the incident, reflected and refracted plane waves are each defined by a single vector perpendicular to the corresponding equal phase planes. A first relation independent of the incident wave's polarization can be obtained by assuming that the components v_i, v_r, v_t of the phase velocities of the three waves (incident, reflected and transmitted) along the separation surface are the same. Using the notation given in Fig. 2.3:

$$\frac{v_i}{\sin\theta_i} = \frac{v_r}{\sin\theta_r} = \frac{v_t}{\sin\theta_t}, \tag{2.6}$$

but if n_1 and n_2 denote the refractive indices of the first and second media, respectively:

$$v_i = \frac{c}{n_1} \qquad v_r = \frac{c}{n_1} \qquad v_t = \frac{c}{n_2}. \tag{2.7}$$

Which therefore gives:

$$\theta_i = \theta_r \tag{2.8}$$

law of reflection

$$n_1 \sin\theta_i = n_2 \sin\theta_t : \tag{2.9}$$

law of refraction (Snell's law)

Other relations independent of polarization can be drawn from the fact that the waves are plane. For example, for each medium:

$$\frac{E}{H} = Z_0 \sqrt{\frac{\mu}{\varepsilon}}, \tag{2.10}$$

where Z_0 is the intrinsic impedance of a vacuum (120π ohms) and ε and σ are the relative permittivity and permeability of the medium in question. As μ is generally equal to 1, ε is equal to the square of the refractive index n of the medium:

$$\mu = 1 \qquad \sqrt{\varepsilon} = n. \tag{2.11}$$

The relative amplitudes of the incident, reflected and transmitted waves can be derived from these continuity equations. As they depend on polarization, it is

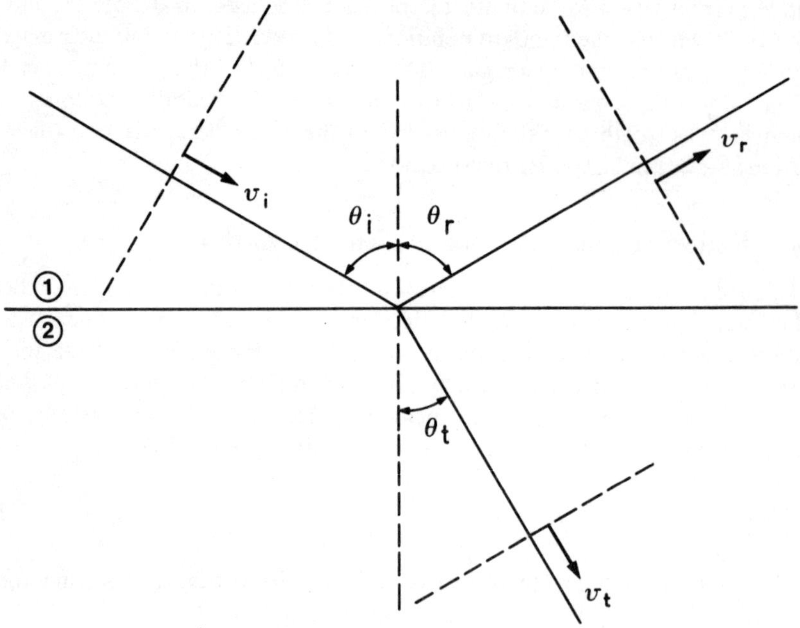

Fig. 2.3. Reflection and refraction of a plane wave.

traditional to study two particular cases which cover most real cases, so-called horizontal and so-called vertical polarization. The plane separating the two media is assumed to be horizontal.

2.4.1.1 Horizontal polarization

The electric field is horizontal, i.e. parallel to the plane of separation. From the continuity relations:

$$E_i + E_r = E_t , \tag{2.12}$$

$$(H_i - H_r) \cos \theta_1 = H_t \cos \theta_2 , \tag{2.13}$$

$$\mu_1 (H_i + H_r) \sin \theta_1 = \mu_2 H_t \sin \theta_2 . \tag{2.14}$$

If the relations, Equations (2.9) and (2.10), which are independent of polarization are taken into account the reflection factor (factor is better than cofficient for dimensionless quantities) for horizontal polarization can be calculated:

$$R_H = \frac{E_r}{E_i} = \frac{H_r}{H_i} . \tag{2.15}$$

As in general μ_1 and μ_2 are equal, we can use the refractive indices n_1 and n_2. Then:

$$R_H = \frac{n_1 \cos \theta_1 - n_2 \cos \theta_2}{n_1 \cos \theta_1 + n_2 \cos \theta_2} \tag{2.16}$$

which can be rewritten taking Equation (2.9) into account:

$$R_H = \frac{\sin (\theta_2 - \theta_1)}{\sin (\theta_2 + \theta_1)} \tag{2.17}$$

or again, using the angle of reflection φ which is the complement of the angle of incidence θ_1:

$$R_H = \frac{\sin \varphi - \sqrt{\left(\frac{n_2}{n_1}\right)^2 - \cos^2 \varphi}}{\sin \varphi + \sqrt{\left(\frac{n_2}{n_1}\right)^2 - \cos^2 \varphi}} \tag{2.18}$$

2.4.1.2. Vertical polarization

The magnetic field is horizontal, i.e. parallel to the plane of separation (the electric field is not vertical, although this might be the impression given by the term used). Continuity relations give:

$$(E_i - E_r) \cos \theta_1 = E_t \cos \theta_2 \tag{2.19}$$

$$H_i + H_r = H_t \tag{2.20}$$

$$\varepsilon_1 (E_i + E_r) \sin \theta_1 = \varepsilon_2 E_t \sin \theta_2 \,. \tag{2.21}$$

If Equations (2.9) and (2.10) are used, the following equations are obtained:

$$R_V = \frac{n_2 \cos \theta_1 - n_1 \cos \theta_2}{n_2 \cos \theta_1 + n_1 \cos \theta_2}, \tag{2.22}$$

$$R_V = \frac{\tan (\theta_1 - \theta_2)}{\tan (\theta_1 + \theta_2)}, \tag{2.23}$$

$$R_V = \frac{\left(\frac{n_2}{n_1}\right)^2 \sin \varphi - \sqrt{\left(\frac{n_2}{n_1}\right)^2 - \cos^2 \varphi}}{\left(\frac{n_2}{n_1}\right)^2 \sin \varphi + \sqrt{\left(\frac{n_2}{n_1}\right)^2 - \cos^2 \varphi}} \,. \tag{2.24}$$

Vertical polarization has a particular property derived from Equation (2.23): the reflection factor vanishes for a value of the angle of incidence given by:

$$\theta_1 + \theta_2 = \frac{\pi}{2} \,. \tag{2.25}$$

If, in addition, we take into account Snell's law, Equation (2.9):

$$\tan \theta_1 = \frac{n_2}{n_1} \tag{2.26}$$

Which defines Brewster's angle. We then talk of Brewster's angle of incidence. In the case of Brewster's angle of incidence the refracted ray is perpendicular to the reflected ray (which has a zero amplitude).

N.B. In Equations (2.18) and (2.24), which give the reflection factors, it can be seen that the refractive indices of the two media only appear in the square of their quotient, which can therefore be replaced by the quotient of permittivities. In cases where the relative permittivity of the first medium is equal to 1 the classic equations can be obtained:

$$R_H = \frac{\sin \varphi - \sqrt{\varepsilon - \cos^2 \varphi}}{\sin \varphi + \sqrt{\varepsilon - \cos^2 \varphi}}, \tag{2.18'}$$

$$R_V = \frac{\sin \varphi - \dfrac{\sqrt{\varepsilon - \cos^2 \varphi}}{\varepsilon}}{\sin \varphi + \dfrac{\sqrt{\varepsilon - \cos^2 \varphi}}{\varepsilon}}. \tag{2.24'}$$

2.4.1.3 Case of grazing incidence

If φ is close to zero, Equations (2.18) and (2.24) can be replaced by a limited expansion. If the ratio n_2/n_1 (the ratio of the refractive index of the second medium to the refractive of the first) is by denoted n:

$$R_H = -1 + \frac{2\varphi}{\sqrt{n^2 - 1}}, \tag{2.27}$$

$$R_V = -1 + \frac{2n^2\varphi}{\sqrt{n^2 - 1}}. \tag{2.28}$$

If furthermore n is very much greater than 1, which is frequently the case:

$$R_H = -1 + \frac{2\varphi}{n}, \tag{2.29}$$

$$R_V = -1 + 2n\varphi. \tag{2.30}$$

2.4.1.4. Case of two media with only slightly different refractive indices

Under certain conditions an approximately plane horizontal surface may form in the atmosphere where its refractive index undergoes a slight discontinuity, by a positive or negative amount Δn. The reflection factor associated with such a change is now considered.

First, assume vertical polarization. The reflection factor is given by Equation

(2.24). As n_1 and n_2 are close to 1:

$$\left(\frac{n_2}{n_1}\right)^2 \simeq 1 + 2\,\Delta n\,. \tag{2.31}$$

Equation (2.24) then becomes:

$$R_V = \frac{(1 + 2\,\Delta n)\sin\varphi - \sqrt{\sin^2\varphi + 2\,\Delta n}}{(1 + 2\,\Delta n)\sin\varphi + \sqrt{\sin^2\varphi + 2\,\Delta n}}\,. \tag{2.32}$$

Figure 2.4 gives the modulus of the reflection factor R_v as a function of angle φ for various values of Δn (in fact R_v is negative). If Δn is positive (movement into a more highly refractive medium), the curves show no discontinuity. On the other hand, if Δn is negative (movement into a less refractive medium), there is total reflection for any angle less than the limiting angle given by:

$$\varphi_l = \sqrt{-2\,\Delta n}\,. \tag{2.33}$$

For values of R_v less than 0.1 the sign of Δn is not very important and:

$$R_V \simeq \frac{1}{2}\frac{\Delta n}{\varphi^2}\,. \tag{2.34}$$

With horizontal polarization and using the same assumptions:

$$R_H = \frac{\sin\varphi - \sqrt{\sin^2\varphi + 2\,\Delta n}}{\sin\varphi + \sqrt{\sin^2\varphi + 2\,\Delta n}}\,. \tag{2.35}$$

For the very small values of Δ_n and φ considered here we can therefore write, for all practical purposes:

$$R_H = R_V\,, \tag{2.36}$$

and the curves in Fig. 2.4 can be used for either type of polarization.

2.4.1.5. Gradual variation of the refractive index within a thin layer

Assume that the refractive index varies over a small layer of atmosphere of depth Δh between altitudes h_1 and h_2 with a variation of n given by:

$$\frac{\mathrm{d}n}{\mathrm{d}h} = f(h)\,. \tag{2.37}$$

Let us also assume that the reflection factor is less than 0.1, making it possible to use the approximation given in Equation (2.34). The fact that the phase difference between the incident ray and the reflected ray will vary within depth Δh must be taken into account, which gives:

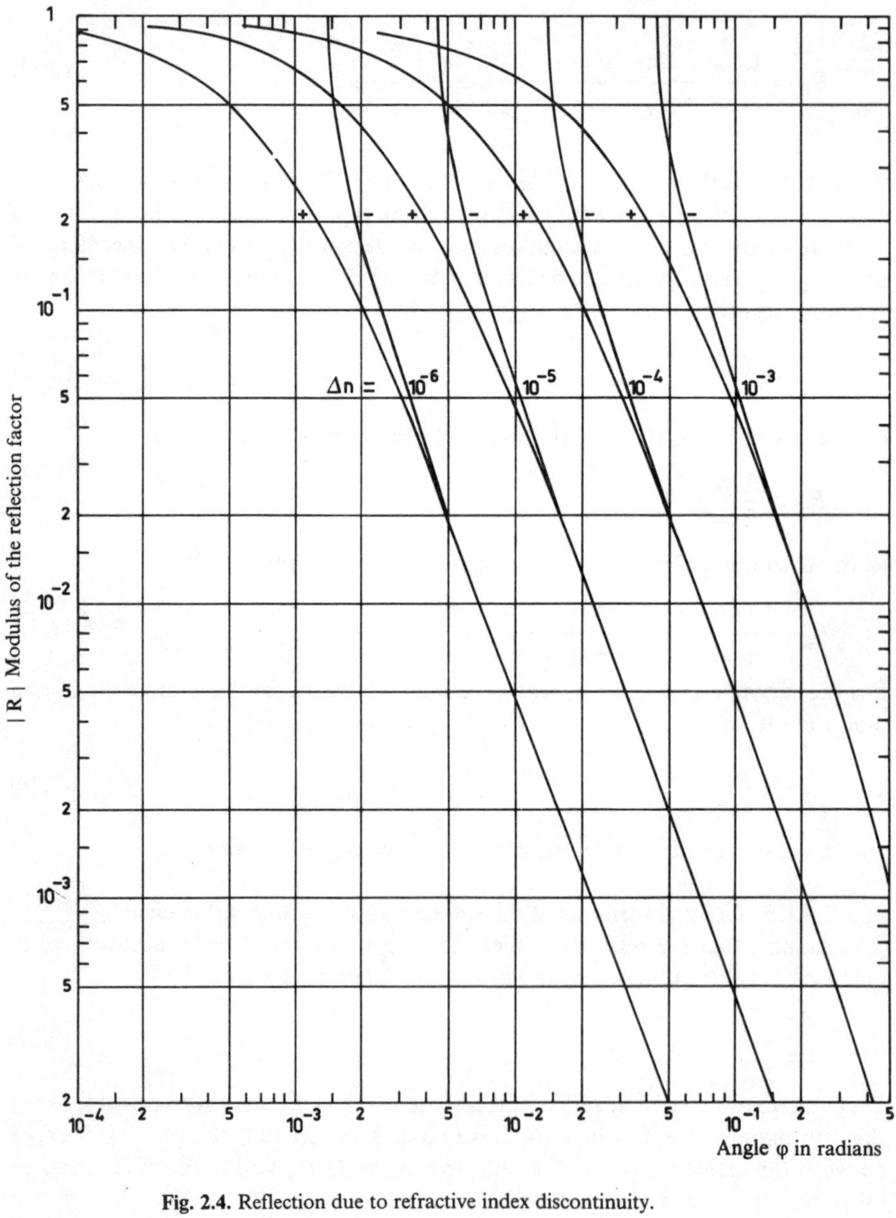

Fig. 2.4. Reflection due to refractive index discontinuity.

$$R = \frac{1}{2\,\varphi^2} \int_{h_1}^{h_1 + \Delta h} f(h) \exp\left(-\,\mathrm{j}\,4\,\pi\,\frac{h}{\lambda}\sin\varphi\right) \mathrm{d}h\,. \tag{2.38}$$

If n is a linear function of h:

$$f(h) = \frac{\Delta n}{\Delta h} = \text{Cte} \tag{2.39}$$

and therefore:

$$|R| = \frac{\Delta n}{2\,\varphi^2} \left| \frac{\sin\left(\dfrac{2\,\pi\,\Delta h}{\lambda}\sin\varphi\right)}{2\,\pi\,\dfrac{\Delta h}{\lambda}\sin\varphi} \right|. \tag{2.40}$$

It is clear that a gradual change in the refractive index throughout a layer gives a reflection factor whose modulus is less than it would be in the case of a sudden transition, which may even vanish under certain conditions or for particular frequencies.

2.4.1.6. Media in which the refractive index varies continuously

If the propagation medium shows no discontinuity but only a continuous variation in refractive index, the paths followed are not straight lines and once again the phenomenon is called refraction. If the equal index surfaces are parallel planes, Snell's law still applies but it has to be expressed as a differential equation. The quantity $n \cos \varphi$ remains constant along the whole of the curved path in question.

If the equal index surfaces are not parallel planes, Snell's law has to be replaced by another which can be derived from Fermat's principle by writing that the value of the integral:

$$\int_a^b n\,\mathrm{d}s \tag{2.41A}$$

is stationary. For example, in cases where the equal index surfaces are concentric spheres, if we use polar coordinates r, θ, the problem becomes one of making the integral:

$$\int_a^b n(r)\sqrt{1 + r^2\,\theta'^2}\,\mathrm{d}r\,. \tag{2.41B}$$

stationary. As the function under the integral sign does not depend on θ, Euler's condition immediately gives:

$$n(r)\,\frac{r^2\,\theta'}{\sqrt{1 + r^2\,\theta'^2}} = \text{Cte} \tag{2.42A}$$

which can also be written as:

$$nr \sin V = \text{Cte} \tag{2.42B}$$

where V is the angle between the tangent to the path and the radial vector (a more elementary demonstration is given in Chapter 4).

2.4.2. Reflection and refraction between a dielectric medium and an 'imperfect' conducting medium

This is a very common case in propagation, since the Earth's surface separates two types of medium. The atmosphere can be regarded for all practical purposes as a vacuum, and the Earth is characterized by its relative permittivity ε and its conductivity σ. It is well known that if we introduce a complex permittivity:

$$\eta = \varepsilon - \mathrm{j}\, 60\, \sigma\lambda \tag{2.43}$$

the equations for reflection and refraction take the same form as in the previous case, but the reflection factor and the angle of refraction take complex values which have to be interpreted physically.

The complex nature of the reflection factor means that reflection introduces a certain phase difference other than 0^0 or 180^0. Calculation of this factor therefore gives both the amplitude and the relative phase of the reflected waves. If the incident wave shows rectilinear polarization, so will the reflected wave. In vertical polarization the modulus of the reflection factor goes through a non-zero minimum for a certain angle of incidence. By analogy with the dielectric medium case, this is called the pseudo-Brewster angle of incidence.

The fact that the angle of refraction is complex may be interpreted as follows. It is known that a plane wave in a dielectric medium may be characterized by a vector $\mathbf{q}$ perpendicular to the equal phase surfaces (called the wave fronts), and θ_2 then represents the angle that this vector makes with the normal to the separation surface between the two media. If the conductivity of the medium is non-zero, there also exists a second vector $\mathbf{p}$ perpendicular to the equal amplitude surfaces, which are, in the case of refraction, parallel to the separation surface. The wave is then said to be disassociated or heterogeneous. In general, the two vectors $\mathbf{p}$ and $\mathbf{q}$ are not identical, and θ_2 then represents the angle formed with the normal by the complex vector:

$$\mathbf{c} = \mathbf{q} + \mathrm{j}\mathbf{p}\,. \tag{2.44}$$

The real part of this vector gives the phase velocity v:

$$v = \frac{\omega}{q} \tag{2.45}$$

and the imaginary part gives the attenuation α perpendicular to the surface:

$$\alpha = e^{-pz}. \tag{2.46}$$

If the incident wave is perpendicular to the separation surface, the two vectors **p** and **q** are collinear, and the refractive wave has the same structure as a plane wave, except that fields **E** and **H** are not in phase and there is a gradual attenuation. If ν and χ denote the real and imaginary parts of the refractive index, i.e.:

$$\sqrt{\eta} = \nu + j\chi, \tag{2.47}$$

we obtain:

$$q = \frac{2\pi}{\lambda}\nu \tag{2.48}$$

$$p = \frac{2\pi}{\lambda}\chi \tag{2.49}$$

where λ is the wavelength in the vacuum.

If the incident wave is no longer perpendicular to the surface but ε is large, it can be shown that Equations (2.48) and (2.49) are still approximately true, since there is only a small angle between vectors **p** and **q**. In other words, if ε is large, the refracted wave enters the conducting medium at almost right angles to the surface. This is the case in particular for refraction of a radio wave by water, for which ε is close to 80 (it is approximately 2 for light waves).

2.4.3. Depth of penetration and wavelength in the conducting medium

Consider Equations (2.48) and (2.49) once again, which are exactly true for normal incidence and approximately true for any incidence when ε is relatively large, two other quantities can be defined. On the one hand, the wavelength λ_m in the conducting medium:

$$\frac{\lambda_m}{\lambda} = \frac{1}{\nu} \tag{2.50}$$

and on the other hand the depth of penetration, i.e the distance l_p at the end of which the amplitude is reduced by a factor *e*:

$$\frac{l_p}{\lambda} = \frac{1}{2\pi\chi}. \tag{2.51}$$

In Equation (2.43) suppose that:

$$\frac{60\,\sigma\lambda}{\varepsilon} = \tan\delta, \tag{2.52}$$

where Δ can be treated as the medium's loss angle. A simple calculation allows us to obtain:

$$\nu - j\chi = \sqrt{\frac{\varepsilon}{\cos\delta}} \exp\left(-j\frac{\delta}{2}\right) \tag{2.53}$$

and therefore:

$$\nu = \sqrt{\frac{\varepsilon}{\cos\delta}} \quad \cos\frac{\delta}{2} \tag{2.54}$$

$$\chi = \sqrt{\frac{\varepsilon}{\cos\delta}} \quad \sin\frac{\delta}{2}. \tag{2.55}$$

Equations (2.50) and (2.15) allow λ_m and l_p to be deduced. It is worth calculating the limiting values of these expressions for cases where the wavelength λ becomes very large. In this case δ is close to π/2, and therefore ν and the χ have approximately equal very large values. Consequently, λ_m and l_p are very small. It is easy to show that (in MKSA units)

$$\lambda_m \simeq \sqrt{\frac{\lambda}{30\,\sigma}} \tag{2.56}$$

$$l_p \simeq \frac{1}{2\,\pi}\sqrt{\frac{\lambda}{30\,\sigma}} \tag{2.57}$$

and therefore:

$$l_p = \frac{\lambda_m}{2\,\pi}. \tag{2.58}$$

We can see that permittivity ε is no longer a factor for very long waves. This can also be concluded directly from Equation (2.43) which shows that any medium behaves as a conductor for very long wavelengths.

2.5. Diffraction

If a wave being propagated in one medium meets a surface separating separating that medium from another, continuity is only preserved by the appearance on the surface of conduction or displacement currents in that surface. These currents in turn create radiation in addition to the incident radiation.

We have seen that, in the case of a plane and infinitely large separation surface, the waves created in this way (by reflection or refraction) obey very simple laws and may be studied using the concept of a ray. If the separation surface is small with respect to the wavelength or if it contains sharp edges or other irregularities, the secondary waves created by the surface no longer obey such simple laws. In order to obtain the laws which apply, we have, in principle, to establish the boundary conditions, which can only be found analytically for a small number of surfaces. The solution obtained, called the diffracted wave, exists throughout space and in particular in the region 'behind' the obstacle, in which the methods of optics would give a zero field. This has given rise to the use of the term diffracted wave only for those waves which cannot be explained by optics.

As it is difficult to calculate diffracted waves, an attempt has only been made in certain cases. In the field of wave propagation two cases are particularly interesting, that of the half plane and that of the sphere. Diffraction by half planes is studied in optics by purely scalar methods without taking polarization into account, which leads to the use of Fresnel integrals and Cornu's spiral. The approximate solutions obtained are generally sufficiently good even for radio waves (see Chapter 7).

Diffraction by a sphere is studied above all in two special cases, that of a point source close to a sphere of very large radius (e.g. diffraction of a wave by the Earth), and that of a plane wave falling on a sphere whose radius is of the same order of magnitude as the wavelength, or smaller (e.g. diffraction by raindrops).

These various cases are examined when they arise in this study of propagation. Here we simply comment on Babinet's transformation because of its usefulness in resolving certain diffraction problems. Babinet's transformation involves replacing a plane screen including full and empty areas by the complementary screen in which the full and empty areas have been interchanged. The fundamental theorem is as follows.

2.5.1. Babinet's theorem

Babinet established the principle in 1837 that at any point the sum of light waves diffracted by two complementary screens whose full areas are perfectly absorbing is equal to the light wave which would have existed in the absence of the screen. This principle was of great value to optics, but when it was attempted to deduce it from Maxwell's equations, several difficulties emerged:

1. Babinet's theorem only concerns scalar quantities, whereas Maxwell's equations use vector quantities.
2. The concept of a perfectly absorbing screen, i.e. a 'black body', is a physical approximation, but cannot be expressed mathematically by boundary conditions on a surface.
3. It is reasonable to look for a relation involving a screen which is a perfect conductor of electricity, as such a screen may be built to within an excellent approximation, but in this case, the complementary screen would have to be perfectly conducting to magnetism, a condition which can be expressed mathematically but cannot be reproduced physically.

All these difficulties have been overcome and it is possible to state Babinet's theorem, deduced from Maxwell's equations, which involves only screens that are infinite conductors of electricity, on condition that two permutations are carried out simultaneously. At the same time as we replace the screen by its complement, we replace the incident electromagnetic sources and fields by conjugate sources and fields, i.e. such that the electric and magnetic fields are reversed. Under these conditions the field $E_2 H_2$ diffracted by the second screen can be deduced from the field $E_1 H_1$ diffracted from the first, using the relations:

$$E_2 = \pm \sqrt{\frac{\mu}{\varepsilon}} H_1 \qquad H_2 = \mp \sqrt{\frac{\varepsilon}{\mu}} E_1 , \tag{2.59}$$

with the change in sign taking place on crossing the screen.

2.5.2. Application: reflection by a limited part of a surface

This theorem can be used to transform one problem into another, which is either simpler or for which a solution is already known. For example the radiation from a slit is the same as that from a dipole, if the electric and magnetic fields are inverted. In the same way, we can deduce the field diffracted from a hole from that diffracted by a disk.

Babinet's theorem can be applied even if reflection takes place, on condition that the reflected path is 'rectified'. For example, diffraction produced by an area of the surface whose size is not very great with respect to the wavelength may be deduced from the diffraction through a hole replacing exactly that section of surface. This property can be applied to the study of reflection of waves by a limited reflecting zone on the Earth's surface (see Section 3.6).

2.6. Scatter

In certain cases the path of the wave includes obstacles whose dimensions are of the same order of magnitude or smaller than the wavelength, but for which the number per unit volume is very great. We can therefore assume that each of these obstacles creates a diffracted wave, and we have to obtain the sum of all those waves. Nevertheless, if the number of these obstacles is sufficiently great, their effects may be considered globally and statistically. This phenomenon is termed scatter. It was studied in optics by Lord Rayleigh for dielectric spheres of very small diameter, and then for radio waves by Mie using homogeneous spheres of finite conductivity, and also by Oguchi for ellipsoids (raindrops) (see Chapter 4).

Given the small size of the obstacles, the concept of a shadow zone does not apply, and scattering is studied throughout space, including in the backward direction, e.g. for radar echoes.

Scattering may be produced not only by obstacles with clearly defined surfaces but also by irregularities in the refractive index of the medium. This is the case of tropospheric and ionospheric scatter which are described in Chapter 8.

Another approach to the problem is possible if the 'obstacles' are particles that are extremely small with respect to the wavelength, and whose number within a cube with a side one wavelength long is extremely large. The medium can then be treated as homogeneous, but with properties depending on the particles present. This, for example, is the case for the ionosphere described in Chapter 5.

2.7. Specular reflection and diffuse reflection: Rayleigh's criterion

Reflection as studied at the beginning of this chapter assumed the existence of two media separated by a surface whose dimensions are very large and whose irregularities are very small with respect to the wavelength. This type of reflection is often termed specular reflection.

If the surface separating the two media has random irregularities, i.e. it is 'unpolished', there is still an energy maximum in the direction of the reflected ray, but there is also energy in nearby directions. This phenomenon is termed diffuse reflection. If the irregularities are not random but show a regular distribution, a further effect is obtained, and there are several directions in which energy maxima appear.

Irregularities in the surface can be taken into account as follows. If Δh denotes the height of an irregularity in the reflection zone and φ denotes the angle between the incident ray and the mean surface (Fig. 2.5), the difference from the path introduced by irregularities is equal to $2\Delta h \sin \varphi$, and this quantity must be compared with the wavelength to calculate phase differences.

We therefore consider the quantity:

$$g = \frac{4\pi \Delta h \sin \varphi}{\lambda} \tag{2.60}$$

The heights of irregularities are generally distributed according to a statistical law, and it is usually accepted that distribution is Gaussian. This means that $4\pi\Delta h/\lambda$ is then defined as the standard deviation of the distribution of irregularity heights

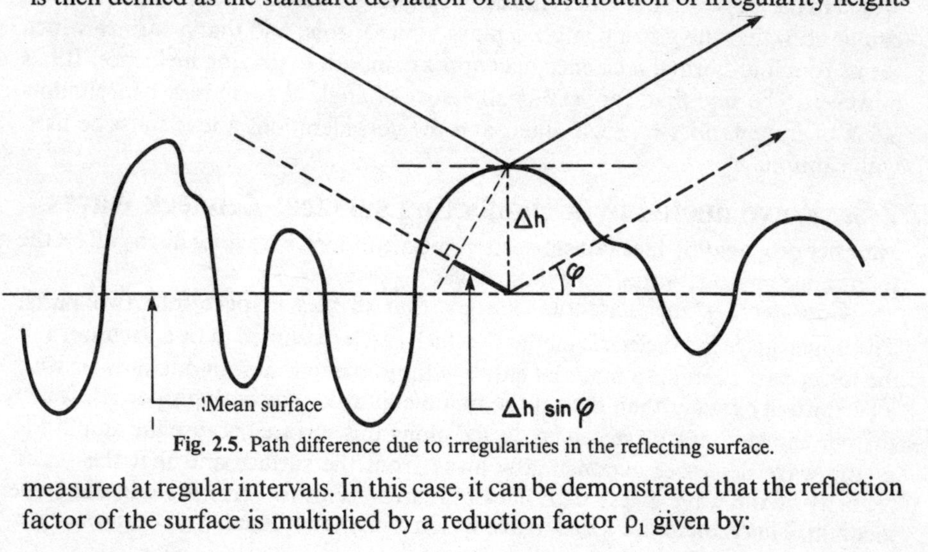

Fig. 2.5. Path difference due to irregularities in the reflecting surface.

measured at regular intervals. In this case, it can be demonstrated that the reflection factor of the surface is multiplied by a reduction factor ρ_1 given by:

$$\rho_1 = \exp(-\tfrac{1}{2}g^2) \tag{2.61}$$

The values calculated in this way conform to observed values if g is very small, but they are too low if g is greater than about 0.1 especially for paths over the sea. Better modelling gives the expression:

$$\rho_2 = \exp(\tfrac{1}{2}g^2).\ \mathrm{I}_0(\tfrac{1}{2}g^2) \tag{2.62}$$

where I_0 is a modified Bessel function of order zero.

In practice this expression may be replaced by the approximation:

$$\rho_2 = \sqrt{\frac{1 + (\tfrac{1}{2}g^2)}{1 + 2.35(\tfrac{1}{2}g^2) + 2\pi(\tfrac{1}{2}g^2)^2}} \tag{2.63}$$

We then obtain the following values which can only be considered as approximate values since the heights of irregularities do not always follow the model chosen.

$\dfrac{\Delta h \sin\varphi}{\lambda}$	ρ_1	ρ_2
$^1/_{100} = 0.010$	0.992	0.995
$^1/_{32} = 0.031$	0.926	0.942
$^1/_{16} = 0.063$	0.735	0.753
$^1/_8 = 0.125$	0.291	0.408
$^3/_{16} = 0.188$	0.0623	0.260
$^1/_4 = 0.25$	0.0072	0.189

Table 2.2. Relationship between the heights of the irregularities and the reduction factor.

A comparison of the two quantities $\Delta h \sin\varphi$ and λ gives Rayleigh's criterion which makes it possible to assess the degree of irregularity of a reflecting surface. It can be seen that the grazing angle φ plays a major role, and that a surface which seems rough to normal incidence may appear smooth to grazing incidence. It has however to be said that, for very small values of angle φ, there may be a shadow effect of irregularities on each other, and the considerations above must be used with caution.

2.8. Wave ducted by a conducting surface: Zenneck waves

Another propagation mechanism needs to be mentioned here since it may affect the long range propagation of very long waves.

Consider a plane horizontal surface that divides a space into two parts. The upper space is a dielectric medium, which can be assumed to be a vacuum, and the lower part contains a material either with non-zero finite conductivity or with a permittivity greater than that of the first medium, or has both properties. It can then be shown that a wave can be ducted along this surface because the amplitude of the wave decreases exponentially away from the surface and that the phase velocity of the wave is less than it would have been for a free wave in the same medium. This is therefore a slow wave which remains confined close to the surface and is called a surface wave.

Surface waves have many applications in link antenna technology, where ducting is provided by materials with the lowest possible degree of loss. In the propagation of waves close to the Earth, there may also be a surface wave along the Earth's surface, due principally to the finite conductivity of the Earth. Consequently, the wave shows exponential attenuation not only in a direction perpendicular to the surface (because of ducting) but also along the surface (because of losses). The theory behind these effects was developed by Zenneck in 1907. It should be noted that the properties of the surface wave are very different from those of the wave due to diffraction around the Earth, often called the ground wave, which is neither a slow wave or a guided wave (see Chapter 7).

The so-called Zenneck wave is characterized by a horizontal magnetic field (parallel to the surface) and a vertical electric field in the plane perpendicular to the magnetic field. It can be shown that the electic field has a vertical component E_z and a horizontal component E_x in the direction of propagation. The ratio of the amplitudes of these two components is given by:

$$\frac{E_z}{E_x} = \sqrt{\varepsilon - \mathrm{j}\,60\,\sigma\lambda} = n\,. \tag{2.64}$$

Since this ratio is complex, the electric field polarization is elliptical.

If the loss angle δ is introduced into the surface, Equation (2.52), the components E_x and E_z are given by:

$$E_x = E_0 \sqrt{\frac{\cos\delta}{\varepsilon}} \cos\left(\omega t + \frac{\delta}{2}\right) \tag{2.65}$$

$$E_z = E_0 \cos\omega t\,. \tag{2.66}$$

The horizontal component leads the vertical component in phase, and the polarization ellipse is generally very elongated vertically, and inclined in the direction of propagation.

The propagation coefficients are equal to:

along the surface

$$\frac{2\,\pi\mathrm{j}}{\lambda} \sqrt{\frac{n^2}{1+n^2}} \tag{2.67}$$

perpendicular to the surface

$$\frac{2\,\pi\mathrm{j}}{\lambda} \frac{1}{\sqrt{1+n^2}} \tag{2.68}$$

It can also be demonstrated that the equal phase planes are orthogonal to the equal amplitude planes.

Advantage is taken of the existence of a horizontal component to the electric field for the reception of long waves using the 'Beverage antenna'. This antenna consists of a very long horizontal wire a few metres above the ground and digitized in the direction of the transmitter.

Chapter 3

The effect of the Earth's surface on propagation: line of sight paths

The Earth's surface is involved in wave propagation by either reflection in the case of line of sight links or diffraction phenomena. Wave propagation within the Earth's surface (e.g. communications with submarines or mine working) is also important. Diffraction phenomena are studied in Chapter 7.

3.1. Electric characteristics of the Earth's surface

As already described in Chapter 2 the characteristics of a non-magnetic medium are defined by its complex permittivity:

$$\eta = \varepsilon - \mathrm{j}\, 60\, \sigma\lambda\,. \tag{3.1}$$

In the case of the Earth's surface ε and σ depend almost entirely on the degree of humidity at the surface: in the radio wave range, ε is of the order of a few units for most rock types but it is about 80 for water. The value of σ is generally very low for rock. For pure water conductivity is roughly proportional to frequency (dipole conductivity). However, the presence of humidity in the ground produces solutions of the salts which can considerably increase conductivity at low frequencies (ionic conductivity). Beyond about 10GHz the total conductivity of water has reached a limit, whereas permittivity decreases rapidly. This information is given in Fig. 3.1 (CCIR).

It can be seen from Equation (3.1) that for any given type of ground there is a transition frequency for which the real and imaginary parts of η are equal. Below this frequency, the ground behaves for all practical purposes like a conductor, while above it, it behaves like a dielectric medium. The transition frequency is shown in the curves in Fig. 3.1, and is 30GHz for pure water and 1GHz for sea water. For any non-maritime surface it is between 0.6 and 6MHz and for ice it is around 100kHz.

3.2. Surface reflectivity

If we assume that the Earth's surface is plane and homogeneous, its reflectivity will be as defined and calculated in Chapter 2. We know that it depends on the wave's polarization and on the angle of reflection and can be expressed by:

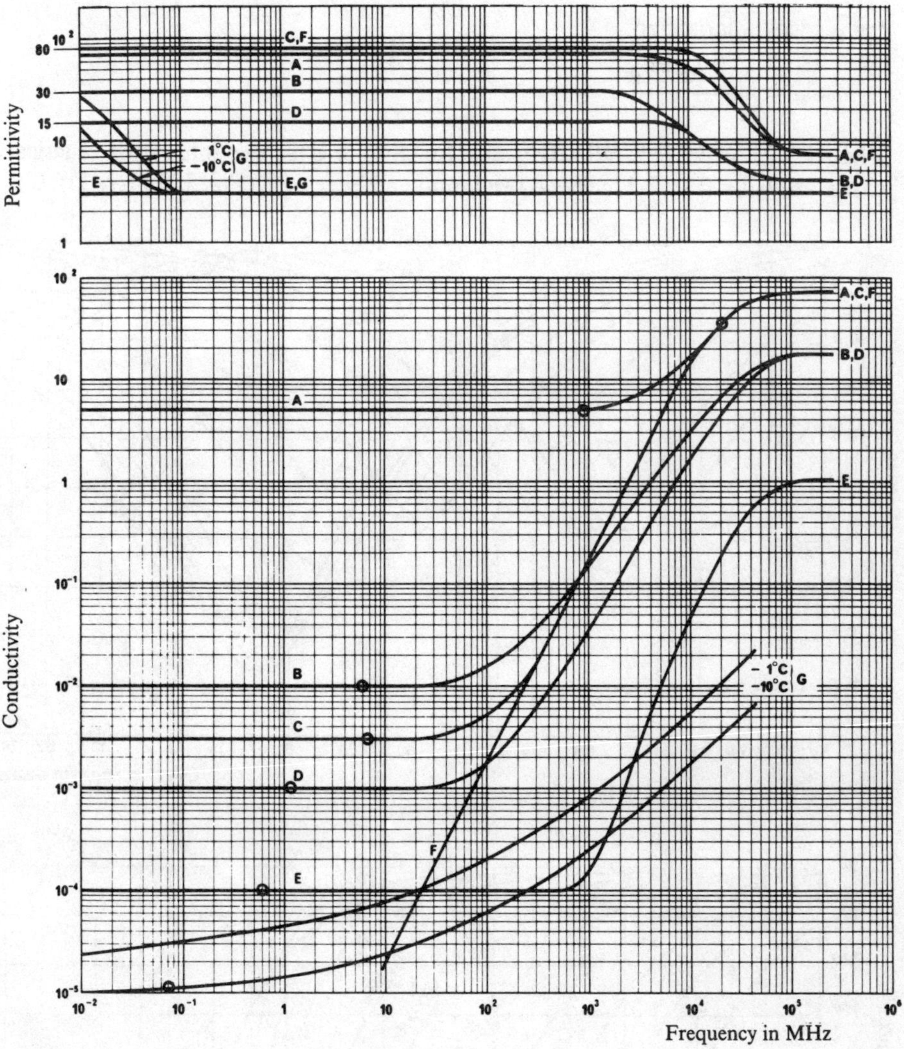

Fig. 3.1 Permittivity ε and conductivity σ of the Earth's surface: A: sea water (mean salinity), 20°C; B: humid soil; C: fresh water, 20°C; D: moderately dry soil; E: very dry soil; F: pure water, 20°C; G: ice (fresh water); ring: transition frequency.

$$R = \frac{\sin\varphi - \sqrt{C}}{\sin\varphi + \sqrt{C}} \tag{3.2}$$

if we introduce a complex function C depending on polarization, which has the two following values:
for horizontal polarization

$$C_H = \eta - \cos^2\varphi\,, \tag{3.3}$$

for vertical polarization

$$C_V = \frac{\eta - \cos^2 \varphi}{\eta^2}. \tag{3.4}$$

Figures 3.2 and 3.3 give reflectivity as a function of φ for two interesting practical cases, sea water (Fig. 3.1, curve A) and moderately dry soil (Fig. 3.1, curve D). It is

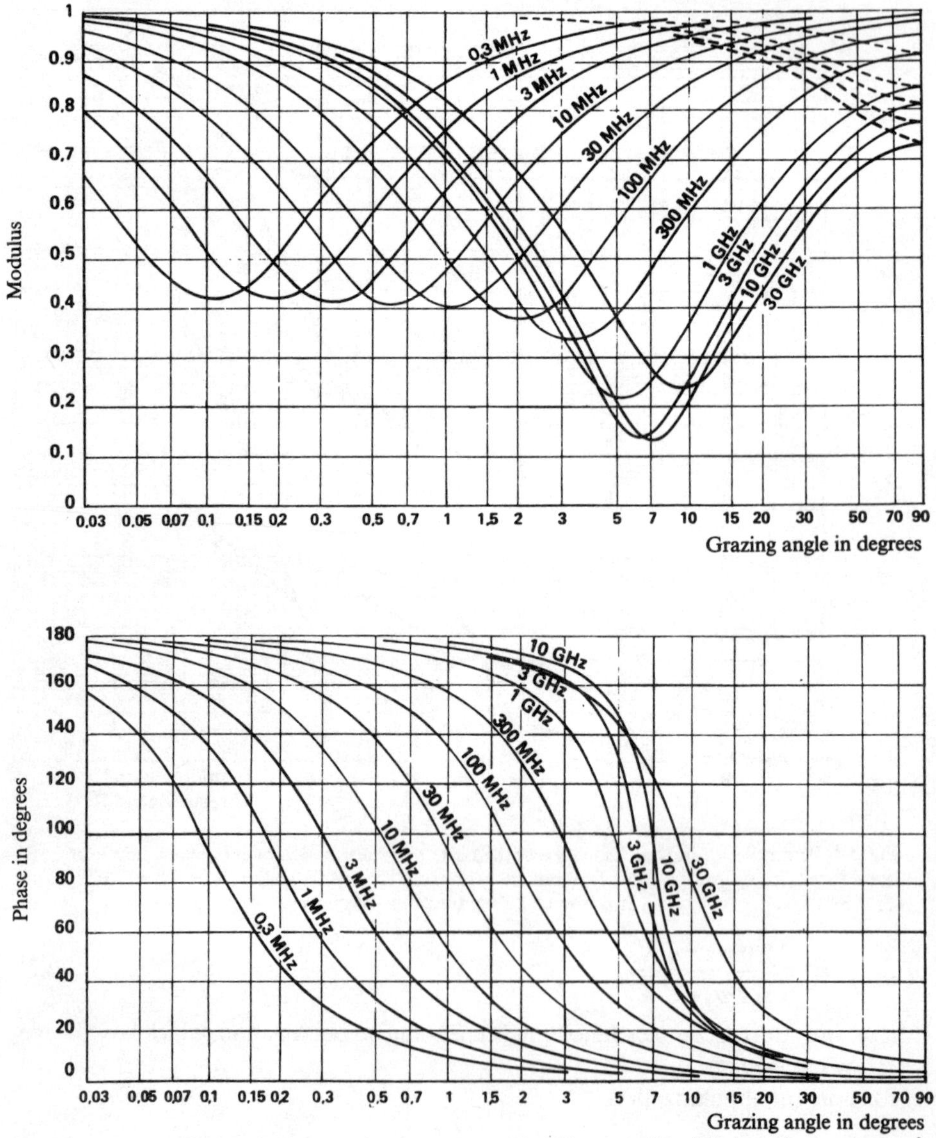

Fig. 3.2. Sea water reflectivity: – horizontal; —— vertical polarization.

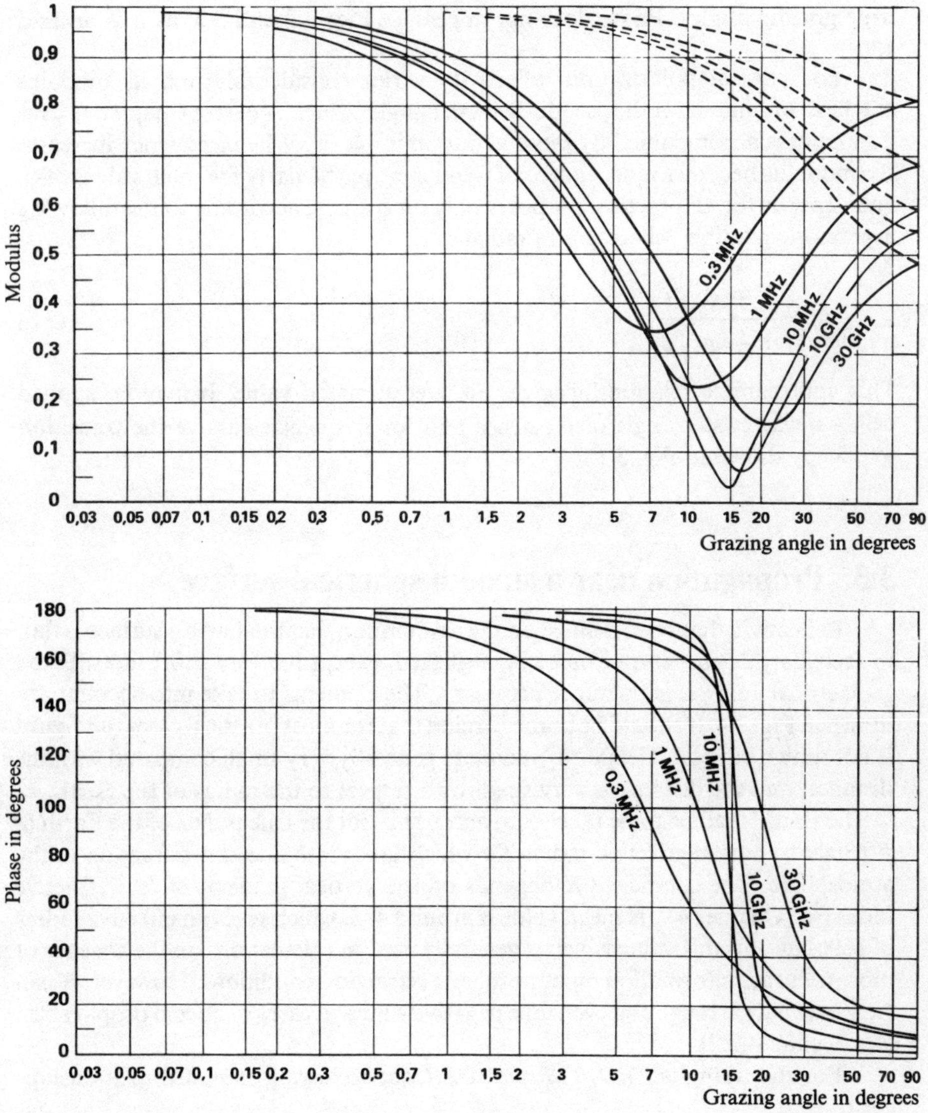

Fig. 3.3. Soil reflectivity (moderately dry soil): – horizontal polarization; —— vertical polarization.

clear that for very small values of φ, the reflectivity is around –1 (modulus 1, phase 180°) whatever the degree of polarization or the nature of the surface. In other words, when φ approaches zero, reflection involves phase inversion.

(The reflection factor is associated with a plane wave, and not, for example, with radiation from an antenna vertical to the ground. See Equation (3.34) and the note at the of Section 3.5)

For horizontal polarization reflectivity remains around –1 even for relatively large grazing angles. Phase does not appear in Figs 3.2 and 3.3, as it is around 180°.

For vertical polarization reflectivity varies considerably and its modulus reaches a minimum for the pseudo-Brewster angle of incidence (see Chapter 2). This minimum is accompanied by rapid variation in phase. When frequency increases the modulus becomes independent of frequency, particularly for small values of φ, and depends for all practical purposes only on φ and ε, according to the following asymptotic relation, where φ is in radians:

$$R \simeq \frac{1 - \varphi\sqrt{\varepsilon}}{1 + \varphi\sqrt{\varepsilon}}. \tag{3.5}$$

This approximate relationship gives an overestimated value. It may be applied below the Brewster angle of incidence and for frequencies above the transition frequency shown in Fig. 3.1.

3.3. Propagation near a smooth spherical surface

Reflectivity has been defined on the assumption that the Earth's surface is flat. In fact, Earth's curvature cannot be neglected, except for very short links. Some geometrical analysis is therefore necessary. The elements to take into account are shown in Fig. 3.4. It should be borne in mind that for most practical cases this figure is not drawn to scale. Heights h_1 and $h_2$are generally very small compared with the distance, and the distance is very small with respect to the radius of the Earth.

The Earth's radius to be taken into account is not the true radius of the Earth (a = 6360km) but an effective radius Ka which takes into account refraction in the atmosphere. The coefficient K depends on the vertical gradient of the refractive index (see Chapter 4). Its mean value is around 4/3, which gives an effective radius of 8500km. It is this value which is generally used in calculations, in the absence of more accurate information on atmospheric refraction conditions. However, K can vary over a wide range and even take negative values in cases of ducted propagation (see Section 4.2.2).

For the triangles TPO, TPR and OPR, the following standard relationships exist:

$$\frac{l_1}{\sin \alpha_1} = \frac{Ka + h_1}{\cos \varphi} = \frac{Ka}{\cos(\varphi + \alpha_1)}, \tag{3.6}$$

$$\frac{l_2}{\sin \alpha_2} = \frac{Ka + h_2}{\cos \varphi} = \frac{Ka}{\cos(\varphi + \alpha_2)}, \tag{3.7}$$

$$d_1 = Ka\alpha_1, \tag{3.8}$$

$$d_2 = Ka\alpha_2, \tag{3.9}$$

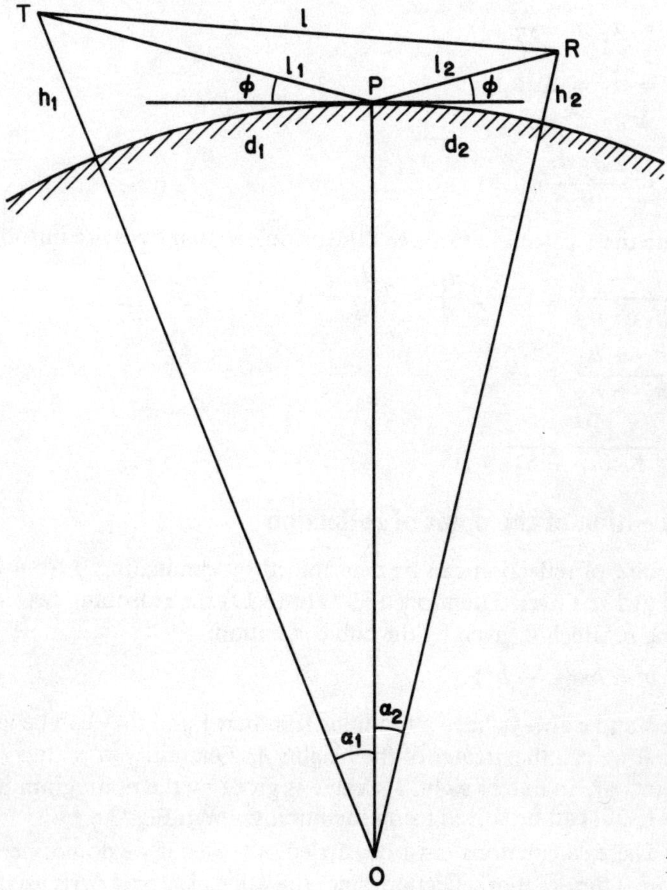

Fig. 3.4. Geometric factors affecting a reflected ray.

$$l^2 = l_1^2 + l_2^2 + 2\, l_1\, l_2 \cos 2\,\varphi\,. \tag{3.10}$$

Equation (3.10) can be rewritten to include a term showing the path difference Δ between the direct and the reflected ray:

$$\Delta = l_1 + l_2 - l = \frac{4\, l_1\, l_2 \sin^2 \varphi}{l_1 + l_2 + l}. \tag{3.11}$$

Let us take the general case where α_1, α_2, and φ are small angles. The geometric relations then become:

$$\frac{h_1}{Ka} = \alpha_1\, \varphi + \frac{\alpha_1^2}{2}, \tag{3.12}$$

$$\frac{h_2}{Ka} = \alpha_2 \varphi + \frac{\alpha_2^2}{2}, \tag{3.13}$$

$$l_1 = d_1 = Ka\alpha_1 , \tag{3.14}$$

$$l_2 = d_2 = Ka\alpha_2 , \tag{3.15}$$

$$\Delta = \frac{2\, d_1\, d_2}{d} \varphi^2 . \tag{3.16}$$

To complete these calculations three dimensionless quantities are introduced:

$$b = \frac{d_1 - d_2}{d_1 + d_2} = 1 - 2\frac{d_2}{d} = 2\frac{d_1}{d} - 1 , \tag{3.17}$$

$$c = \frac{h_1 - h_2}{h_1 + h_2}, \tag{3.18}$$

$$m = \frac{d^2}{4\, Ka(h_1 + h_2)} . \tag{3.19}$$

3.3.1. Location of the point of reflection

The point of reflection can be determined by eliminating φ from Equations (3.12) and (3.13). Given Equations (3.17) to (3.19) the reasoning here reduces to determining b, which is given by the cubic equation:

$$c = b + bm(1 - b^2) . \tag{3.20}$$

Quantities b and c always have a modulus less than 1 and they can be assumed to be positive if we call the greater of the heights h_1. Quantitity m is always less than 1 if the points are in line of sight. Its value is given by the nomogram in Fig. 3.5. Equations (3.20) can be solved using the nomogram in Fig. 3.6.*

N.B. These calculations must be carried out, even if we do not need to know the position of the point of reflection, since the value of b (or d_1/d) is used in all later calculations.

3.3.2. Angle of incidence

This angle is given by the relation:

$$\varphi = \frac{h_1 + h_2}{d} [1 - m(1 + b^2)] . \tag{3.21}$$

The first term in this expression is the value of angle φ_0 corresponding to a flat Earth, and can be obtained from the nomogram in Fig. 3.5. The second term is therefore a reduction factor Z, due to the curvature of the Earth, given by nomogram in Fig. 3.7. If the distance between transmitter and receiver is sufficiently short, only the first term needs to be taken into account.

* This equation may also be solved by the following expression, valid for m positive:

$$b = 2\sqrt{\frac{m+1}{3\,m}} \cos\left[\frac{\pi}{3} + \frac{1}{3} \text{Arc cos}\left(\frac{3\,c}{2}\sqrt{\frac{3\,m}{(m+1)^3}}\right)\right]$$

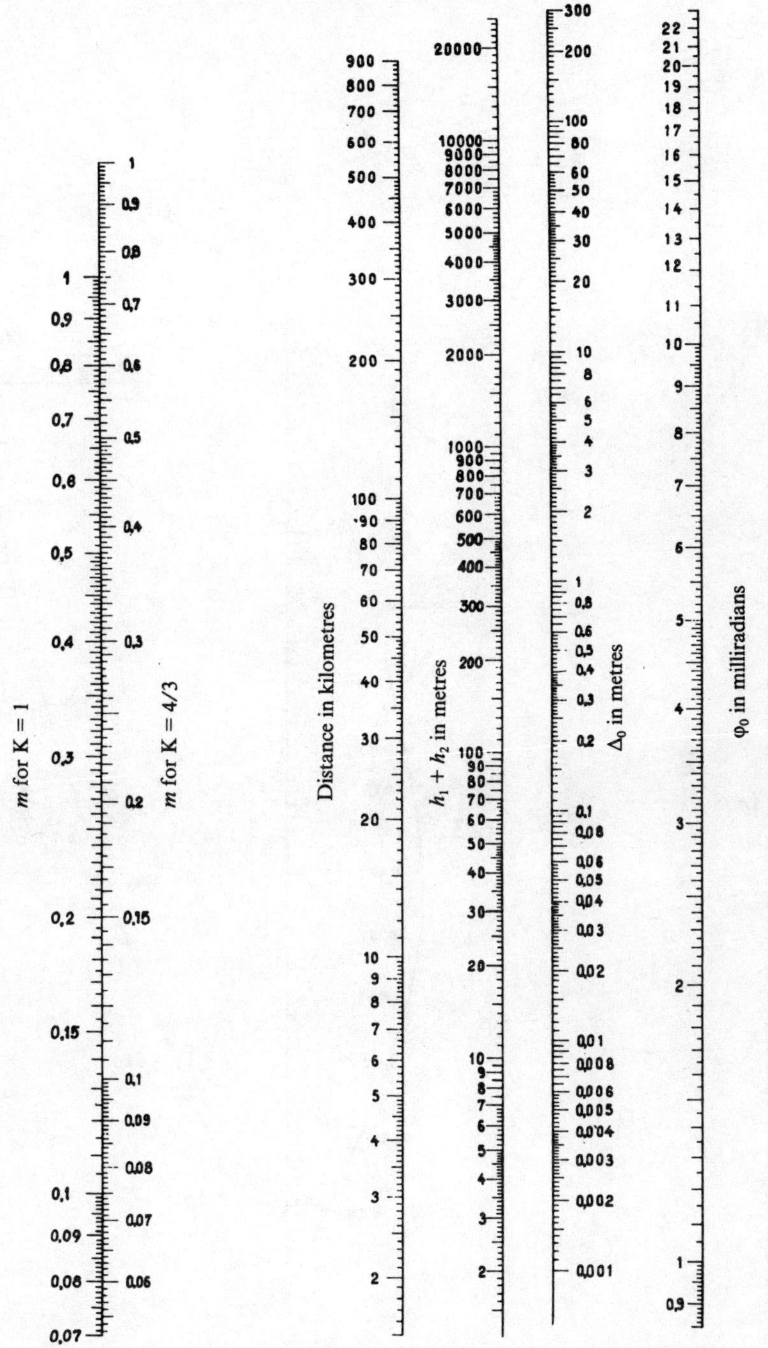

Fig. 3.5. Calculation of m, Δ_0 and φ_0.

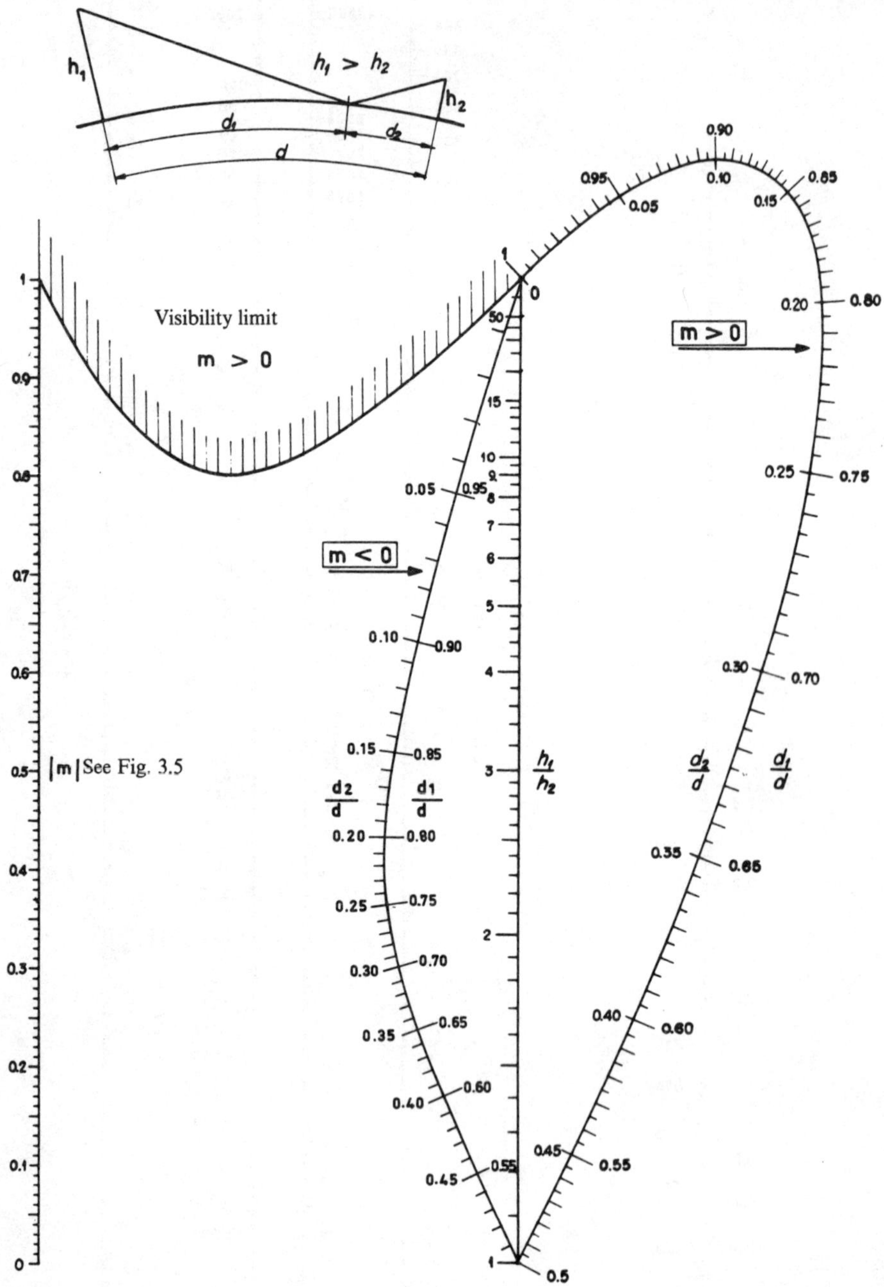

Fig. 3.6. Determining the point of reflection.

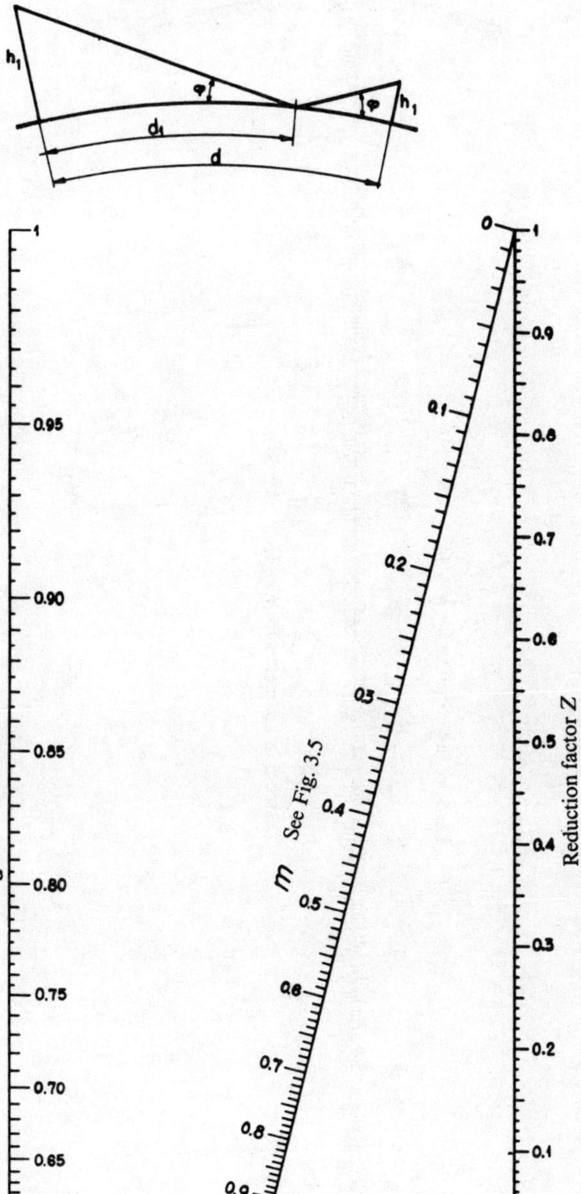

Fig. 3.7. Grazing angle reduction factor on a spherical Earth.

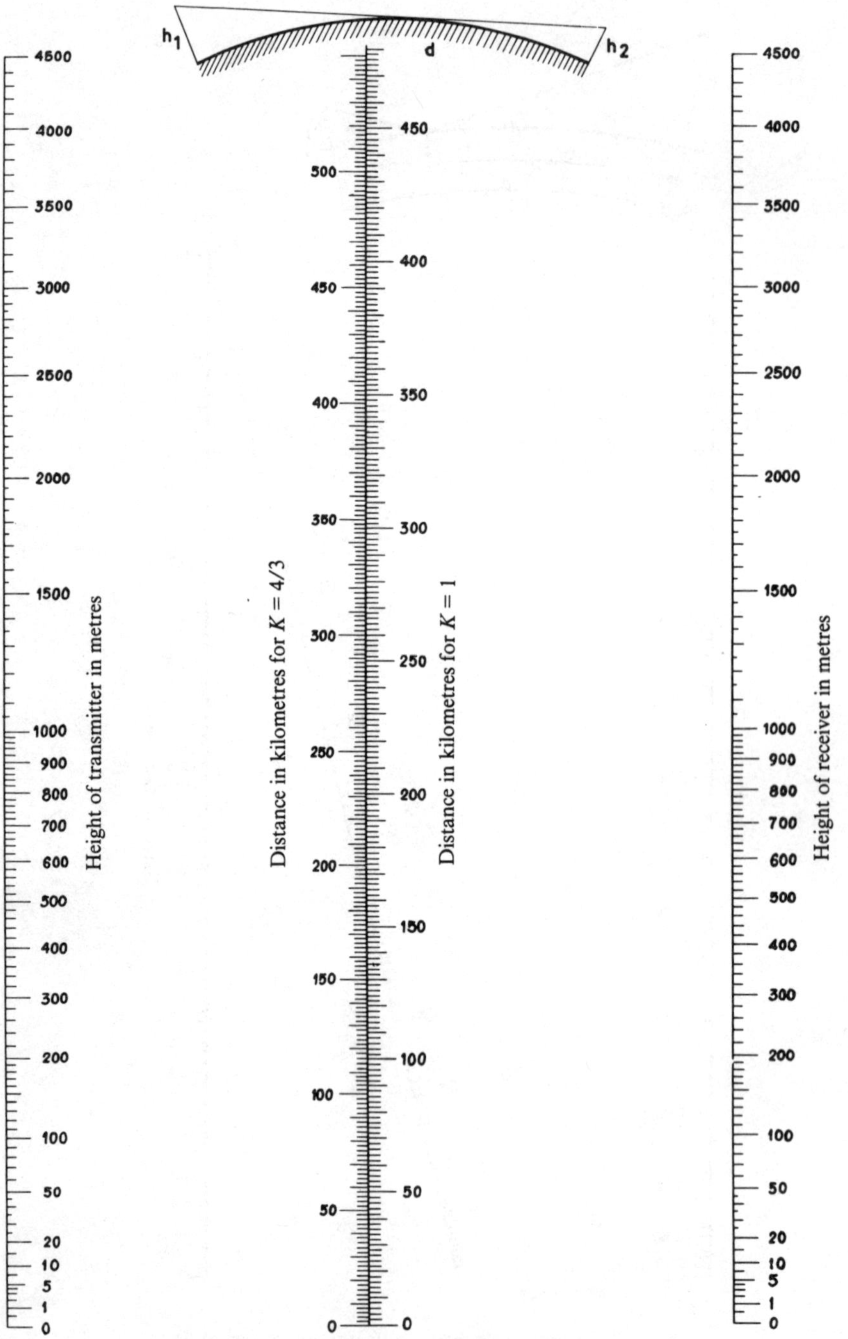

Fig. 3.8. Maximum line of sight distance.

3.3.3. Distance giving a tangent link

This distance is obtained when angle φ becomes zero:

$$m(1 + b^2) = 1 . \tag{3.22}$$

If geometric quantities are considered again, we obtain the standard equation, which can also be deduced directly:

$$d_l = \sqrt{2\,Ka}\left(\sqrt{h_1} + \sqrt{h_2}\right). \tag{3.23}$$

If heights h_1 and h_2 are expressed in metres and the distance d in kilometres, and if the atmospheric conditions are assumed to be normal, i.e. $Ka = 8500$km, we obtain the following practical formula, which can be calculated by the nomogram in Fig. 3.8:

$$d_{l(\mathrm{km})} = 4.1\left[\sqrt{h_{1\,(\mathrm{m})}} + \sqrt{h_{2\,(\mathrm{m})}}\right]. \tag{3.24}$$

3.3.4. Path difference between the direct and reflected rays

Path differences are given by Equation (3.16), which can be re-written:

$$\varDelta = 2\,d\,\frac{d_1}{d}\,\frac{d_2}{d}\,\varphi^2 , \tag{3.25}$$

this expression allows us to calculate the difference once we know φ and the position of the point of reflection. To calculate the difference by nomogram we write:

$$\varDelta = \frac{(h_1 + h_2)^2}{2\,d}(1 - b^2)\left[1 - m(1 + b^2)\right]^2 = \varDelta_0\, q \tag{3.26}$$

with:

$$\varDelta_0 = \frac{(h_1 + h_2)^2}{2\,d} \quad \text{and} \quad q = (1 - b^2)\left[1 - m(1 + b^2)\right]^2 \tag{3.27}$$

Δ_0 is given by Fig. 3.5 and q by Fig. 3.9.

If the Earth can be considered as flat, i.e. if K is very large or d is sufficiently small:

$$\varDelta \simeq \frac{2\,h_1\,h_2}{d}. \tag{3.28}$$

The path difference varies if we move vertically or horizontally, and consequently the reflected ray will sometimes reinforce and sometimes diminish the direct ray. Therefore this is an interference phenomenom which is particularly apparent if reflection takes place from the surface of the sea.

1. If the receiver moves vertically, a series of regularly spaced maxima and minima is traced (Fig. 3.10). The distance between two consecutive maxima

(or minima) is equal to:

$$\delta h_2 = \frac{\lambda d}{2\, d_1\, \varphi} = \frac{\lambda d}{2\, h_1} \cdot \frac{1 + m(1 - b^2)}{1 - m(1 + b^2)}. \tag{3.29}$$

If the distance is short enough, the Earth can be assumed to be flat, and Equation (3.29) reduces to:

$$\delta h_2 = \frac{\lambda d}{2\, h_1}. \tag{3.30}$$

2. If the receiver moves horizontally, as would be the case in an aircraft,

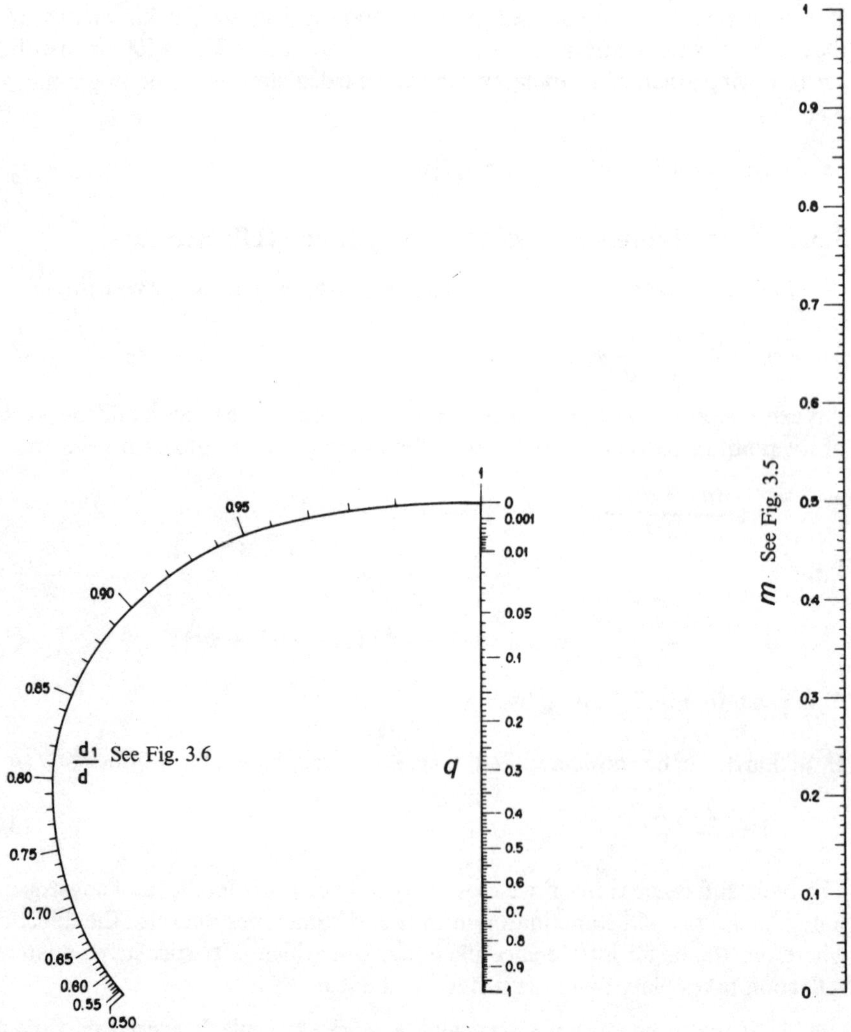

Fig. 3.9. Path difference between the direct and the reflected rays.

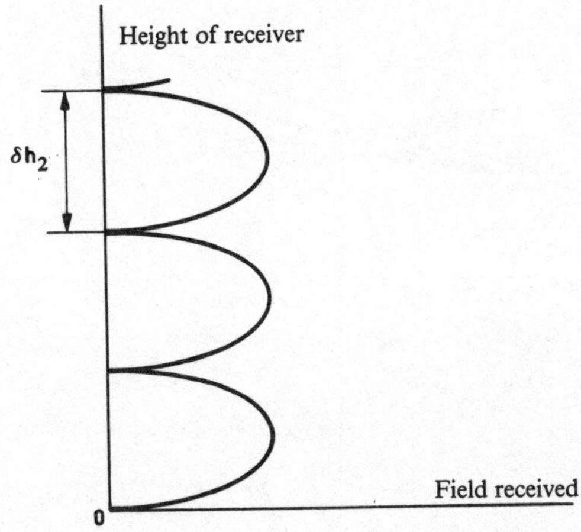

Fig. 3.10. Interference lobes.

maxima and minima are no longer regularly spaced, but close up as the transmitter is approached. The distance between two consecutive maxima or minima is approximately equal, assuming the Earth to be flat, to:

$$\delta d = \frac{\lambda d^2}{2\, h_1\, h_2}. \tag{3.31}$$

This relation is obviously inadequate as the horizon is approached, since it only applies on a flat Earth. On a spherical Earth the last maximum appears a little before the horizon, after which the field decreases very rapidly. But the method derived from optics, which we are using here, then ceases to apply. To go further the diffraction theory described in Chapter 7 must be used.

3.3.5. Divergence factor

If the area of the earth's surface involved is so large that it can no longer be regarded as flat, reflectivity has to be corrected. This is because a beam broadens after reflection from a convex spherical surface, leading to an apparent reduction in reflectivity. This means that reflectivity must be multiplied by a coefficient less than 1 called the divergence factor.

A beam of solid angle Ω_1 produces a beam of solid angle Ω_2 greater than Ω_1 after reflection (Fig. 3.11). The two beams are of area Σ_1 and Σ_2 in the plane of the receiver. As power per unit area is inversely proportional to area, the received field is multiplied by a factor D less than 1 called the divergence factor.

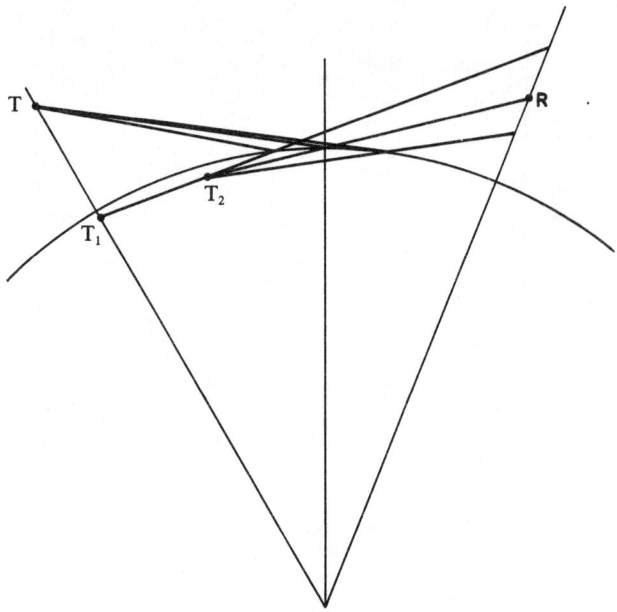

Fig. 3.11. Divergence of a beam of rays.

$$D = \sqrt{\frac{\Sigma_1}{\Sigma_2}} = \left(1 + \frac{2}{Ka \sin \varphi} \frac{l_1 l_2}{l_1 + l_2}\right)^{-1/2}. \tag{3.32}$$

Using the quantities already calculated the divergence factor is given by the following expressions:

$$D = \frac{1}{\sqrt{1 + \frac{2d}{Ka} \frac{d_1}{d} \frac{d_2}{d} \frac{1}{\varphi}}} = \sqrt{\frac{1 - m(1 + b^2)}{1 + m(1 - 3b^2)}} \tag{3.33}$$

and the factor can be either calculated directly, or derived from the nomogram in Fig. 3.12. It can be seen in particular for $m = 0.5$ that D takes the value $1/\surd 3$ independently of b. This expression, which vanishes when the grazing angle becomes zero, is not however applicable to grazing incidence, as it does not take diffraction phenomena into account. The minimum grazing angle below which calculation by optics methods is no longer possible is given by the approximate relation:

$$\varphi_{(\mathrm{mrad})} = \left(\frac{2\,100}{F_{(\mathrm{MHz})}}\right)^{1/3} = (7\,\lambda_{(\mathrm{m})})^{1/3} \tag{3.34}$$

which, by way of example, gives a limiting angle of 2.76mrad at 100MHz and 0.59mrad at 10GHz.

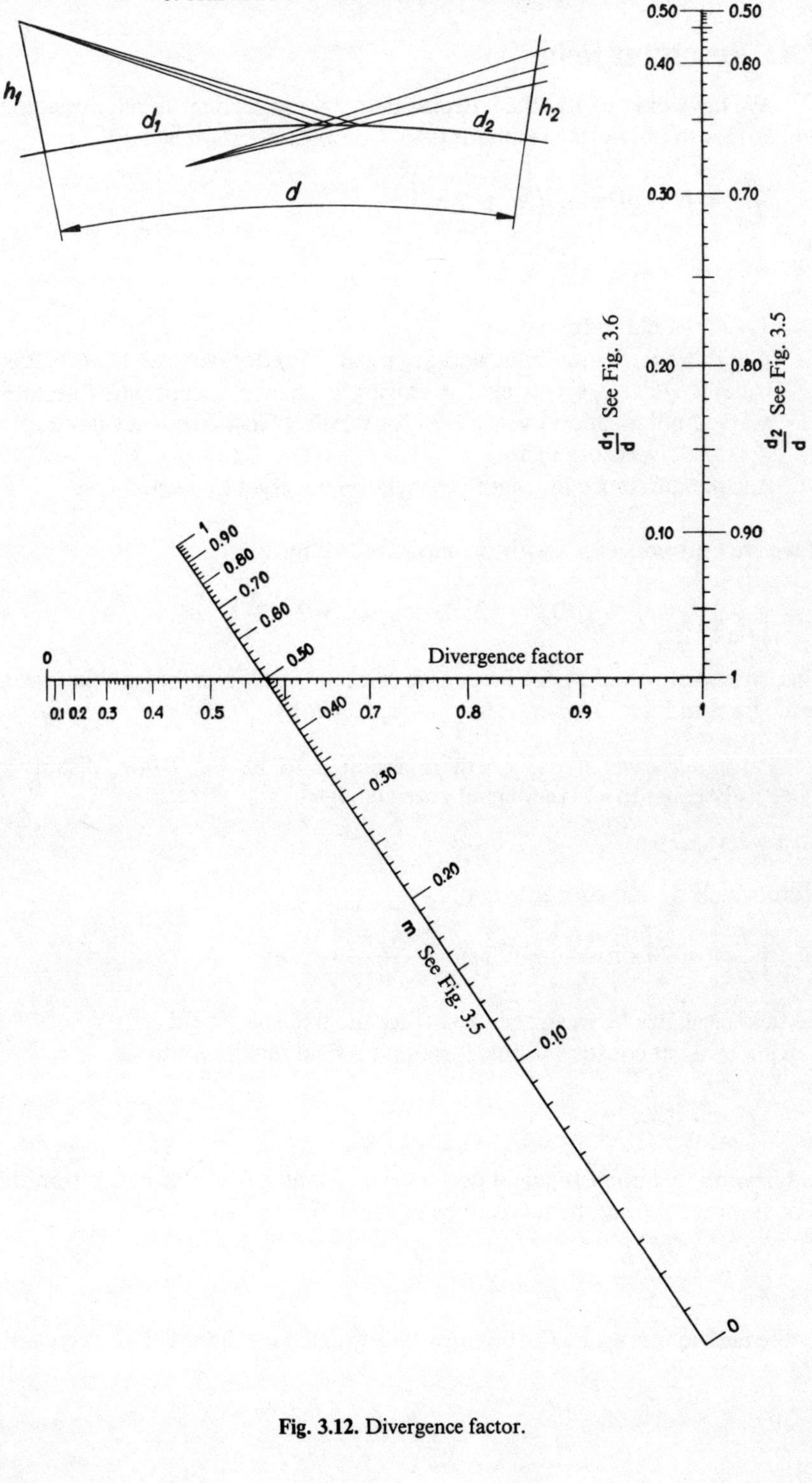

Fig. 3.12. Divergence factor.

3.4. Resultant field

We have only to add the reflected field, taking account of relative phases and amplitudes to obtain the resultant field. Complex notation is used:

$$\frac{E}{E_0} = 1 + \rho D \exp \mathrm{j}\left(\psi + 2\pi\frac{\Delta}{\lambda}\right), \tag{3.35}$$

where:

E_0 is the field in free space;
ρ and ψ are the modulus and argument of reflectivity, and can be taken to equal 1 and π respectively for grazing incidence, except when dealing with vertical polarization over the sea for relatively low frequency (see Fig. 3.2);
D is the divergence factor;
Δ is the difference in length between the direct and reflected rays.

If we are only concerned with the modulus, we have:

$$\left|\frac{E}{E_0}\right| = \sqrt{1 + (\rho D)^2 + 2\rho D \cos\left(\psi + 2\pi\frac{\Delta}{\lambda}\right)}. \tag{3.36}$$

This expression, all of whose factors we can calculate, gives the resultant field. The nomogram in Fig. 3.13 can be used to replace R by ρD and φ by $\psi + 2\pi\Delta/\lambda$.

Particular case: if the Earth is assumed to be flat (short distance) and reflectively equal to −1 (very small grazing angle):

$D=1$ $\rho=1$ $\psi=\pi$

From which we can conclude that:

$$\left|\frac{E}{E_0}\right| = 2\left|\sin\frac{\pi\Delta}{\lambda}\right| = 2\left|\sin\frac{2\pi h_1 h_2}{\lambda d}\right|. \tag{3.37}$$

the field oscillates between zero and twice the free space field.
Moving away at constant altitude, we find a final maximum for:

$$d = \frac{4 h_1 h_2}{\lambda}, \tag{3.38}$$

and beyond this point the field decreases constantly. Very far away from the last maximum the arc can be taken to be equal to the sine and:

$$\frac{E}{E_0} \simeq \frac{4\pi h_1 h_2}{\lambda d}. \tag{3.39}$$

As the field in free space E_0 decreases with $1/d$, the resultant field E decreases with $1/d^2$.

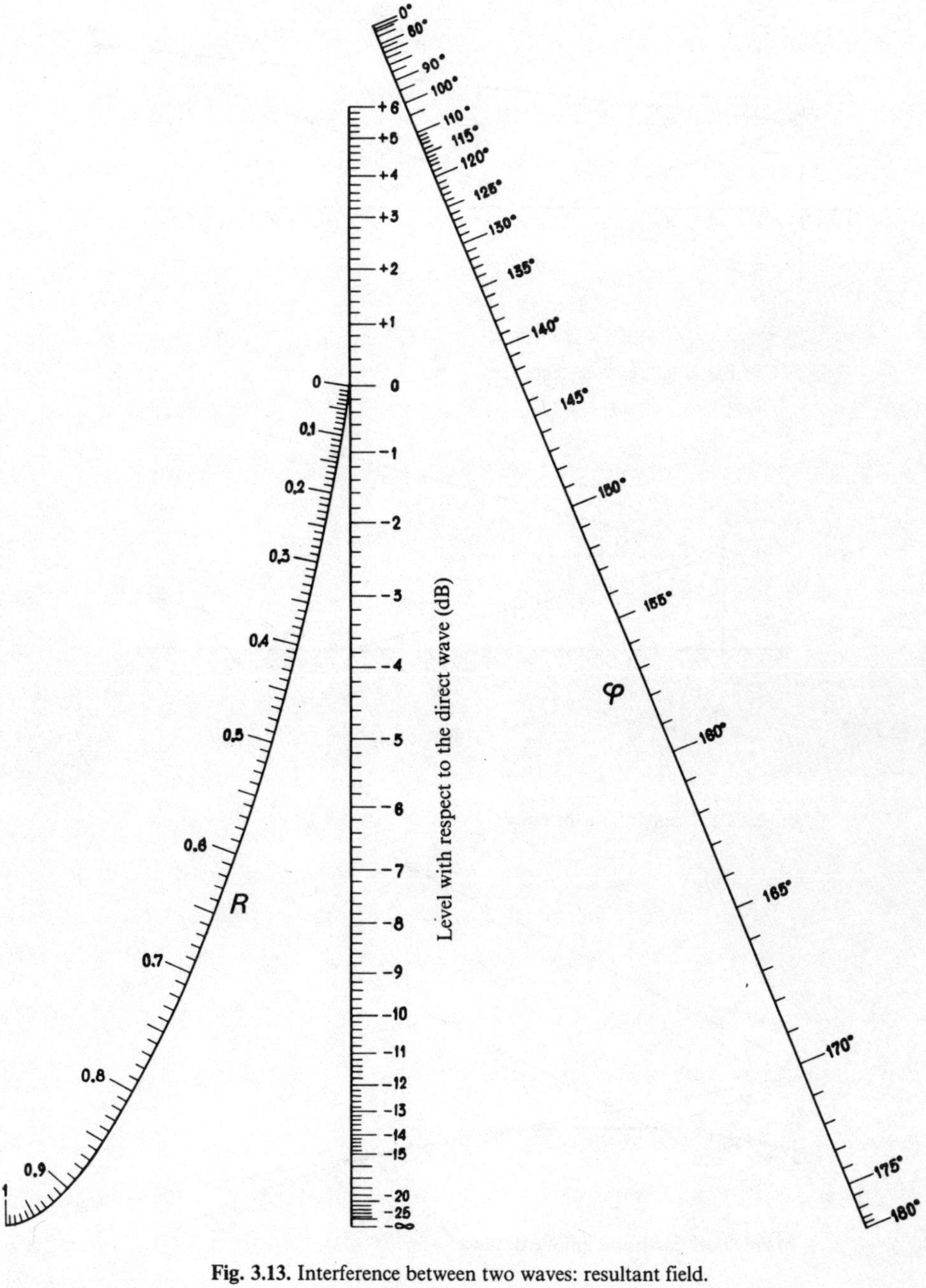

Fig. 3.13. Interference between two waves: resultant field.

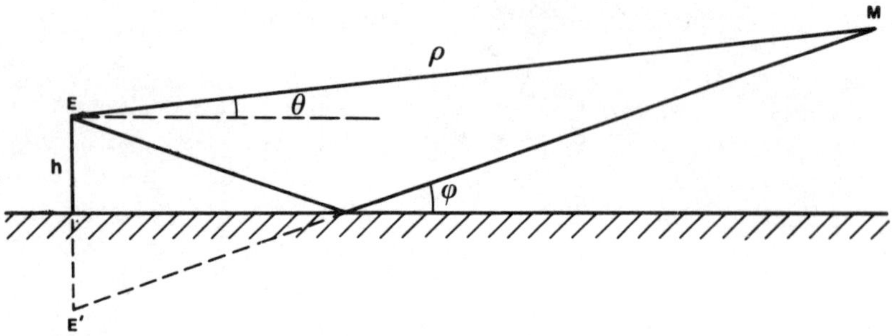

a) Flat Earth and finite distance.

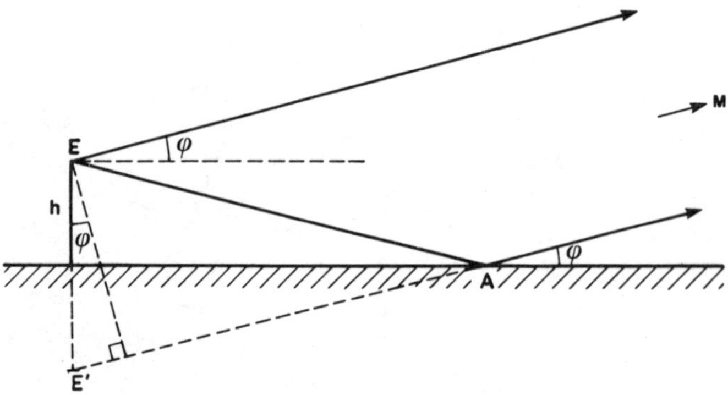

b) Flat Earth and infinite distance.

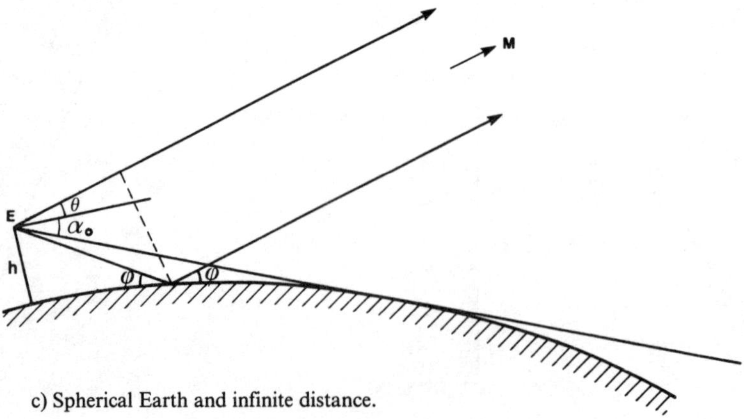

c) Spherical Earth and infinite distance.

Fig. 3.14. Influence of the Earth on a link with a moving aircraft.

3.5. Apparent radiation pattern of an antenna taking the Earth's surface into account: links with aircraft or spacecraft, target detection using radar

Reflection from the Earth's surface profoundly alters antenna patterns, which can have extremely important consequences, e.g. for links with aircraft or for target detection using radar. The effect of the Earth's surface can be calculated, where appropriate, in different ways. We consider the case of radar, but the same calculations can be applied to other radio links (Fig. 3.14).

a) In general, the results obtained previously can be used (Section 3.3), but the calculations are relatively long.

b) If the Earth's surface is treated as a plane, i.e. in particular if the height of the radar above the ground is relatively small, the image method can be used. This is based on the fact that the antenna pattern in the presence of the Earth's surface is the same as the pattern that would be obtained for the real antenna, with an image antenna symmetric to it with respect to the ground, if the ground itself is ignored. Note that currents in the image antenna have the same amplitude as in the real antenna, the same direction for components normal to the Earth's surface and opposite directions for components parallel to the Earth's surface (the Earth is assumed to be perfectly conducting).

If the height of the radar above the plane is equal to h and if ρ and θ denote the polar coordinates of a target in a vertical plane with the radar at its origin, the path difference between the direct and reflected rays can be obtained using a simple geometric calculation:

$$\Delta = \sqrt{\rho^2 + 4\,h\rho\sin\theta + 4\,h^2} - \rho \tag{3.40}$$

and the angle of incidence φ on the ground is given by:

$$\tan\,\varphi = \frac{2\,h + \rho\sin\theta}{\rho\cos\theta}. \tag{3.41}$$

If the target is at an infinite distant from the radar, Δ and φ tend towards their limiting values, which can then be obtained directly:

$$\varphi \to \theta \qquad \Delta \to 2\,h\sin\theta\,. \tag{3.42}$$

This gives the pattern at infinity of the radar antenna in the presence of the Earth's surface, which is assumed to be flat. If the reflectivity of the Earth's surface, for the angle considered, is expressed by:

$$R = |\,R\,|\,\mathrm{e}^{\mathrm{j}\delta} \tag{3.43}$$

the ratio of the resultant field E to field E_0 in the absence of reflection is given by:

$$\frac{E}{E_0} = 1 + |\,R\,|\,\mathrm{e}^{\mathrm{j}\left(2\pi\frac{\Delta}{\lambda} + \delta\right)}. \tag{3.44}$$

If we now assume that the grazing angle φ is relatively small, we can make a first approximation:

$$|R| = 1 \qquad \delta = \pi \tag{3.45}$$

and under these conditions, the modulus of the resultant field is given by:

$$\frac{E}{E_0} = 2\left|\sin\frac{2\,\pi h\varphi}{\lambda}\right|. \tag{3.46}$$

If the antenna used has a radiation pattern represented by function $f(\varphi)$ in the vertical plane, the situation is the same as if following an antenna whose pattern is represented by the following function was used:

$$2\,f(\varphi)\left|\sin\frac{2\,\pi h\varphi}{\lambda}\right|. \tag{3.47}$$

At very high frequencies the quantity h/λ is very large and the function $\sin\frac{2\,\pi h\varphi}{\lambda}$ oscillates very rapidly. Each lobe of the antenna diagram is then broken down into a certain number of very narrow lobes and there is a zero in the direction of the horizon, which explains the difficulty of detecting low targets.

c) In certain applications terrestrial curvature must be taken into account but we are still interested in the antenna pattern at infinity. Calculations can be carried out as described in Section 3.3 but there lengths l_2 and h_2 are infinite. Equation (3.11) becomes:

$$\Delta = 2\,l_1\,\sin^2\varphi\,. \tag{3.48}$$

Relatively simple geometrical calculation makes it possible to express the path length differences Δ and the divergence factor D as a function of the grazing angle incidence φ and height h:

$$\Delta = \frac{4\,h\sin^2\varphi}{\sqrt{\sin^2\varphi + \dfrac{2\,h}{Ka}} + \sin\varphi} \tag{3.49}$$

$$\frac{1}{D^2} = 2\sqrt{1 + \frac{2\,h}{Ka\sin^2\varphi}} - 1\,. \tag{3.50}$$

It is, however, preferable to express these quantities as a function of the angle of elevation θ of the target seen by the radar. Unfortunately, the calculations become far more complicated if we do, and we only consider the case where the height h of a radar is relatively small, and where we are only concerned with small values of θ. If we write α_0 or the absolute value of the (negative) angle of elevation of the horizon seen from the radar:

$$\alpha_0^2 = \frac{2\,h}{Ka}, \tag{3.51}$$

we obtain the following expressions, where angles are expressed in radians:

$$\Delta = \frac{4\,h\alpha_0}{27}\left[\left(\left(\frac{\theta}{\alpha_0}\right)^2 + 3\right)^{3/2} - \frac{\theta}{\alpha_0}\left(\left(\frac{\theta}{\alpha_0}\right)^2 - 9\right)\right], \tag{3.52}$$

$$D^2 = \frac{1}{3}\left[1 + \frac{2\dfrac{\theta}{\alpha_0}}{\sqrt{\left(\dfrac{\theta}{\alpha_0}\right)^2 + 3}}\right]. \tag{3.53}$$

Having calculated the values of Δ and D, and using the approach described in Section 3.3.5, we can calculate the effective antenna pattern, taking into account the free space pattern. An example is given in Fig. 3.15, where we only consider the main lobe of the antenna. It can be seen that for very low elevation angles the divergence factor plays a major role.

N.B. In the hectometre wave range (MF) and longer waves for antennas on the Earth's surface the radiation characteristics of the antennas can only be defined taking the presence of the Earth's surface into account. But in this case we can no longer talk of line of sight paths (see Section 2.3).

3.6. The influence of the size and position of a reflecting area of the Earth's surface

All the calculations above are based on the assumption that there is no limitation to the extent of the reflecting area. They are therefore directly applicable to paths over the sea. For terrestrial paths it is often the case that only part of the surface between the transmitter and the receiver can be regarded as reflecting. If that

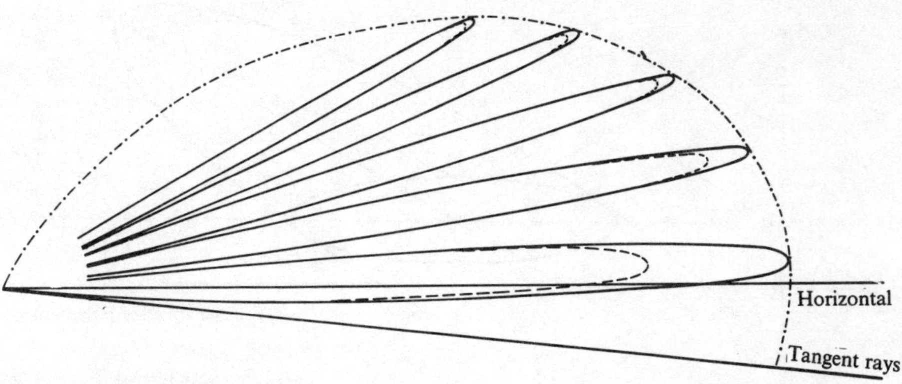

Fig. 3.15. Influence of the Earth's surface on the antenna radiation pattern: Height of antenna 20m. Gain 40dB; λ = 10cm. Heights scale multiplied by 50. —— not taking into account the divergence factor. - taking into account the divergence factor. -.-.- main lobe of the antenna in free space.

part is too small, given the grazing angle in question, diffraction phenomena cannot be neglected.

It is therefore necessary first of all to determine which part of the Earth's surface reflects most of the transmitted energy. To answer this question, we only need to use Babinet's theorem (see Chapter 2), which here means replacing the reflecting part by a transparent opening cut out of an opaque surface, with the transmitter and receiver on either side of the surface. We can then use the results discussed in Chapter 2 concerning Fresnel ellipsoids which lead to the definition of Fresnel zones for reflection. For a plane, the Fresnel zones are obtained by taking the intersection with this plane of the Fresnel ellipsoids defined by one end of the link and the symmetrical equivalent of the other end with respect to the plane. In the more general case, the Fresnel zone of rank n is the set of points within the reflecting surface for which the sum of distances to the transmitting and receiving antennas is given by the relation (Fig. 3.16):

$$TM + MR = TI + IR + k\frac{\lambda}{2} \tag{3.54}$$

where:

$$n - 1 < k < n. \tag{3.55}$$

The first Fresnel zone is an ellipsoid which is always very elongated in the direction of propagation. Its centre is relatively close to the geometric point of reflection, but does not generally coincide with that point. Its shorter axis equal, to within a good approximation, to the equatorial diameter of the corresponding first Fresnel

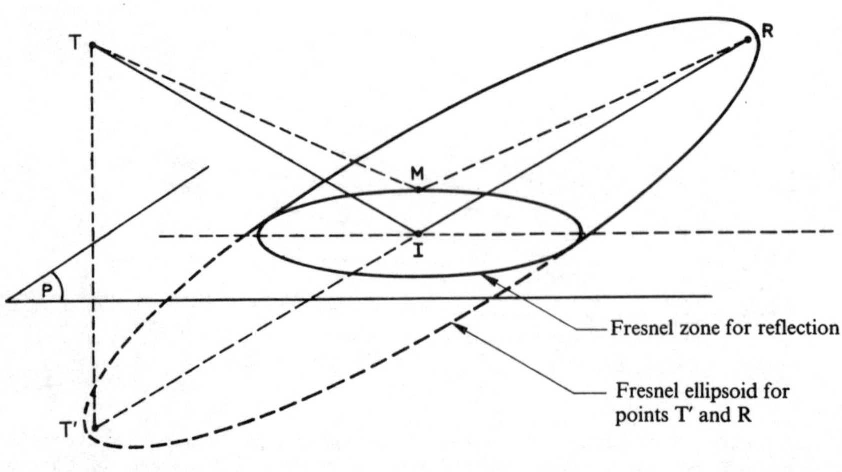

Fig. 3.16. Relation between a Fresnel ellipsoid and a Fresnel zone for reflection.

ellipsoid, i.e. $\sqrt{\lambda d}$. On the other hand, its major axis can vary considerably in size depending on the heights of the antennas above the plane tangent to the reflecting surface at the point of reflection. If h_1' and h_2' denote these heights, d is the length of the link and λ is the wavelength, the length of the Fresnel zone is given by:

$$L = d\,\frac{\sqrt{1 + \dfrac{4h_1' h_2'}{\lambda d}}}{1 + \dfrac{(h_1' + h_2')^2}{\lambda d}}. \tag{3.56}$$

For example, for a 6GHz link over 50km between antennas 75m high the axes of the first Fresnel zone are some 16km and 50m, respectively, if it is assumed that the Earth's surface is flat. The length increases to 24km if terrestrial curvature is taken into account.

Given the very elongated shape of the first Fresnel zone, the reflecting areas associated with terrestrial paths generally cover the whole width of the zone, but only part of its length. Under these conditions, the relative level of the reflected ray with respect to the direct ray depends on the size and position of the reflecting area.

In general, in terrestrial links, the angle of reflection is always very small, typically less than 1°. Consequently, a reflecting zone on the ground of any form will be seen from either end of the link as a narrow band. This then makes it possible, by using Babinet's theorem, to reduce the problem to the diffraction of a wave by a slit. This is a well known phenomenon whose analysis involves using Fresnel integrals. We know that the ratio between the field in the presence of the slit and that which would exist in free space is given by:

$$\frac{E}{E_0} = \frac{1}{1+\mathrm{j}} \int_{v_a}^{v_b} \mathrm{e}^{\mathrm{j}\pi \frac{t^2}{2}}\, \mathrm{d}t. \tag{3.57}$$

The dimensionless quantities v_a and v_b are functions only of the geometric elements defining the respective positions of the edges a and b of the slit. In particular, if we assume that the reflecting area of the Earth's surface is horizontal and if we define (Fig. 3.17):

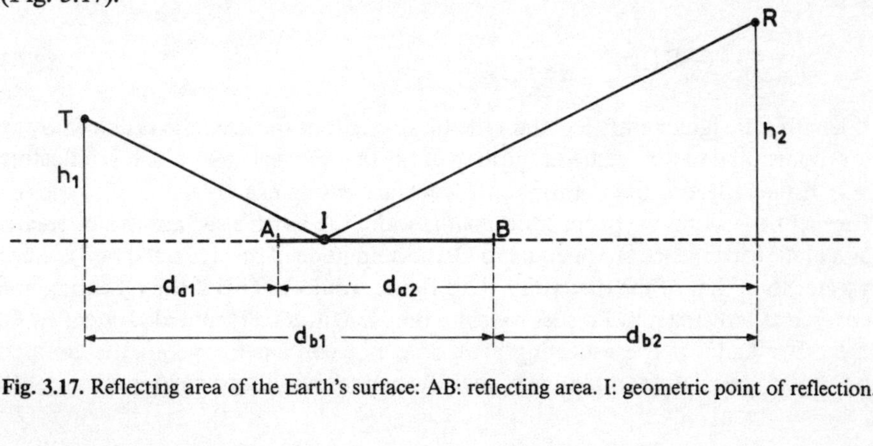

Fig. 3.17. Reflecting area of the Earth's surface: AB: reflecting area. I: geometric point of reflection.

d for the length of the path;
h_1 and h_2 for the heights of the antennas above the reflecting area;
d_{a_1} and d_{a_2} for the distance from edge A to the ends of the path;
d_{b_1} and d_{b_2} for the distances from edge B to the ends of the path;
λ for the wavelength;

we have:

$$v_a = \sqrt{\frac{2}{\lambda d}} \cdot \frac{h_1 d_{a_2} - h_2 d_{a_1}}{\sqrt{d_{a_1} d_{a_2}}}, \tag{3.58}$$

$$v_b = \sqrt{\frac{2}{\lambda d}} \cdot \frac{h_1 d_{b_2} - h_2 d_{b_1}}{\sqrt{d_{b_1} d_{b_2}}}. \tag{3.59}$$

The quantities v are zero when the corresponding edge is at the geometric point of reflection. If neither edge is very far from the point of reflection, it can be shown that the ratio of received powers, due to reflection on the one hand and in free space on the other, is given by the approximate expression:

$$\frac{P}{P_0} = \frac{1}{2}(v_b - v_a)^2 \tag{3.60}$$

and in this case the quantity $v_b - v_a$ can be expressed as a function of the width Δx of the reflecting area, independently of the exact position of the edges:

$$\frac{P}{P_0} = \frac{(h_1 + h_2)^4 (\Delta x)^2}{h_1 h_2 \lambda d^3}. \tag{3.61}$$

If the link is symmetric, i.e. if:

$$h_1 = h_2 = h, \tag{3.62}$$

the preceding ratio can be expressed as a function of the length L of the first Fresnel zone:

$$\frac{P}{P_0} \simeq (2\,\Delta x)^2 \left(\frac{1}{L^2} - \frac{1}{d^2}\right). \tag{3.63}$$

If length L is significantly less than the total length of the link, the ratio of powers is proportional to the square of the area of the first Fresnel zone which is reflecting.

If the reflecting area of the Earth's surface covers exactly the area of the first Fresnel zone along the propagation path, and if the Earth's surface can be treated as a plane (divergence factor equal to 1), the amplitude of the 'reflected ray' is 2.6dB greater than that of the direct ray. This figure would be 6dB if the reflecting area covered exactly the first Fresnel zone not only longitudinally, but also laterally. On the other hand, if the reflecting area does not contain the geometric point of reflection, the relative amplitude of the reflected ray will not be greater than –3.4dB, and if the reflecting area is completely outside the first Fresnel zone, the amplitude of the reflected ray will be less than −11.5dB.

3.7. Effect of irregularities in the Earth's surface

As suggested in Chapter 2, the surface can only be considered to be perfectly reflecting if its irregularities are small, taking into account the wavelength and the angle of reflection. This is indeed the case for sea, particularly for vertical polarization, but on land reflectivity is generally well below what one would expect from the characteristics of the ground alone. The remarks made in Section 2.7 have to be taken into account.

For a link between two points near the Earth's surface angles of reflection are generally between 1 and 10mrad. Irregularities which influence propagation consequently have heights of the order of 100 to 1000 wavelengths. In the centimetre wave range this corresponds to altitudes between a few metres and tens of metres.

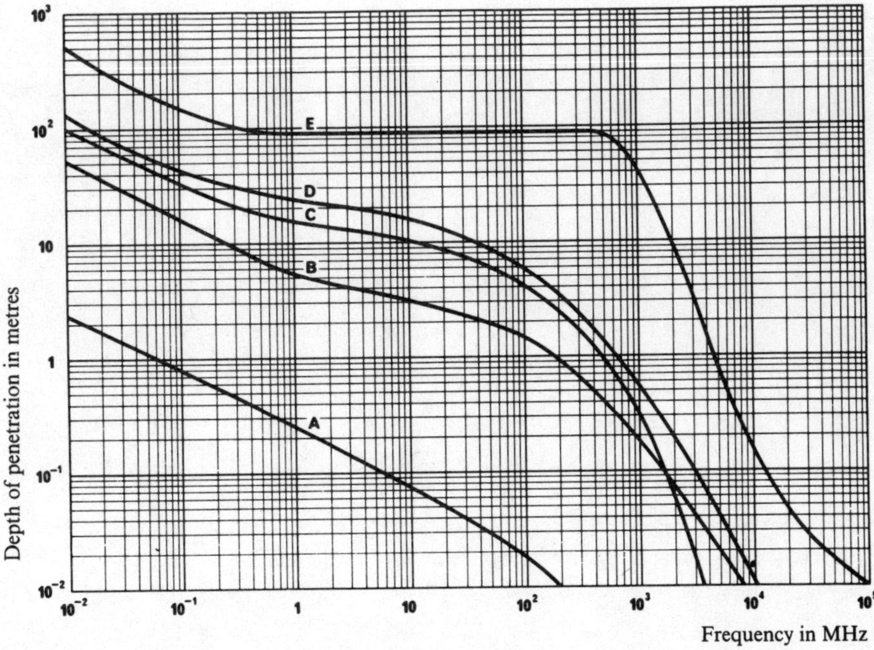

Fig. 3.18. Depth of penetration of waves into the Earth's surface: A: sea water; B: humid soil; C: fresh water; D: moderately dry soil; E: very dry soil.

3.8. Wavelength and depth of penetration into the Earth's surface

In certain cases we need to know how far waves will penetrate into the Earth's surface. For this, we have only to use the results in Chapter 2, taking the permittivity and conductivity of the ground as given in Fig. 3.1. This gives the curves in Fig. 3.18. It is clear that in sea water the depth of penetration is very small, except for very low frequencies. For this reason, the kilometre wave range (LF) or even the VLF wave range is used to communicate with submerged submarines equipped with an antenna a few metres below the surface.

Chapter 4

Influence of the troposphere: refraction effects

If we ignore special cases such as links between satellites, all radio paths travel through the atmosphere. The influence of a homogeneous medium on propagation is entirely determined by its refractive index. We therefore briefly consider the refractive index of clear atmosphere. We then proceed to analyse the effects of refraction on propagation. Other effects due to tropospheric scatter are studied in Chapter 8.

4.1. Refractive index of the troposphere

For a mixture of gases the extent by which the refractive index is greater than 1 can be expressed as follows:

$$n - 1 = \sum m_i\left(A_i + \frac{B_i}{T}\right) \tag{4.1}$$

where m_i is the density, T is the absolute temperature, A_i and B_i are two constants of which the second is only non-zero for gases whose molecules have an electric dipole moment, e.g. water vapour. If Mariotte's law applies, m_i can be replaced by P_i/T, where p_i is the partial pressure of the gas.

As n–1 is very small, we generally use a quantity N which is 10^6 times greater and is called the refractivity. When dealing with radio waves the refractivity of air is given by the following approximate formula, which is adopted throughout the world:

$$N = (n - 1)\,10^6 = \frac{77.6}{T}\left(p + 4\,810\,\frac{e}{T}\right) \tag{4.2}$$

where p is the total pressure in millibars;
e is the partial pressure of the water vapour in millibars;
T is the absolute temperature in degrees Kelvin.

At sea level the value of N is around 300.

Atmospheric humidity can be expressed in various other ways, which give various different forms of the expression for N. We can in particular use:

1. Specific humidity, which is the ratio between the mass of water vapour mixed with the air and the mass of humid air.

2. The mixing ratio, which is the ratio between the mass of water vapour mixed with the atmosphere and the corresponding mass of dry air.
3. Relative humidity, which is the ratio between the water vapour contents actually observed and the saturated water vapour content at the same temperature, or, in other words, the ratio between the existing mixture ratio and the saturated mixture ratio.

In general (except at very high temperatures), the amount of water which air can support is sufficiently low for us not to have to distinguish between specific humidity and the mixture ratio; the two quantities are given by the following approximate relation:

$$s = 622\frac{e}{p} \tag{4.3}$$

where s is expressed in grams of water per kilogram of air. Under these conditions N takes the form:

$$N = p\left(\frac{77.6}{T} + 600\frac{s}{T^2}\right). \tag{4.4}$$

The first term, which does not depend on humidity, is called the 'dry term', while the second is called the 'wet term'.

Meteorological services tend to present humidity information in the form of relative humidity figures, but tables are available to deduce the saturated mixture ratio from them. In the range of temperatures normally met we can calculate the saturated mixture ratio from the approximate relation:

$$\log_{10} s_0 = 12.22 - \log_{10} p - \frac{2\,360}{T}. \tag{4.5}$$

The nomograms in Figs 4.1 and 4.2 allow the dry and wet terms to be calculated separately. The wet term is derived from relative humidity figures.

4.1.1 Refractive index at ground level (N_s) and sea level (N_0)

The refractive index near the ground plays a particular role since it is easy to measure and certain methods of predicting propagation are based on its use.

However, as meteorological stations can be located at different heights, it is preferable to take sea level as the reference altitude than to calculate N_0 for these altitudes.

To do so, certain assumptions concerning the variation of refractive index with altitude have to be made. It is generally supposed that index N (Fig. 4.3) is an exponential function of altitude, but this can only be regarded as a fairly rough approximation near the ground.

4.1.2. Reference atmosphere and the real atmosphere

As altitude increases, the refractive index n generally decreases and tends

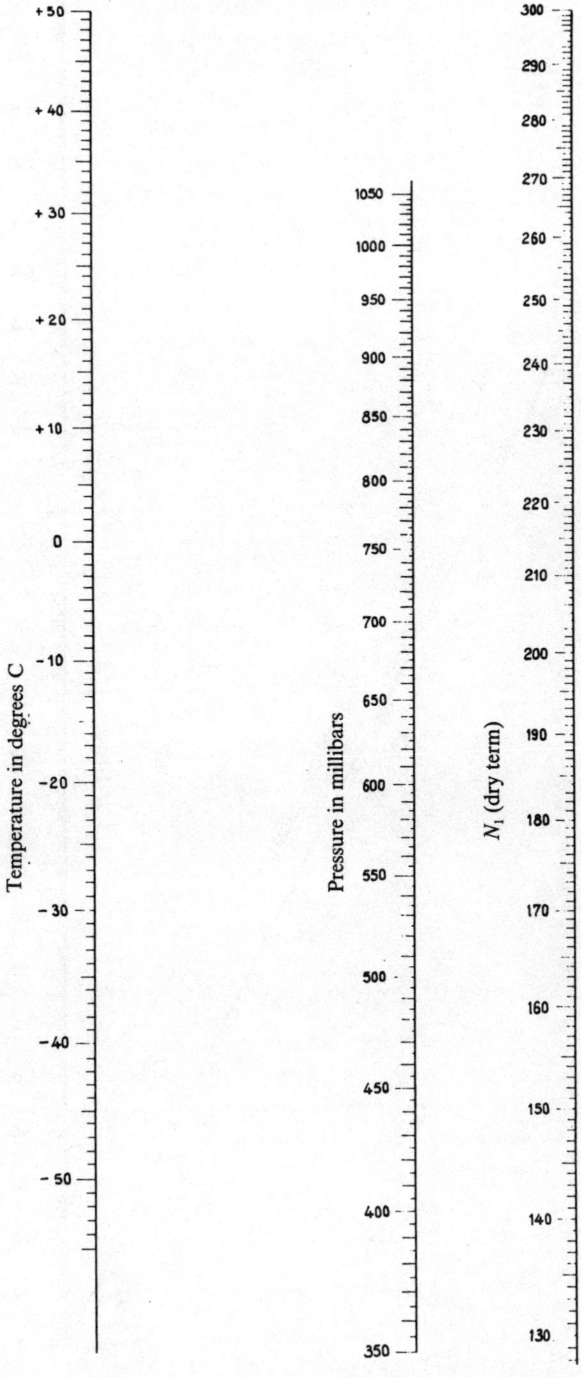

Fig. 4.1. Refractive index of air N_1 (dry term).

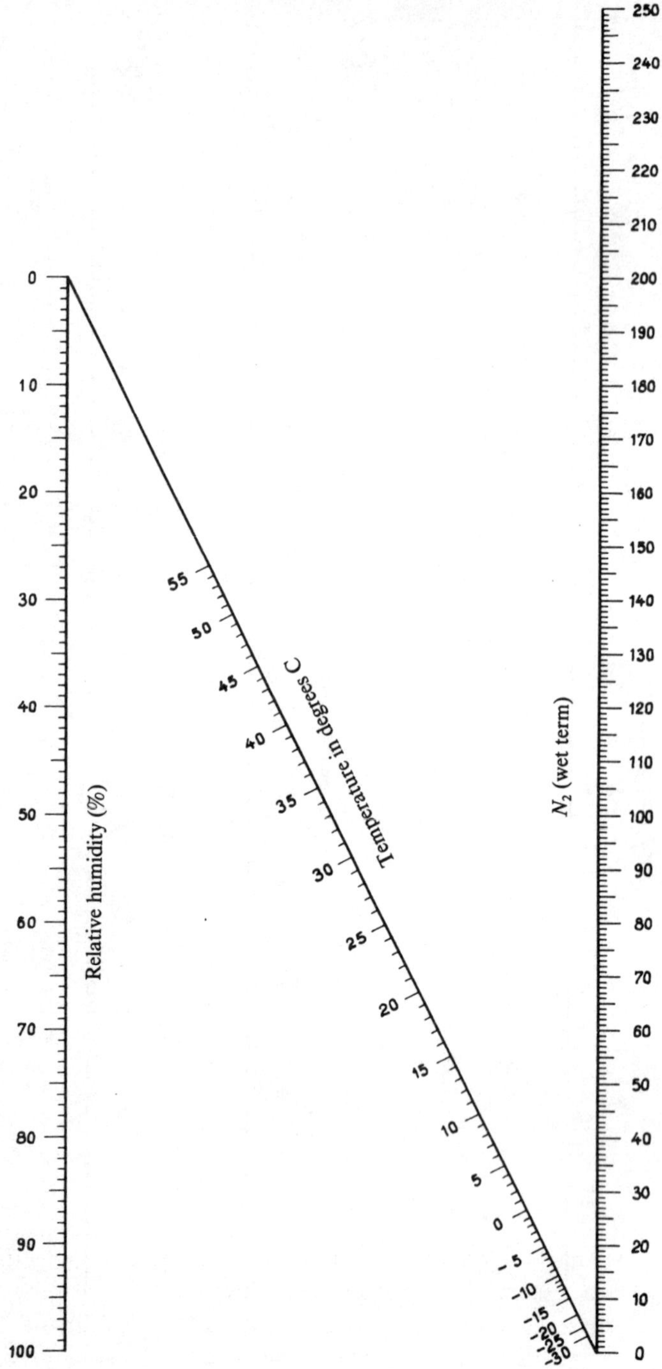

Fig. 4.2. Refractive index of air N_2 (wet term)

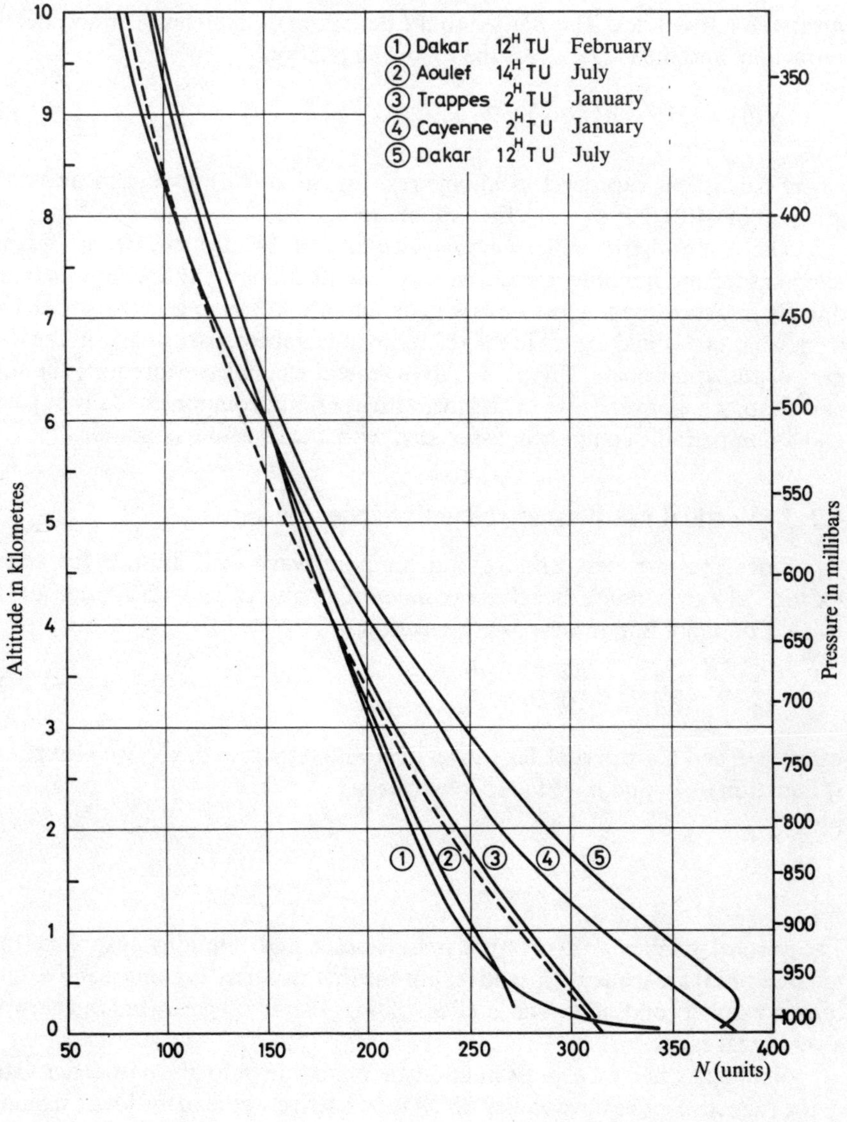

Fig. 4.3. Examples of refractive index curves (montly means): ---- reference atmosphere for refraction (CCIR).

towards 1; and consequently N tends towards zero. However, this overall decrease is generally subject to considerable variations in the lower parts of the atmosphere. In simple terms, we can say that the overall decreasing trend is determined by reducing pressure, whereas variations about the mean value are due to fluctuations in temperature and humidity.

Since pressure decreases exponentially with altitude, the same is true for the mean refractive index. This has led to the definition of a reference atmosphere for refraction, and the CCIR chose the following relation:

$$N(h) = 315 \exp(-\ 0.136\, h)\,, \tag{4.6}$$

where height h is expressed in kilometres. Equation (4.6) gives a mean vertical gradient of –40N/km over the first kilometre.

The atmospheric index curve, since it can be deduced from pressure, temperature and humidity measurements as a function of altitude, or as it can be directly measured by a refractometer in an aircraft, differs in general considerably from what one would expect in the reference atmosphere, particularly in the lower part of the atmosphere. Figure 4.3 gives several examples. Moreover, the index curve shows major variations with time, with a periodic component (daily or yearly) and a non-periodic component associated with local weather conditions.

4.1.3. Vertical gradient of refractive index of air

Since pressure, temperature and humidity vary with altitude (at a given location at a given time), the refractive index of air also varies with altitude and this value is of major importance to propagation:

$$\frac{\mathrm{d}N}{\mathrm{d}h} = A\frac{\mathrm{d}p}{\mathrm{d}h} + B\frac{\mathrm{d}T}{\mathrm{d}h} + C\frac{\mathrm{d}s}{\mathrm{d}h}\,; \tag{4.7}$$

where A, B and C are partial derivatives of N with respect to three variables, p, t and s, Equation (4.4), and have the following signs:

$$A > 0$$
$$B < 0$$
$$C > 0\,.$$

The vertical gradients of pressure, temperature and humidity (expressed as a mixture ratio) are generally negative, but the first two may become positive under certain weather conditions. The gradient of N is generally negative but can vary over a very great range.

Note that when we express humidity as a mixture ratio, the partial derivatives of the refractive index have values which vary relatively little in the lower regions of the atmosphere, and we can sometimes use the following approximation:

$$\frac{\mathrm{d}N}{\mathrm{d}h} = 0.35\frac{\mathrm{d}p}{\mathrm{d}h} - 1.3\frac{\mathrm{d}T}{\mathrm{d}h} + 7\frac{\mathrm{d}s}{\mathrm{d}h} \tag{4.8}$$

in which pressure is expressed in millibars, temperature in degrees Celsius and humidity in grams of water vapour per kilogram of dry air.

Figure 4.4 gives examples of the statistical distribution of the mean gradient over the first 100 metres, for three points at different latitudes. It can be seen that the range of variation is very large. These variations would be considerably smaller

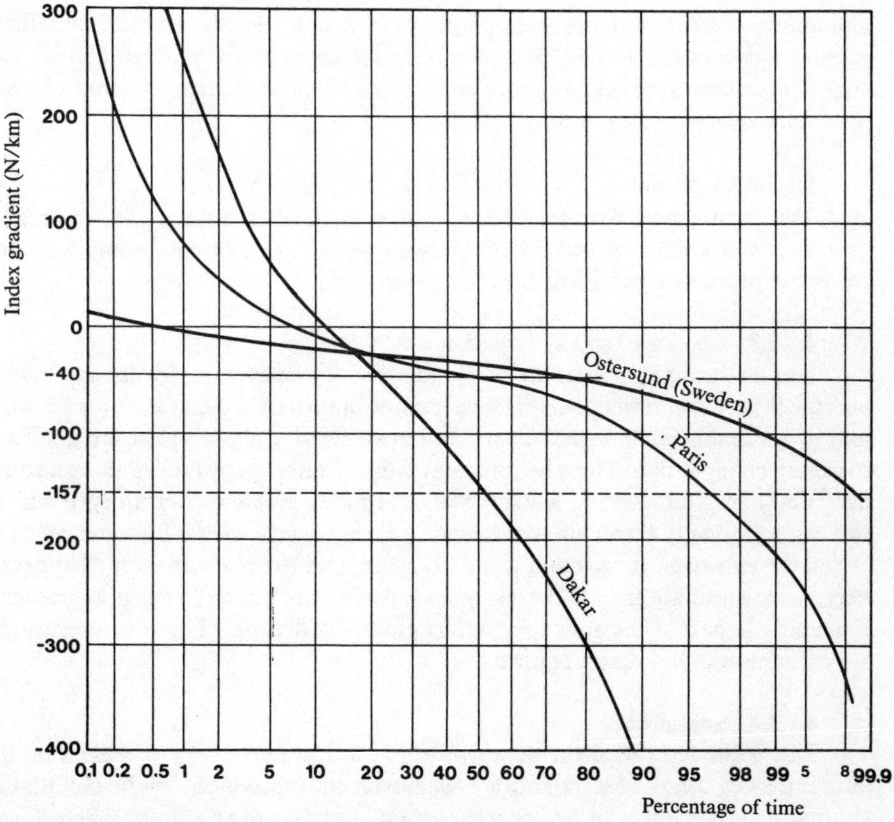

Fig. 4.4. Statistical distribution of the refractive index gradient in the first 100 metres of the atmosphere.

if we considered a greater depth of atmosphere, or a region of the atmosphere at a greater altitude.

Since large negative values (–157N/km) of the index gradient play a particularly important role in propagation, we briefly study the weather conditions which can create them. Note first of all that, given the signs of A, B and C, large negative values of the vertical gradient will be obtained when at least one of the following conditions applies:

$\mathrm{d}T/\mathrm{d}h$ is positive (temperature inversion);

$\mathrm{d}s/\mathrm{d}h$ is very large and negative (dry air layer above a humid air layer).

The weather conditions producing large negative refractive index gradients are given below.

4.1.3.1. Evaporation

Above humid soil or sea water heated by the sun, there is a layer of atmosphere containing water vapour, in very large quantities near the Earth's surface and falling rapidly with altitude, producing a very steep refractive index gradient. Above sea such a layer, called the ocean surface limit layer, is in general 10 to 15m thick. It can exist for a relatively long time.

4.1.3.2. Advection

This term is used for the horizontal movement of air masses with respect to each other. If a mass of hot dry air passes over a region of cold humid air, the refractive index gradient is steep over a depth of some tens of metres.

4.1.3.3. Cooling of soil by radiation

During the night, if there are no clouds, the ground cools rapidly by radiation, and the layer of air immediately above it cools in turn by contact with the ground and by radiation, while higher layers cool more slowly. The temperature gradient therefore changes sign. This phenomenon is (quite improperly) called temperature inversion and it can lead to a steep refractive index gradient over a depth which increases gradually throughout the night and can reach several hundred metres. When the sun rises, the ground heats up again, and the phenomenon disappears very quickly near the ground but less quickly further up. The steep negative gradient remaining in part of the atmosphere then gradually declines. This phenomenon is very common in hot desert regions.

4.1.3.4. Subsidence

This is the term used for the downward vertical movement of masses of air (and is the opposite of ascendance). This movement is produced by the fact that a volume of air previously heated near the ground has risen to an altitude where it has cooled slightly and may have become dryer by precipitation of some of its water. It then moves downwards again slowly (a few centimetres a second) on to a mass of colder air which diverges horizontally, producing a layer of atmosphere with a steep negative refractive index where the two regions meet. Such a phenomenon, which is always associated with the presence of an anticyclone (high barometric pressures), can last several days.

It is clear that all the phenomena producing steep negative refractive index gradients have a major property in common. They correspond to stable arrangements of the air masses concerned, as the densest layers of air are lower than the others. They can therefore last hours or even some days and extend over large areas, as much as several hundred kilometres across or even several thousand kilometres above the sea.

4.2 Wave refraction in the troposphere

From the point of view of radio waves, a so-called clear terrestrial atmosphere (i.e. a terrestrial atmosphere which does not contain liquid or solid particles) is characterized by its refractive index, which depends on meteorological conditions

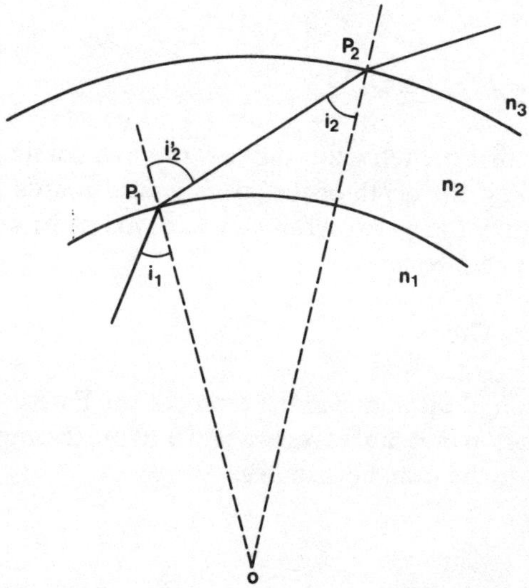

Fig. 4.5. Refraction in a medium with spherical stratification.

(pressure, temperature, humidity) and consequently is a function of space and time. Given the complexity of this function and the difficulties involved in working it out, the study of these phenomena has to be broken down into several stages each of which is analysed using simplifying assumptions.

The first stage assumes that the refractive index is a function of altitude alone, with large scale space and time variations. In other words, we are only concerned with 'mean' characteristics of refraction. Having established general formulas, we study first terrestrial links and then Earth-satellite links.

The second stage of study will involve small scale variations, in space or time, in the refractive index, i.e. those which give rise to phenomena called tropospheric multi-paths and scintillations.

4.2.1. General equations for tropospheric refraction

We start by establishing the relation which gives the path of a radio wave in an atmosphere with spherical stratification (Fig. 4.5).

At point P_1 Snell's law can be written as:

$$n_1 \sin i_1 = n_2 \sin i_2' . \tag{4.9}$$

Moreover, in triangle $OP_1\,P_2$ the following relation holds:

$$\frac{r_1}{\sin i_2} = \frac{r_2}{\sin i_2'} . \tag{4.10}$$

Eliminating i'_2:

$$n_1 r_1 \sin i_1 = n_2 r_2 \sin i_2 . \tag{4.11}$$

If we now assume that the refractive index varies in a continuous way, Equation (4.11) remains valid if the depth of the layers tends towards zero. Consequently, over the whole path of the wave (in this case a curved path) we have the following relation, giving the path equation:

$$n(r).r.\sin i = \text{Cte} . \tag{4.12}$$

where n is a function of distance r to the centre of the Earth.

In the propagation of radio waves, we prefer to use the angle of elevation φ (or angle of sight) above the local horizon, i.e.

$$\varphi = \frac{\pi}{2} - i . \tag{4.13}$$

The previous relations can therefore be written:

$$n.r.\cos \varphi = \text{Cte} . \tag{4.14}$$

This fundamental equation is, in fact, the differential equation of the paths, but it may also be considered as a first integral of a second order equation, since it contains a constant and therefore represents an invariant function over the entire path. The calculations become reasonably simple if we work in polar coordinates, with the origin at the centre of the Earth and the polar axis passing through the origin of the path. In this case, angle V for the tangent of the vector ray is merely the angle of incidence i, i.e. the complement of φ (Fig. 4.6). The following standard relations ought to be borne in mind:

$$\cos V = \frac{\mathrm{d}r}{\mathrm{d}s}; \quad \sin V = \frac{r\,\mathrm{d}\theta}{\mathrm{d}s}; \quad \tan V = \frac{r\,\mathrm{d}\theta}{\mathrm{d}r} \tag{4.15}$$

where $\mathrm{d}s$ is an element of the arc of the path.

Equation (4.14) can then be rewritten:

$$nr \sin V = n_0\, a \cos \varphi_0 \tag{4.16}$$

and using the relations in Equation (4.15) the integral is obtained:

$$\theta = \int_a^{a+h} \frac{\mathrm{d}r}{r\sqrt{\left(\frac{n}{n_0}\right)^2 \left(\frac{r}{a}\right)^2 \frac{1}{\cos^2 \varphi_0} - 1}}, \tag{4.17}$$

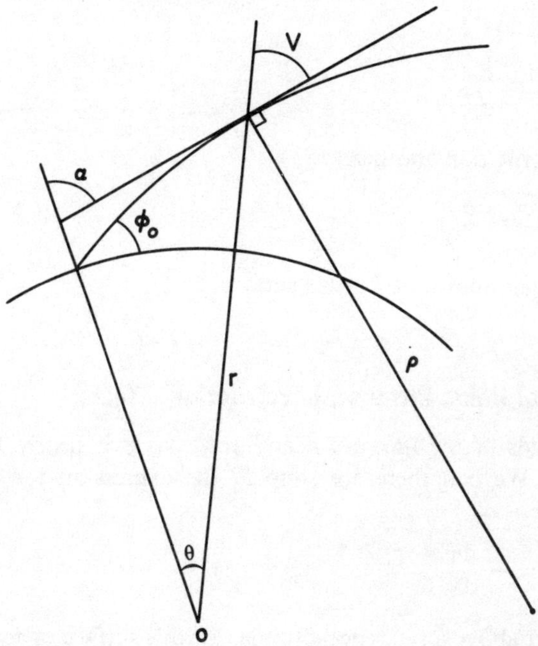

Fig. 4.6. Refraction in a medium with constant variation in the refractive index.

which gives the path equation, as long as $n(r)$, i.e. the 'refractive index profile' of the atmosphere, is known.

It is worth knowing the radius of curvature of the paths. To obtain this, differentiate Equation (4.14) taking Equation (4.15) into account:

$$\frac{\mathrm{d}n}{n} + \frac{\mathrm{d}r}{r} + \frac{\mathrm{d}V}{\tan V} = 0\,. \tag{4.18}$$

If we furthermore take Equation (4.7) into account as well as relation:

$$\alpha = \theta + V \tag{4.19}$$

Equation (4.18) takes the form:

$$\frac{\mathrm{d}n}{n} + \frac{\mathrm{d}\alpha}{\tan V} = 0. \tag{4.20}$$

Equation (4.18) can also be rewritten to show the radius of curvature $\mathrm{d}s/\mathrm{d}\alpha = \rho$:

$$\frac{\mathrm{d}n}{n} + \frac{\mathrm{d}\alpha}{\mathrm{d}s} \cdot \frac{\mathrm{d}s}{r\,\mathrm{d}\theta} \cdot \mathrm{d}r = 0 \tag{4.21}$$

i.e.

$$\frac{1}{\rho} = -\frac{\sin V}{n} \cdot \frac{dn}{dr} \tag{4.22}$$

which is usually written in another way:

$$\frac{1}{\rho} = -\frac{\cos \varphi}{n} \cdot \frac{dn}{dh} \tag{4.23}$$

where h is the height above the Earth's surface.

4.2.2. Terrestrial links: large scale refraction effects

If the two ends of the link are near Earth, cos φ is nearly 1; n is always approximately 1. We can therefore simplify the expression for the radius of curvature:

$$\frac{1}{\rho} \simeq -\frac{dn}{dr} = -\frac{dn}{dh} \tag{4.24}$$

calling h the height above a reference altitude (Earth's surface or sea level). It can be seen that apart from the sign, the curvature of the path is equal to the vertical gradient of the refractive index. If the vertical gradient is constant, paths are circular arcs.

4.2.2.1. Modified refractive index

We know that n is very close to 1 at sea level. We can therefore write:

$$n = 1 + N\,10^{-6}. \tag{4.25}$$

Furthermore, if the two ends of the link are in the atmosphere, the height n above the Earth's surface of any point on the path is small compared with the radius of the Earth a and we can write:

$$r = a + h \quad \text{where } h \ll a. \tag{4.26}$$

Finally, if angle φ is small:

$$\cos \varphi = 1 - \frac{\varphi^2}{2}. \tag{4.27}$$

Under these conditions, the fundamental relation for propagation in a spherical stratified atmosphere, Equation (4.14), can be written, accurate to the second order:

$$N\,10^{-6} + \frac{h}{a} - \frac{\varphi^2}{2} = \text{Cte}. \tag{4.28}$$

Adopting the following practical dimensions where:

altitude = h metres, refractive index gradient = N units/km, earth radius = a × 10^3km (i.e. a = 6.37 and 1/a = 0.157) and the ray angle = φ milliradians.

Equation (4.28) can be written in the form:

$$N + \frac{h}{a} - \frac{\varphi^2}{2} = \text{Cte}\,. \qquad (4.29)$$

The quantity $M = N + h/a$ is called the modified refractive index. It can be used to deduce propagation in an atmosphere of refractive index M above a flat Earth (a is infinite). The path equation then becomes:

$$M - \frac{\varphi^2}{2} = \text{Cte}\,. \qquad (4.30)$$

4.2.2.2. Effective radius of the Earth

There is another modification which is often usefully applied. It employs an effective radius of the Earth chosen in such a way that by measuring altitudes above this hypothetical terrestrial surface, paths become straight lines. The new terrestrial surface is therefore assumed to be in an atmosphere of constant refractive index. If we now differentiate Equation (4.29) with respect to h:

$$\frac{\mathrm{d}N}{\mathrm{d}h} + \frac{1}{a} - \varphi\,\frac{\mathrm{d}\varphi}{\mathrm{d}h} = 0\,. \qquad (4.31)$$

If we now let:

$$\frac{\mathrm{d}N}{\mathrm{d}h} + \frac{1}{a} = \frac{1}{Ka} \qquad (4.32)$$

allowing us to define a sphere of radius Ka for which paths will be straight lines because:

$$\frac{1}{Ka} - \varphi\,\frac{\mathrm{d}\varphi}{\mathrm{d}h} = 0 \quad \text{or} \quad \frac{h}{Ka} - \frac{\varphi^2}{2} = \text{Cte}\,, \qquad (4.33)$$

is the path equation in an atmosphere of constant refractive index. Coefficient K defined in this way is called the multiplying coefficient of the radius of the Earth:

$$K = \frac{1}{1 + a\,\dfrac{\mathrm{d}N}{\mathrm{d}h}}\,. \qquad (4.34)$$

As K varies with time, paths ought to be drawn for various values of K. Generally, however, just one chosen to suit the application intended is used. Paths are then only straight lines for that particular type of atmosphere. This procedure is very commonly used for determining the heights of antennas used in radio relay links (see Section 6.3.1).

4.2.2.3. Normal atmosphere: subrefraction and super-refraction

A normal atmosphere is a spherical stratified atmosphere with a constant vertical refractive index gradient whose value is:

$$dN/dh = -39 \text{ units/km.} \tag{4.35}$$

For such an atmosphere we therefore find that the following approximation applies:

$$K \simeq 4/3 \text{ or } Ka \simeq 8500\text{km.} \tag{4.36}$$

This value of the gradient is a mean value over the first kilometre of atmosphere in a temperate climate.

If the algebraic value of the gradient is greater than –39N/km, we say that subrefraction takes place, paths are less curved and the effective radius of the Earth is less than 8500km. In particular, if the value of the gradient is zero, the atmosphere is linear, and paths are straight lines. If the gradient is positive, the effective radius of the Earth is less than its real radius.

In particular, if the gradient is less than –157N/km, K takes a negative value (note that K changes sign not by passing through zero, but by passing through infinity). Path curvature is then greater than that of the Earth's surface. The paths therefore turn downwards, and may be reflected by the Earth's surface. Guided propagation or ducted propagation takes place (see Section 4.2.2.4).

These different refraction conditions can easily be shown on a graph representing either M or N as a function of height, by considering the gradient of the

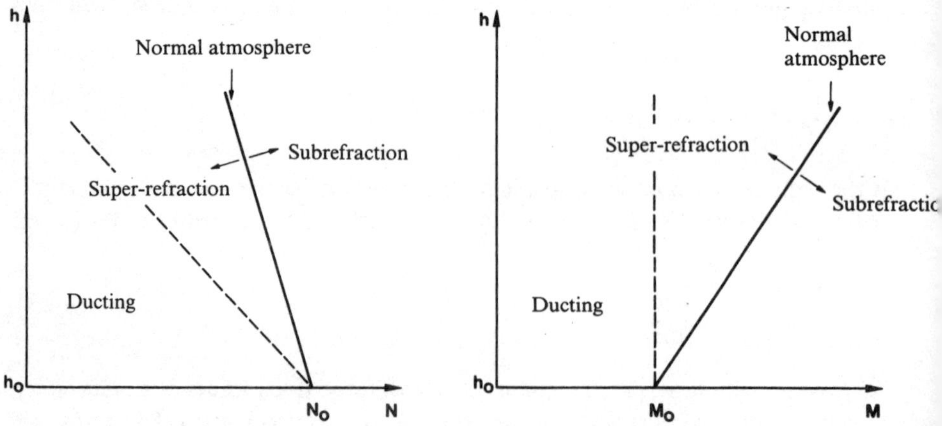

Fig. 4.7. The different refraction conditions in the atmosphere.

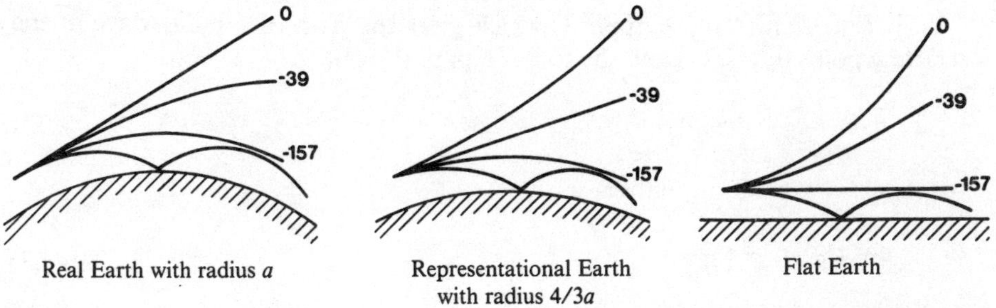

Fig. 4.8. Various representations of a path starting in a horizontal direction. Parameter: three particular values of the refractive index gradient.

tangent to the curve $N(h)$ or $M(h)$ near a point at altitude h_0 (Fig. 4.7).

Figure 4.8 shows three possible path representations depending on whether we are considering the real Earth, a representational Earth with a radius of 8500km ($K = 4/3$) or a flat Earth. The vertical gradient which gives a straight line path is obviously not the same in each of the three cases.

If we compare vertical soundings carried out simultaneously at two points a few tens of kilometres or a few hundred kilometres apart, we find that super-refraction conditions extend over very large areas, as they correspond to stable atmospheric structures (the densest air layers are the lowest). On the other hand, subrefraction conditions extend only a short way, as they correspond to unstable atmospheric structures; consequently, for a given path low values of K will not appear simultaneously at all points on the path. This means that there is a damping effect to variations in K in the low value region, and the longer the path, the greater this effect. A curve deduced from experimental results and showing this effect is given in Chapter 6; it gives the minimum value of K for a path.

4.2.2.4. Ducted propagation in the atmosphere

4.2.2.4.1. Necessary conditions

We have seen that if the vertical refractive index gradient is less than −157N/km, radio waves are more curved than the Earth's surface. An atmospheric zone in which such refractive conditions exist is called a ducting layer. If it is low enough, a path which starts by travelling upwards will take the form of an arc which will meet the Earth again at another point. If, at this point, the Earth's surface is reflecting, a second arc will form, and so on. This gives a groundbase duct. If the atmospheric zone in which these refraction conditions exist is sufficiently high, paths will not return to Earth, but will take various different curvatures, alternating between higher values than the Earth's curvature in the duct zone, and less than the Earth's curvature in the region of the atmosphere below it. They therefore remain between two fixed altitudes defining an elevated duct. In all these cases, we speak of ducted propagation (Fig. 4.9). Conditions creating such situations were outlined in Section 4.1.3.

If Equation (4.14) is applied, which gives the path of a radio wave at the starting point (n_0, r_0, φ_0) and the top of a path (n, r, 0):

$$n_0\, r_0 \cos \varphi_0 = nr \tag{4.37}$$

where:

$$\cos \varphi_0 = \frac{n}{n_0} \cdot \frac{r}{r_0}. \tag{4.38}$$

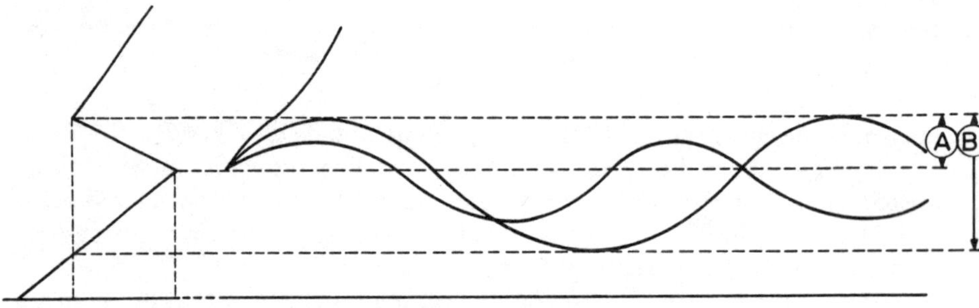

a) Radio duct above the Earth's surface, propagation without reflection from the Earth

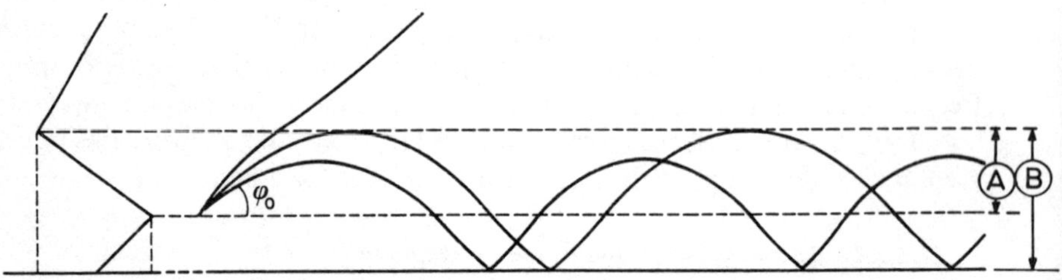

b) Radio duct above the Earth's surface, propagation with reflection from the Earth

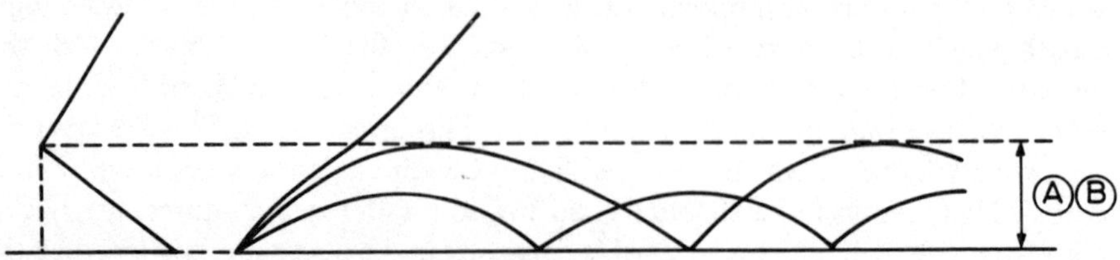

c) Radio duct at the Earth's surface

Fig. 4.9. Ducted tropospheric propagation: A: ducting layer; B: tropospheric radio duct.

As an atmospheric layer with a very high refractive index gradient must necessarily be of limited thickness, the two quantities n/n_0 and r/r_0 must be near 1, and therefore φ_0 is small. The only waves that can be ducted are therefore those which begin at a very low angle of elevation, generally less than 1° and consequently in the following cos φ can be replaced by a limited expansion.

As, in ducted propagation, the effective radius of the Earth can take widely varying values, it is best to use a flat Earth representation, i.e. to use the modified refractive index M:

$$M = N + \frac{h}{a}. \tag{4.39}$$

Under these conditions, we have seen that the path equation can be written in the form:

$$\varphi^2 - 2\,M = \text{Cte}\,. \tag{4.40}$$

Now write this relation for a starting point (M_1, φ_0) and the point at which the tangent becomes horizontal (M_2, 0):

$$\varphi_0^2 = 2(M_1 - M_2)\,. \tag{4.41}$$

It is clear that ducting will only take place if within an atmospheric layer ($M_1 - M_2$) is positive. The maximum angle φ_0 is obtained if the point of departure is at the height where M is maximum. If we use straight lines as a schematic representation of the variations of M as a function of altitude, we obtain the situation shown in Fig. 4.9. It is clear that the elevated duct lies between two altitudes at which we find the same value for M, the other altitude being that at which ducting conditions end. In cases of a duct with reflection by Earth, which is the most frequently met, the lower part of the curve for M may be missing. In this case, M decreases from the ground, and path curvature is always downwards.

As shown in Section 4.1.3 the thickness of a duct is seldom greater than a few hundred metres, but it can extend a very long way horizontally (several hundred kilometres) particularly above sea because of the homogeneous conditions at the surface. It also needs to be borne in mind that there is sometimes an evaporation duct at the sea's surface which may be of thickness between a few metres and a few tens of metres, and which can exist for a long time.

4.2.2.4.2. Theory of propagation in a duct

There are two ways of approaching the study of propagation in a duct: geometric theory and mode theory. We examine both briefly.

(a) Geometric theory

The geometric method is based on tracing paths emerging from the transmitting antenna, for regularly increasing angles of elevation, e.g. for every milliradian.

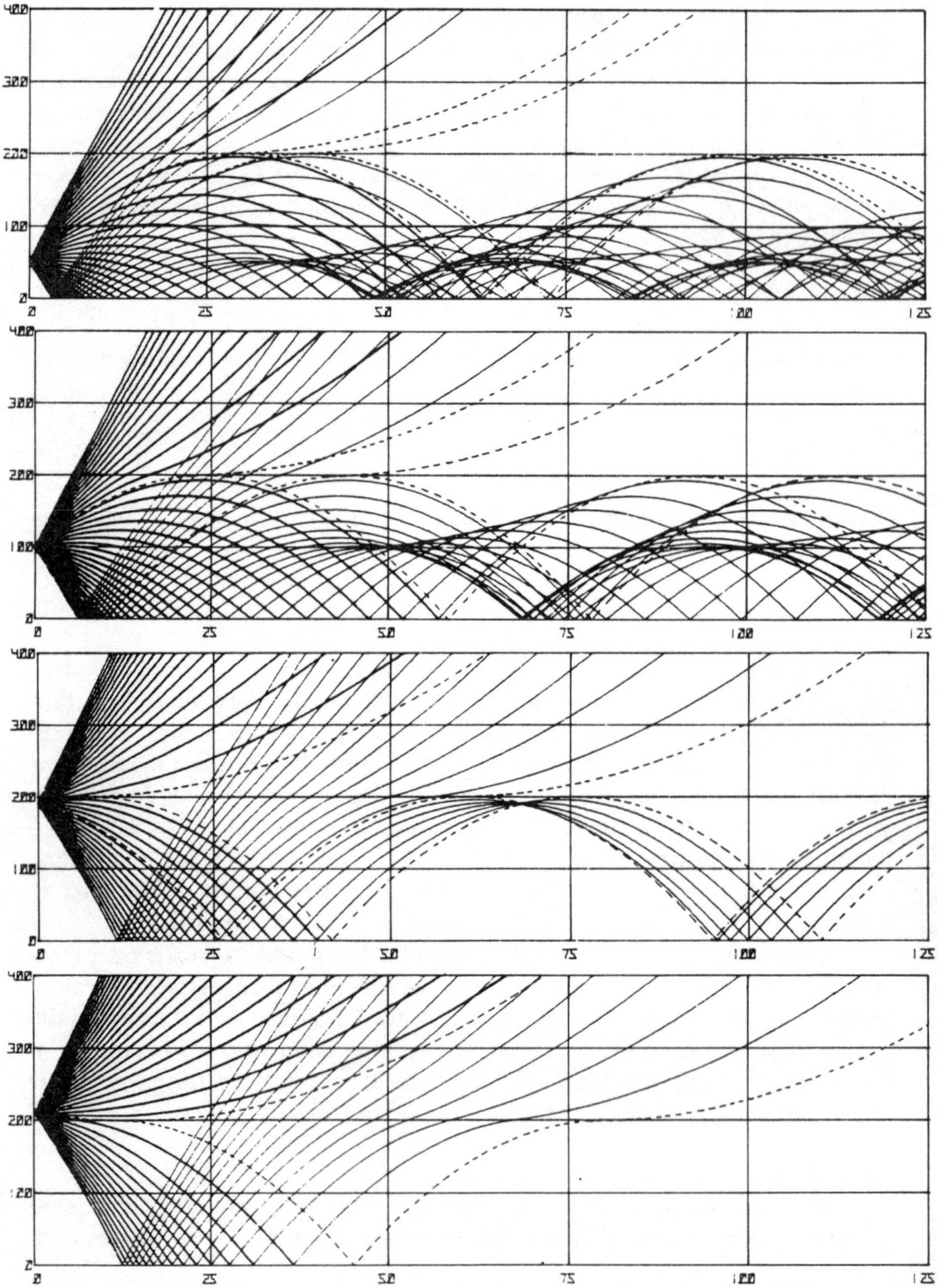

Fig. 4.10. Path forms: Refractive index gradient in the duct: –500N/km. Thickness of the duct: 200m. Height of transmitter: 50m; 100m; 190m; 210m.

Examination of all these paths shows that energy is concentrated along certain lines (caustics) and that it is impossible to reach certain regions even within a duct (radio holes). Outside these regions and caustics, the concentration of paths gives an idea of the energy concentration (Fig. 4.10).

How to calculate these paths is shown in Appendix 4.1 for a duct at ground level. Each path is then made up of a series of arcs whose maximum lengths between two points of reflection depend on the height H of the duct, and the refractive index gradient $\mathrm{d}N/\mathrm{d}h$ within the duct as specified by the relation:

$$L = 2\sqrt{\frac{2\,000\,H}{-(\mathrm{d}N/\mathrm{d}h + 157)}}, \tag{4.42}$$

where H is given in metres, L in kilometres and $\mathrm{d}N/\mathrm{d}h$ in N/km ($\mathrm{d}N/\mathrm{d}h < -157$).

Table 4.1 gives L in kilometres for various typical values of H and $\mathrm{d}N/\mathrm{d}h$. Lengths can be very great, and on a path including ducting effects, the number of arcs is often very small.

Appendix 4.1 also shows that if the transmitter and receiver are at the same height within the duct and if the Earth's surface is reflecting, there will always be at least one possible path between them if the height is less than 3/4 of the thickness of the duct. This is not true for greater heights.

(b) Mode theory

As with metallic ducts, a wave can be propagated within a tropospheric duct in a number of modes. That number depends on the ratio of the thickness of the duct to the wavelength. Assuming linear decrease of the modified index M with altitude the theory suggests the following relation for the minimum depth H for the n order mode (see Appendix 4.1):

$$\frac{H}{\lambda} = \frac{A_n}{\sqrt{\Delta M}}. \tag{4.43}$$

ΔM is the total decrease in M in the duct layer and A_n is a number depending on the

Table 4.1. Values of L for values of H.

$\frac{\mathrm{d}N}{\mathrm{d}h}$ (N/km) \ H (m)	100	200	300
− 160	516	730	894
− 170	248	351	430
− 180	186	264	324
− 200	136	192	236
− 250	93	131	161
− 300	75	106	129
− 500	48	68	84
− 1 000	31	44	53

rank of the mode considered. For a duct at ground level the following approximate values exist:

$$A_1 = 400 \qquad A_2 = 930 \qquad A_3 = 1\,460\,. \tag{4.44}$$

Instead of taking the value of ΔM in the duct layer as a parameter, we can use the refractive index gradient or even the (negative) value of the multiplying coefficient of the Earth's radius, taking account of the following relation (note that ΔM which represents a decrease in M has the opposite sign to the gradient of M):

$$-\frac{\mathrm{d}M}{\mathrm{d}h} = \frac{\Delta M}{H}, \tag{4.45}$$

we obtain the following expression, in which $\mathrm{d}M/\mathrm{d}h$ is expressed in units of M/m:

$$\frac{H^{3/2}}{\lambda} = A_n/\sqrt{-\,\mathrm{d}M/\mathrm{d}h}\,. \tag{4.46}$$

Moreover:

$$\frac{\mathrm{d}M}{\mathrm{d}H} = \frac{\mathrm{d}N}{\mathrm{d}h} + \frac{1}{a} = \frac{1}{Ka}, \tag{4.47}$$

we can also express the minimal height as a function of coefficient K:

$$\frac{H^{3/2}}{\lambda} = A_n\sqrt{|\,K\,|\,a} = C_n\sqrt{|\,K\,|} \tag{4.48}$$

with:

$$C_1 = 1\,004 \qquad C_2 = 2\,342 \qquad C_3 = 3\,685 \tag{4.49A}$$

where H and λ are expressed in metres.

Example: Let us suppose that near the Earth there is an atmospheric layer 100m thick with a refractive index gradient of 300N/km, and normal atmosphere above it. Within this layer the decrease in M is equal to:

$$\Delta M = 30 - \frac{100}{6.37} = 14.30 \text{ M units} \tag{4.49B}$$

which gives for the first mode from Equation (4.43):

$$\frac{H}{\lambda} = \frac{400}{14.30} = 105.8\,. \tag{4.49C}$$

The longest wavelength that can be ducted is therefore:

$$\lambda = \frac{100}{105.8} = 0.945 \text{ m} \quad \text{or } 317\text{MHz} \tag{4.49D}$$

We can see from this example that a tropospheric duct whose thickness is of the order of a few hundred metres can duct metre range waves and shorter waves (VHF and above); the number of ducted modes is in single figures for the metre wave range and in double figures for the decimetre wave range. If the number of possible modes is very high, it is best to use geometric optical methods to calculate paths.

The cut-off frequency defined in this way is not as clean-cut a boundary as in the case of metallic wave ducts. It would be more accurate to say that once the frequency falls below the cut-off frequency, the duct lets more and more energy escape. Moreover, fluctuations in the refractive index near the top of the duct will make the cut-off frequency even less definite.

4.2.2.4.3. Propagation attenuation in a duct

Neither the geometric theory nor the mode theory provide reliable values for attenuation in the duct, even if the structure of the duct is stable and well understood, though in fact, it is never either. Moreover, the field is not homogeneous throughout the duct. If we consider energy alone, however, and assume that the ducting distance is very long, we can use the following reasoning.

Energy emerging from the transmitter remains confined between two horizontal parallel planes. The power flux density across a cylindrical surface with a vertical axis and radius d diminishes as we move further away from the transmitter, on the one hand as a function of $1/d$ as the total area of the cylinder over which power is being distributed increases with distance, and on the other hand exponentially, because of losses due to irregularities in the lower and upper surfaces of the duct. Power received is therefore proportional to:

$$\frac{1}{d}\,\mathrm{e}^{-\alpha d}\,. \tag{4.50}$$

If we compare this power with what would have been received in free space, which falls off by $1/d^2$, it is clear that attenuation with respect to free space varies, beyond a certain distance, as a function of distance given by:

$$d.\mathrm{e}^{-\alpha d}\,. \tag{4.51}$$

It is therefore not impossible to obtain received levels significantly higher than those in free space if α is sufficiently small, particularly above sea. Over land, the surface does not generally reflect enough for there to be authentic ducted propagation with reflection from the Earth's surface. In this case, the path is generally reduced to a single 'arc'. Obviously, there can be elevated ducts for which reflection from the surface is not necessary, but such ducts have no major role to play unless both ends of the link are close to them.

4.2.2.4.4. Role of ducting

Although ducting can provide extremely high received levels, it is almost impossible to use because of its sporadic occurrence. Its role is almost entirely harmful both for line of sight links, in that it can lead to deep fades, and for paths beyond the horizon, because it causes interference. These two aspects are studied in Chapters 6 and 10, respectively.

4.2.3. Links with satellites: large scale refraction effects

When the path passes through a large part of the atmosphere at a high angle of elevation (links with aircrafts or rockets) or if it passes through the entire atmosphere (links with a satellite), the approximations made above are no longer valid. In particular, the concept of an effective radius of the Earth can no longer be used, and we have to start again from general equations. Since these links are often established with moving bodies or are intended for measurements, certain refraction effects must be accurately known in order to make the necessary corrections (angle of arrival, propagation delay, Doppler effect, etc.).

4.2.3.1. Directional variations

Because of the curvature of paths due to refraction, the radio direction of a satellite is not exactly the same as its geometric direction (Fig. 4.11). The difference between the two is expressed as the difference: $\varphi_0 - \varphi_1$. If the satellite is a long way off, it can be seen from the figure that the difference is almost exactly equal to the difference: $\alpha - \alpha_0$. Returning to Equation (4.20), and integrating over the entire depth of the atmosphere:

$$\alpha - \alpha_0 = \int_1^{n_0} \tan V \frac{\mathrm{d}n}{n} . \tag{4.52}$$

If we calculate tan V taking Equation (4.16) as our starting point we obtain:

$$\alpha - \alpha_0 = \int_1^{n_0} \frac{\cos \varphi_0 (a/r)\, \mathrm{d}(n_0/n)}{\sqrt{1 - (a/r)^2 (n_0/n)^2 \cos^2 \varphi_0}} . \tag{4.53}$$

For a real refractive index pattern and even for most reasonable refractive index profile models, it is only possible to carry out these calculations using numerical methods. An upper limit for $\alpha - \alpha_0$ can however easily be obtained. If we return again to Equation (4.52) and integrate by parts:

$$\alpha - \alpha_0 = \int_1^{n_0} \tan V \frac{\mathrm{d}n}{n} = [\ln n \tan V]_{n=1}^{n=n_0} - \int_{V_\infty}^{V_{\mathrm{sol}}} \frac{\ln n}{\cos^2 V} \mathrm{d}V . \tag{4.54}$$

Where the integrated part is equal to:

$\ln n_0 \tan \alpha_0$.

As n_0 has a value close to 1:

$$n_0 = 1 + (n_0 - 1) . \tag{4.55}$$

If we now consider the series expansion of the logarithm:

$$\ln n_0 = (n_0 - 1) - A . \tag{4.56}$$

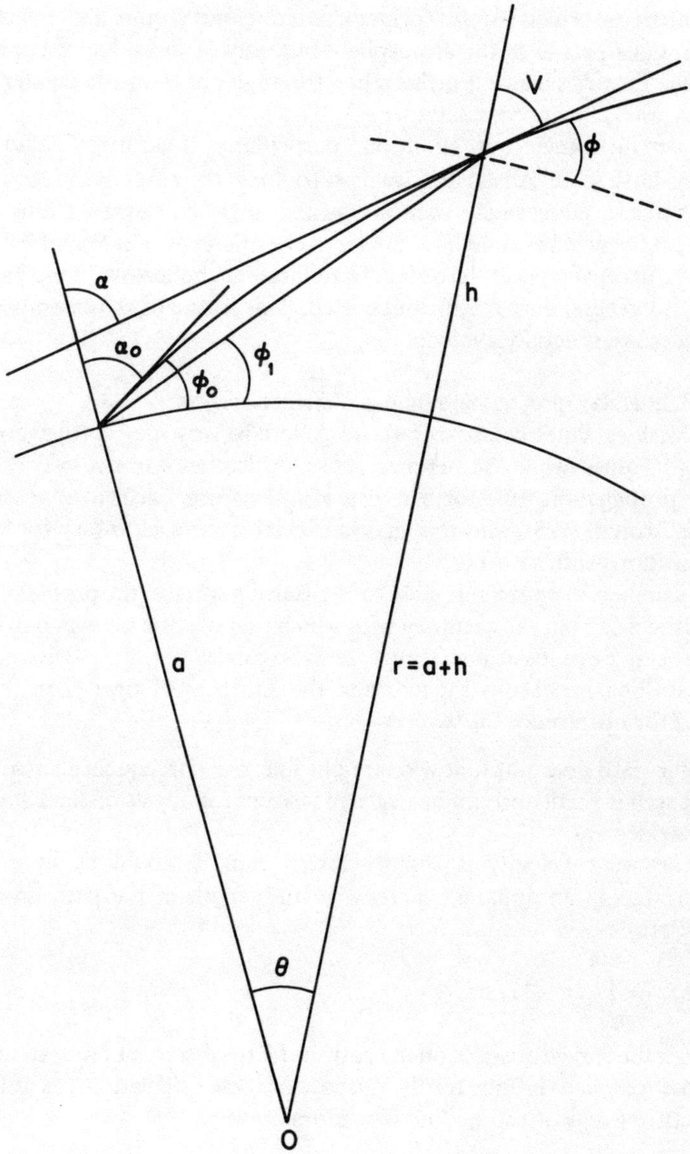

Fig. 4.11. Path between the Earth's surface and a point in space.

Where A is a positive quantity which is small with respect to $n_0 - 1$ and the remaining integral is positive. The following approximation therefore gives a value for the upper limit:

$$\alpha - \alpha_0 \simeq \frac{n_0 - 1}{\tan \varphi_0}. \tag{4.57}$$

Very frequent use is made of this formula, as it does not require a knowledge of the refractive index profile of the atmosphere but only of the value of the refractive index at the Earth's surface. Furthermore, for angles of elevation greater than 20° it gives an adequate approximation.

For smaller angles of elevation, particularly if accurate radio ranging operations have to be carried out, we need to know the refractive index profile for the atmosphere, increasingly accurate as the angle of elevation falls. A mean monthly profile may be enough for angles of elevation between 5° and 20°, if we are prepared to accept errors of the order of a milliradian, but below 5° and particularly below 2° no general correction can be used. This is one of the reasons why such small angles are generally avoided.

4.2.3.2. Delay, propagation time and distance errors

As links with satellites or spacecraft are generally very long, propagation times reach high values which sometimes cause difficulties for two-way links. For example, propagation time for a return journey from Earth to a geostationary satellite is around 0.25s, and this makes special devices necessary for telephone communications with satellites.

For satellites or spacecraft close to the Earth's surface, the problem is not the length of the delay but the accuracy with which that delay is known, particularly if remote sensing or geodesic operations are to be carried out. Propagation times for waves travelling between a satellite and the Earth are longer than the figures calculated for open space for two reasons:

1. The path does not follow a straight line: the consequent increase in path length is small and can be neglected except for angles of elevation below a few degrees.
2. The wave velocity is slightly lower than it would be in a vacuum, producing an apparent increase in the length of the path given by the relation:

$$\Delta L = \int_0^R (n - 1)\, \mathrm{d}s\,. \tag{4.58}$$

s being the curved abscissa on the path and R the distance of the satellite, which can be treated as infinite for the purpose of these calculations, as atmospheric effects not only affect the first few kilometres.

As the real path does not deviate very much from a straight line with angle of elevation φ_0 as long as φ_0 is greater than a few degrees:

$$\Delta L = \int_0^\infty \frac{(n - 1)}{\sin \varphi_0}\, \mathrm{d}h \tag{4.59}$$

where h is the vertical altitude. An approximate value of ΔL can be obtained assuming that the atmosphere is exponential:

$$n - 1 = (n_0 - 1)\, \mathrm{e}^{-bh}, \tag{4.60}$$

where b is a scale of altitude associated with n_0 and with its vertical gradient near the Earth's surface by the relation:

$$b = \frac{(\mathrm{d}n/\mathrm{d}h)_0}{n_0 - 1}. \tag{4.61}$$

Under these conditions, Equation (4.59) gives the following result:

$$\Delta L = \frac{(n_0 - 1)^2}{(\mathrm{d}n/\mathrm{d}h)_0 \sin \varphi_0}. \tag{4.62}$$

This relation is a good approximation as long as the variation in n near the ground is not too irregular, and for angles φ_0 greater than about 5°. For example, if the atmosphere has the following characteristics:

$N_0 = 300$ i.e. $n_0 = 1.0003$

$(\mathrm{d}N/\mathrm{d}h)_0 = 40$ i.e. $(\mathrm{d}n/\mathrm{d}h)_0 = 40.10^{-6}$.

We find that ΔL is equal to 2.25m in the direction of zenith. Figure 4.12 gives the mean increase in path length in dry air and for a saturated humid atmosphere.

4.2.4. Effects of small scale variations in refraction

If we study the atmosphere on a smaller scale, we find that locally equal refractive index surfaces are not perfect concentric spheres but constantly varying roughly irregular surfaces. At any moment there can therefore be several possible propagation paths between any two given points.

A comment needs to be made first of all on the spatial scale of fluctuations of this kind. The Earth's surface represents a fundamental discontinuity surface for the atmosphere, with an influence not only on the mean variation on the refractive index but even on small scale fluctuations. For example, we may at certain moments find fluctuations which are much larger horizontally than vertically. Irregularities of this kind are often called tropospheric layers and they produce partial reflections (Section 2.4.1.4). A major consequence of this situation is that, over horizontal links, propagation phenomena sometimes appear different from those encountered in oblique links. In other words, the horizontal link cannot always be treated as an oblique link whose angle of elevation tends towards zero. The various paths which may exist at the same time through the troposphere generally create multiple secondary waves which interfere with the main wave, having followed paths with slightly different lengths. In general, all these additional waves are of small amplitude compared with the main wave, and only cause scintillations about the received level. Under certain conditions, however, one of these waves, or a small number of them can reach an amplitude of the same order of magnitude as the main wave, and lead to very deep fades. We then speak of multi-path propagation.

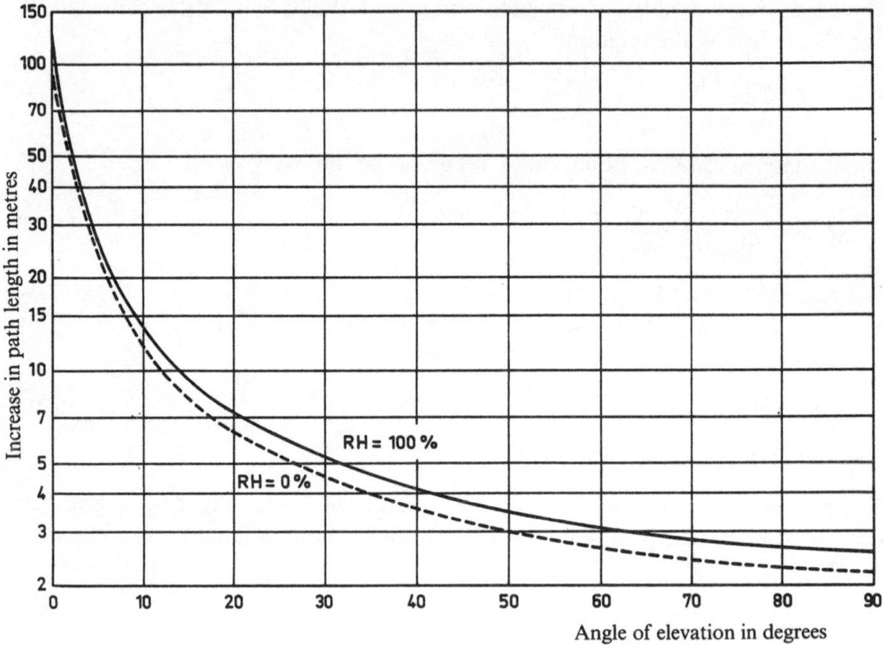

Fig. 4.12. Apparent increase of the path in length due to refraction.

4.2.4.1. Statistical distribution of the amplitude of a wave undergoing propagation by tropospheric multi-paths

If no stable reflection from the Earth's surface takes place and there is no tropospheric duct for an appreciable proportion of the time, the following considerations apply. The wave leaving the transmitting antenna undergoes alteration, as a result of refractive index fluctuations, which increase with distance. These alterations may be considered as due to additional waves created by radiation from irregularities excited by the incident wave. If we ignore losses due to the transmission medium, the sum of the re-radiated power and the power remaining in the initial wave is constant. The total power flux density decreases as a proportion of $1/d^2$ just as though no multi-paths had existed. However, on reception, the various waves interfere with each other and produce fades. From the statistical point of view, the received level may be considered as the resultant of a geometric sum of a determinist vector and a random vector with amplitude following a Rayleigh distribution and uniformly distributed phase. This gives the Nakagami-Rice distribution.

This model gives a distribution which has two characteristics also observed in experimental results:

1. The distribution of deep fades has the same slope as a Rayleigh distribution (10dB for every decade of probability).

2. Lower amplitude fluctuations have a Gaussian distribution.

For this model to be applicable to a given link, we have to know the relations between the path characteristics and those of the statistical distribution. If we assume that small scale fluctuations in refractive index are distributed homogeneously throughout the path, the decrease in the deterministic vector is an exponential function of distance:

$$P_d = P_0 \, e^{-d/\Delta} . \tag{4.63}$$

where P_0 is the power in free space and Δ is the parameter representing a length and which must be experimentally determined for given frequencies and climates. Assume (see Appendix 1) that:

$$e^{-d/\Delta} = \cos^2 \varphi . \tag{4.64}$$

We can then show that for very small fractions of total time the probability that a very low relative level is not reached is given by the asymptotic relation:

$$\mathrm{Prob}\,(X^2 < x^2) \simeq e^{-1/\tan^2\varphi} \, \frac{x^2}{\sin^2 \varphi} , \tag{4.65}$$

where x^2 is the ratio between the power considered and the power received under free space conditions. It is clear that this probability value is proportional to the power received. This relation is valid for attenuation greater than about 15dB ($x^2 < 3{,}10^{-2}$), and for distances greater than 40km. For shorter distances or lower attenuation we have to use the general formula (see Appendix 1):

$$\mathrm{Prob}\,(X^2 < x^2) = e^{-1/\tan^2\varphi} \int_0^{\frac{x^2}{\sin^2\varphi}} e^{-t} \, I_0\!\left(\frac{2\sqrt{t}}{\tan\varphi}\right) dt . \tag{4.66}$$

For levels which are not very different from the level in open space, i.e. for values of x^2 close to 1, it can be shown that the distribution of variable x tends towards a Gaussian distribution with a mean value and standard deviation given by:

$$\begin{aligned} m &= \cos \varphi \\ \sigma &= \frac{\sin \varphi}{\sqrt{2}} . \end{aligned} \tag{4.67}$$

This situation corresponds in particular to short distances, for which the mean value of the field is close to 1 (0dB), and the standard deviation is given by the relation:

$$\sigma \simeq \sqrt{\frac{d}{2\,\Delta}} . \tag{4.68}$$

4.2.4.2. Application to terrestrial communications

The model described above applies very well to terrestrial communications in the frequency range between about 2 and 30GHz. Frequency does not appear explicitly, but as a factor in the value of Δ. Experimentally, we find that the influence of frequency is relatively slight. An acceptable empirical approximation is to assume that Δ is inversely proportional to the fourth power root of frequency. (The function sometimes used, according to which the probability of fading of given depth is proportional to frequency, generally introduces larger corrections. Moreover, such a correction, which is independent of the time percentage and distance, can only be very restricted in its application.)

Regarding the influence of climate on the value of Δ the data obtained by the CCIR show that for North-West Europe and for the least favourable month at 4GHz, a value of 265km should be used. For the United States and Japan different values are needed, but they are relatively difficult to work out, as the data mentioned by the CCIR take into account certain paths involving reflection from the ground or are calculated for the least favourable season rather than the least favourable month. Under these conditions, the range of values of Δ would be around 200km in certain regions of the United States and 350km in Japan.

For communications over distances longer than about 50km, the amplitude of the deterministic vector is no longer predominant, and tends rapidly towards a Rayleigh distribution, which gives a limiting value to attenuation with increasing distance. A further assumption has to be made. We can assume that the mean power associated with Rayleigh's vector is not a fixed number, but has a log-normal distribution, as in cases of communications by tropospheric scatter. We have to take into account that the signal received during the least favourable month at the end of a long line-of-sight link has characteristics relatively similar to those of a tropospheric scatter link. Experience shows that on line of sight links the standard deviation of this log-normal distribution is proportional to distance.

$$\sigma = \frac{d}{d_0}. \tag{4.69}$$

Appendix 1 gives the statistical distribution resulting from the combination of a Rayleigh distribution and a log-normal distribution. With practical applications in mind, it is worth noting that for low levels, i.e. for high proportions of time, the log-normal distribution can be taken into account by subtracting, from the level corresponding to a Rayleigh distribution alone, a number of decibels equal to 0.115 $(d/d_0)^2$.

For North-West Europe, a distance of 25km is generally used for d_0. The correction is negligible for short distances and only becomes significant beyond about 50km. The graphs of distributions obtained with these values, for a certain number of distances are shown in Fig. 4.13. In other parts of the world other values have to be used for Δ and d_0.

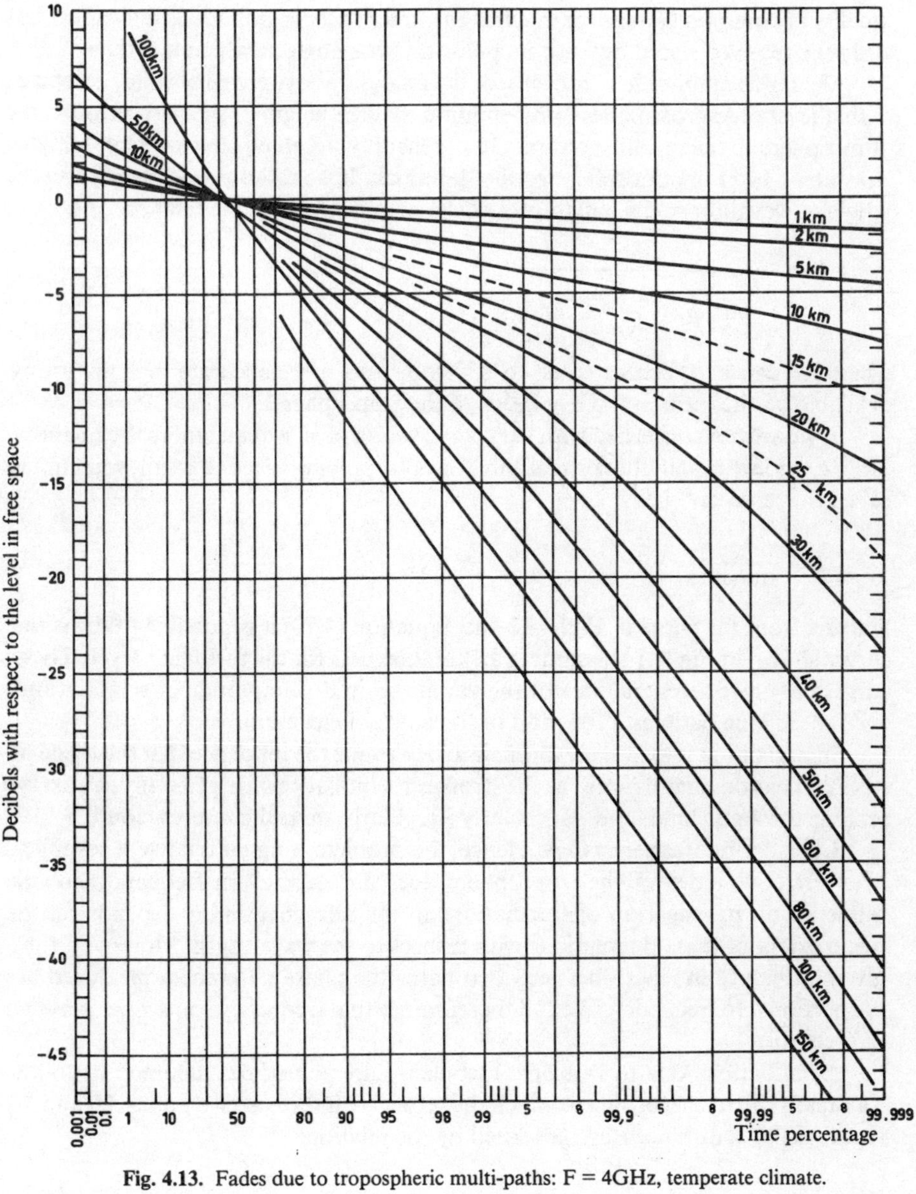

Fig. 4.13. Fades due to tropospheric multi-paths: F = 4GHz, temperate climate.

4.2.4.3. Application to Earth-satellite communications

If the angle of elevation at the point of departure on the Earth's surface is not very small, the length of path in the troposphere is relatively short, and generally does not involve deep fading. Scintillations, on the other hand, may affect the quality of communications for much of the time. These scintillations may affect not only the received signal but also its phase and the angle at which it arrives.

As far as amplitude is concerned, the analysis above remains valid, at least at a first level of approximation, on condition that the length of the path through the atmosphere is taken into account. It is generally accepted that we have only to consider a layer of atmosphere about 1km thick. If we call this thickness h and the angle of elevation φ, the length of l of the path is given by the relation:

$$l = \frac{2\,h}{\sqrt{\sin^2 \varphi + \dfrac{2\,h}{Ka}} + \sin \varphi}, \tag{4.70}$$

where Ka can be taken as equal to 8500km if no more accurate information is available on the mean refractive index of the troposphere.

If we are not concerned with very small values of φ, we can ignore the quantity $2\,h/Ka$, and we obtain the form of this relation applying to a flat Earth, sometimes called the 'cosecant' law:

$$l \simeq \frac{h}{\sin \varphi}. \tag{4.71}$$

Starting from the curve in Fig. 4.12 and Equation (4.70) it is possible to draw the curves shown in Fig. 4.14, assuming a 1km thickness for the turbulent layer. These curves only give a general idea of the variations in the amplitude of scintillations over an oblique path, as a function of the angle of elevation.

In practice, it is extremely difficult to determine the influence that the angle of elevation has on scintillations, as the strongest scintillations are generally caused by passage through cloud, and particularly irregularly shaped cumulus clouds.

Considering frequency dependence, the problem is again relatively complex. The refractive index of the troposphere does not depend on frequency but the influence on propagation of fluctuations in the refractive index depends on the mean extent of these fluctuations with respect to the wavelength. Moreover, for a given difference in length between two paths the phase differences produced are proportional to frequency. We can therefore see that frequency sometimes plays an indirect role.

Scintillations due to isotropic turbulence are sometimes assumed to follow Tatarski's theory, according to which the refractive indices at two points M_1 and M_2 separated by a distance r are governed by the relation:

$$\overline{[n(M_1) - n(M_2)]^2} = C_n^2\, r^{2/3}, \tag{4.72}$$

where Cn^2 is a characteristic of the amplitude of the fluctuations in refractive index. Its value near the Earth's surface is of the order of between 10^{-15} and 10^{-10} $m^{-2/3}$ but it falls off suddenly beyond 1 or 2km. Equation (4.72) is valid for r between two values l_0 and L_0 which are of the order of a few millimetres and a few tens of metres, respectively. It is for wavelengths within this range of values that the most striking effects occur.

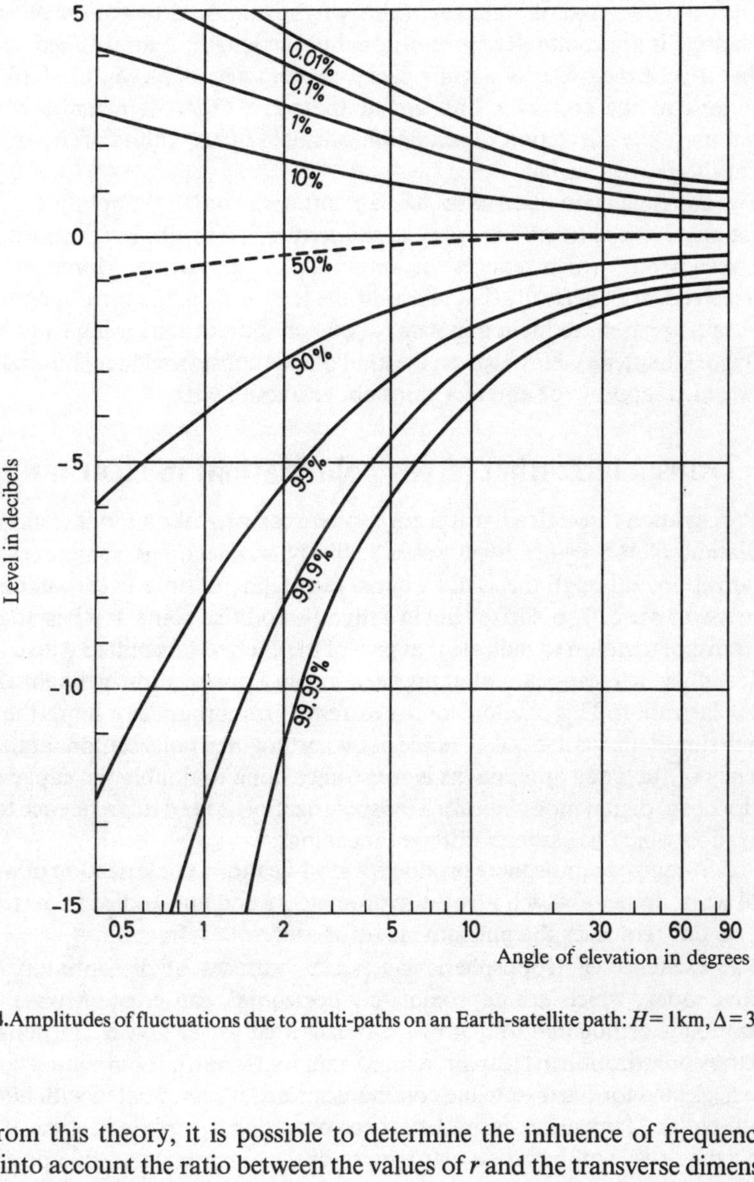

Fig. 4.14. Amplitudes of fluctuations due to multi-paths on an Earth-satellite path: $H = 1$ km, $\Delta = 350$ km.

From this theory, it is possible to determine the influence of frequency by taking into account the ratio between the values of r and the transverse dimensions of the Fresnel ellipsoid in the turbulent zone. In particular, if this ratio is far greater than 1, it can be deduced from theory that the variance of the logarithm of the signal amplitude is proportional to $F^{7/6}$, while, if it is much smaller, the variance is proportional to F^2. It is therefore possible to find both types of frequency dependency within a single path.

Fluctuations in angle of arrival can be important in certain cases. Given the structure of the atmosphere, fluctuations in angle of elevation are very much greater (some ten times greater) than those in the azimuth for low angles of elevation.

Experience shows that the median value of the standard deviation of angular scintillations is approximately proportional to the length of atmosphere crossed, and that it is of the order of 3 millidegrees for an angle of elevation of 10°. Also fluctuations at the end of a link are of the same order of magnitude as the uncertainty in the correction to refraction calculated from values of the refractive index at the Earth's surface.

The antenna gain used also has an influence on the amplitude of these fluctuations. If the antenna has a very large aperture, it smooths out the fluctuations at the wave front, which reduces the amplitude of fluctuation. Moreover, as the wave received is not perfectly flat, phase differences between the various points lead to a certain apparent reduction in gain, as on communications using tropospheric scatter (see Chapter 8). However, in the kind of case considered here, this reduction only begins to appear for antenna gains above about 50dB.

4.3. Transpolarization (cross polarization) in clear air

Propagation through a natural atmosphere can provoke a modification in the polarization of the transmitted waves. This phenomenon is sometimes called depolarization, although this is not what in fact happens, since in certain cases the wave remains perfectly polarized but in a slightly modified plane. It is best to use the word transpolarization to indicate that part of the energy transmitted with a certain polarization which reappears, after propagation, in a polarization orthogonal to the initial polarization. This phenomenon is extremely important, as it limits the extent to which simultaneous use can be made of two orthogonal polarizations at the same frequency of the same antenna, as is sometimes done to double the capacity of a radio link (e.g. digital modulation). Transpolarization is used in preference to cross polarization which has several different meanings.

A non-ionized atmosphere produces a modification in polarization of waves in several ways, some of which involve hydrometeors and are studied in Section 5.4. Here, we only consider the phenomena associated with refraction.

The existence of tropospheric layers, i.e. surfaces of discontinuity in the refractive index, which are approximately horizontal, can create a wave whose polarization is orthogonal to that of the transmitted wave. This effect, sometimes called transpolarization in clear air, is important for terrestrial communications but can be neglected for Earth-satellite communications. It is associated with fades due to multi-paths. Many mechanisms have been proposed to explain it, some of which assign a major role to the crosspolar pattern of the antenna. It would appear that, to reduce the influence of the antenna, a crosspolar pattern which is roughly homothetic to the copolar pattern in the main lobe needs to be used, rather than a crosspolar pattern with a zero in the axis.

It can be shown experimentally for a given level of fading on the parallel polarization, called copolar fading (*CPF*), that the level on orthogonal polarization follows a log-normal distribution. If *XPD* denotes the orthogonal polarization discrimination (i.e. *XPD* is the median relative level expressed in decibels of the orthogonal polarization with respect to the level in open space of the parallel polarization) we obtain the following approximate relation:

$$CPF + XPD = C \tag{4.73}$$

where C takes a value (in decibels) which depends on the path (frequency, distance) and on the characteristics of the antennas, and is of the order of 45 to 55dB, for a distance of some tens of kilometres for UHF wave range.

Appendix 4.1
Ray paths for a duct close to the Earth's surface

Assume that above the Earth's surface, which is regarded as spherical, there is a layer of atmosphere of thickness H where the vertical gradient of the index $(dN/dh)_1$ is less than -157N/km, above which there is an atmosphere of very great depth, where the vertical gradient $(dN/dh)_2$ is assumed to have a value, for example, close to the normal value: -40N/km. Finally assume that a transmitter is in a tropospheric duct at an altitude y_0. We wish to calculate the paths emerging from the transmitter as a function of their initial angle φ_0. For this problem it is natural to use a representation based on a flat Earth. Distances are written x and heights y (Fig. 1).

Equation (4.29) gives the path equation in the units mentioned. If N_0 denotes the refractive index at the Earth's surface, this formula can be written:

$$N_0 + \left(\frac{dN}{dh}\right)_1 y + \frac{y}{a} - \frac{\varphi^2}{2} = N_0 + \left(\frac{dN}{dh}\right)_1 y_0 + \frac{y_0}{a} - \frac{\varphi_0^2}{2}. \tag{1}$$

let:

$$\frac{1}{a} + \left(\frac{dN}{dh}\right)_1 = b_1, \tag{2}$$

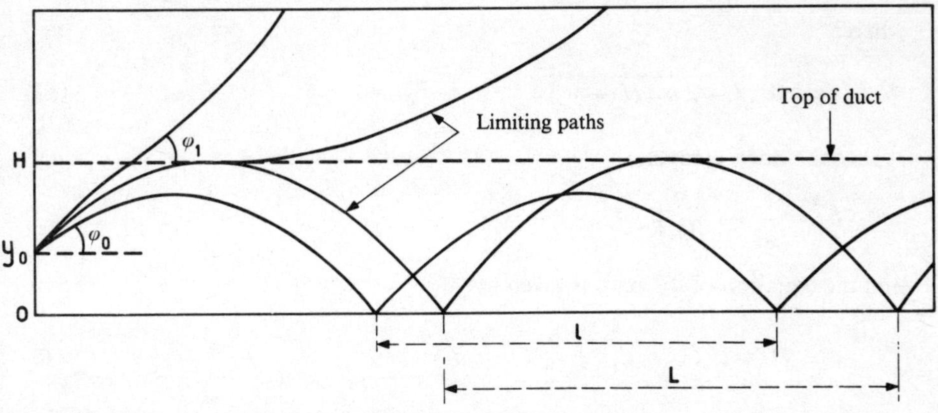

Fig. 1. Path forms (duct at the Earth's surface).

where b_1 is the gradient of the modified index in the duct (b_1 is negative):

$$b_1 y - \frac{\varphi^2}{2} = b_1 y_0 - \frac{\varphi_0^2}{2} \tag{3}$$

which on differentiation, gives:

$$b_1 \, dy - \varphi \, d\varphi = 0 ; \tag{4}$$

furthermore, as angle φ is small:

$$\frac{dy}{dx} = \tan \varphi \simeq \varphi , \tag{5}$$

from which we can deduce an expression in parameter form for the path:

$$b_1 x = \varphi - \varphi_0 , \tag{6}$$

$$b_1(y - y_0) = \frac{\varphi^2}{2} - \frac{\varphi_0^2}{2} , \tag{7}$$

eliminating parameter φ gives the equation in Cartesian coordinates:

$$y = y_0 + \varphi_0 x + \frac{b_1}{2} x^2 . \tag{8}$$

from which it is clear that we are dealing with a parabola which, within the duct, is concave downwards. The parabola must be cut off at the Earth's surface ($y = 0$), but it can also be cut off at the top of the duct ($y = H$) if the starting angle is too large. The path then leaves the duct. If φ_1 denotes the output angle and d_1 for the corresponding output distance:

$$\varphi_1 = \sqrt{\varphi_0^2 + 2 b_1(H - y_0)} \tag{9}$$

$$d_1 = \frac{\varphi_1 - \varphi_0}{b_1} , \tag{10}$$

where:

$$| \varphi_0 | \geqslant \sqrt{- 2 b_1(H - y_0)} . \tag{11}$$

Above the ducting layer:

$$b_2 = \frac{1}{a} + \left(\frac{dN}{dh}\right)_2 \tag{12}$$

and the equation of the path is given by:

$$y = H + \varphi_1(x - d_1) + \frac{b_2}{2}(x - d_1)^2 . \tag{13}$$

Length of 'arcs' in a duct

If angle φ_0 is less than the value calculated above, the path remains in the duct. It will then be reflected from the ground at distance:

$$\frac{\varphi_0 + \sqrt{\varphi_0^2 - 2\,b_1\,y_0}}{-\,b_1}. \tag{14}$$

This will lead to the formation of a series of arcs. The length of each of these is given by the relation:

$$L = \frac{2}{-\,b_1}\sqrt{\varphi_0^2 - 2\,b_1\,y_0}\,. \tag{15}$$

The maximum value of L is obtained when angle $|\varphi_0|$ is equal to the limiting value calculated above:

$$|\,\varphi_0\,| = \sqrt{-\,2\,b_1(H - y_0)}\,. \tag{16}$$

The maximum arc length is then obtained:

$$L_{max} = 2\sqrt{\frac{2\,H}{-\,b_1}}\,. \tag{17}$$

Condition to obtain at least one path reaching the same height as the transmitter within the duct irrespective of distance

A path leaving the transmitter with angle φ_0 will pass again, after reflection, through the same height at distance L calculated before. For the first arc, i.e. the one that passes through the transmitter, the abscissas of the points at height h are, from the path equation:

$$x_1 = 0 \tag{18}$$

$$x_2 = \frac{2\,\varphi_0}{-\,b_1}. \tag{19}$$

The abscissas of subsequent points at the same height form two series:

$$n_1\,L$$

$$\frac{2\,\varphi_0}{-\,b_1} + n_2\,L, \qquad n_1 \text{ and } n_2 \text{ take values } 1, 2, 3, 4 \ldots \tag{20}$$

Limiting paths, tangential to the upper surface of the ducts, are obtained for:

$$\varphi_0 = \pm\sqrt{-\,2\,b_1(H - y_0)} \tag{21}$$

and their length L_{max} is:

$$L_{max} = 2\sqrt{2}\sqrt{\frac{H}{-b_1}}. \tag{22}$$

The abscissas of the first series do not depend on the sign of φ_0 and the corresponding paths cross at height y_0. The abscissas of the second series depend on the sign of φ_0 and consequently the rising and falling paths do not cross at the same height y_0.

There will always be a path at height y_0 irrespective of distance, if there is no interval between rising paths having undergone n reflections and falling paths having undergone $n+1$ reflections. If φ_0 denotes the positive value of the angle, the condition can therefore be written:

$$\frac{2\,\varphi_0}{-b_1} + nL \geqslant \frac{2\,\varphi_0}{b_1} + (n+1)\,L \tag{23}$$

or:

$$\frac{4\,\varphi_0}{-b_1} \geqslant L \tag{24}$$

where:

$$\varphi_0^2 \geqslant \frac{b^2\,L^2}{16}; \tag{25}$$

Taking the expressions for φ_0 and L into account, the condition can finally be written:

$$y_0 \leqslant \frac{3\,H}{4}. \tag{26}$$

The transmitter must be in the bottom three quarters of the duct.

Appendix 4.2
Conditions for the existence of duct modes in a tropospheric duct

As tropospheric ducts are practically parallel to the Earth's surface, we assume that the Earth is flat and replace the real index n by the modified index m, and N by M. Distances are denoted by x and heights by y.

The phase difference due to a small displacement ds with an angle of inclination φ is equal to:

$$\mathrm{d}\Phi = \frac{2\,\pi}{\lambda}\,m\,\mathrm{d}s = \frac{2\,\pi}{\lambda}\,m[\mathrm{d}x\cos\varphi + \mathrm{d}y\sin\varphi] \tag{1}$$

and the phase difference between points (x_0, y_0) and (x_1, y_1) on the path is equal to the integral of the quantity given above.

But according to Snell's law:

$$m \cos \varphi = m_0 \cos \varphi_0 = A \, . \tag{2}$$

Consequently:

$$\Phi_2 - \Phi_1 = \frac{2\pi}{\lambda}\left[A(x_2 - x_1) + \int_{y_1}^{y_2} m \sin \varphi \, dy\right]. \tag{3}$$

Since the duct is limited in the 0y direction, the only possible solution is a set of stationary waves between the upper and lower boundaries of the duct. This means that the term containing the integral must be equal to a multiple of 2π whenever the integral is calculated for a return journey from the bottom to the top of the duct. If the quarter wavelength advance which appears when a wave moves through a focus or a caustic (limiting surface of the duct) is taken into account, we obtain the relation which defines the modes which may exist in a duct between heights h_0 and h_2 above ground:

$$\frac{4\pi}{\lambda}\int_{h_0}^{h_2} m \sin \varphi \, dy - \pi = 2k\pi \, , \tag{4}$$

where k is an integer of any given value.

If Snell's law is considered between any point in a path and the top of that path, i.e.

$$m \cos \varphi = m_2 \, , \tag{5}$$

Equation (4) becomes:

$$\int_{h_0}^{h_2} \sqrt{m^2(y) - m_2^2} \, dy = \left(k + \frac{1}{2}\right)\frac{\lambda}{2} \, . \tag{6}$$

as m is close to 1, M is preferred:

$$m = 1 + M \, 10^{-6} \tag{7}$$

and consequently Equation (5) may be written as:

$$\int_{h_0}^{h_2} \sqrt{M(y) - M_2} \, dy = \left(k + \frac{1}{2}\right)\frac{\lambda}{2\sqrt{2}} 10^{-3} \tag{8}$$

where $M_2 = M(h_2) = M(h_0)$.

To calculate this integral, we need to know the function $M(y)$. Assume that it is a linear function of altitude within the duct (Fig. 1):

$$M(y) - M_2 = \frac{\Delta M}{h_1 - h_0}(y - h_0) \quad \text{when} \quad h_0 < y < h_1 \tag{9}$$

$$M(y) - M_2 = \frac{\Delta M}{h_1 - h_0}(y - h_2) \quad \text{when} \quad h_1 < y < h_2 \tag{10}$$

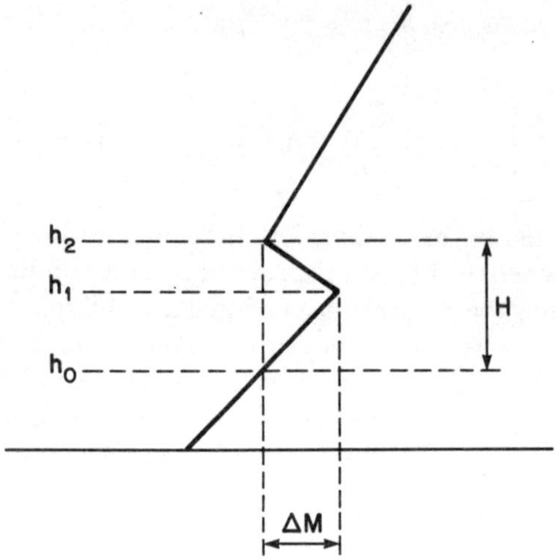

Fig. 1. Vertical pattern of M.

Where ΔM is the total decrease in M within the duct. Integration gives the following relation, where H represents the total height of the duct between the altitudes h_0 and h_2:

$$\frac{H}{\lambda} = \frac{3\left(k + \frac{1}{2}\right)}{4\sqrt{2}} \, 10^3 \, \frac{1}{\sqrt{\Delta M}}. \tag{11}$$

When $k = 0$ (first mode):

$$\frac{H}{\lambda} = 265 \frac{1}{\sqrt{\Delta M}}. \tag{12}$$

When $k = 1, 2, 3$, we obtain 795–1, 325–1, 855, respectively.

Case of a ducting layer at the Earth's surface

If the ducting layer is at the Earth's surface, there is no layer of atmosphere below h_2. Moreover, reflection takes place from the ground with a phase change of π. Equation (4) can then be replaced by:

$$\frac{4\,\pi}{\lambda} \int_0^H m \sin \varphi \, dy - \frac{\pi}{2} - \pi = 2\,k\pi \tag{13}$$

and Equation (11) can be replaced by:

$$\frac{H}{\lambda} = \frac{3\left(k + \frac{1}{4}\right)}{4\sqrt{2}} . 10^3 \frac{1}{\sqrt{\Delta M}} . \tag{14}$$

When $k = 0$ (first mode):

$$\frac{H}{\lambda} = 398 \frac{1}{\sqrt{\Delta M}} . \tag{15}$$

When $k = 1, 2, 3$ we obtain 928–1, 458–1, 988, respectively.

Chapter 5

Influence of the troposphere: attenuation due to gases and hydrometeors

5.1. Introduction

The density of electromagnetic energy from an antenna decreases during propagation because of two completely different phenomena:

1. the wave diverges from the antenna;
2. the wave is progressively absorbed or scattered by the medium.

The first phenomenon occurs even in a medium without losses. Its effect can be countered using a large wave collector (antenna). On the other hand, for the second phenomenon the energy lost is unrecoverable in cases of absorption, or partially recoverable but with a low success rate in cases of scatter.

In the preceding chapters we studied only the first phenomenon. This chapter is concerned with the second, and particularly with the following questions:

1. molecular absorption by atmospheric gases;
2. absorption and scatter by liquid and solid particles in the atmosphere particularly by raindrops.

These effects begin to appear at frequencies above a few Gigahertz, and increase rapidly with increasing frequency. Moreover, particles can lead to a modification in the polarization of the incident wave.

5.2. Attenuation by atmospheric gases

Since nitrogen has no absorption bands in the radio frequency range, molecular absorption is due almost entirely to oxygen and water vapour.

5.2.1. Specific attenuation due to oxygen

At frequencies less than 350GHz oxygen has an isolated absorption line at 118.74GHz and a series of very close lines between about 50 and 70GHz. In the lower part of the atmosphere these lines behave as a continuous absorption band. Furthermore, the oxygen molecule has a magnetic moment which produces a zero frequency line in its absorption spectrum.

Using the Van Vleck and Weisskopf line pattern and taking account of available observation results the specific attenuation in dB/km due to oxygen in the lower part of the atmosphere can be represented by the following approximate expressions:*

$$\gamma_{O_2} = \left(\frac{6.6}{f^2 + 0.33} + \frac{9.8}{(f - 57.5)^2 + 2.2}\right) f^2 . 10^{-3}\,, \quad f < 57.5 \text{ GHz} \quad (5.1)$$

$$\gamma_{O_2} = 14.7 \text{ dB}\,, \qquad 57.5 < f < 62.5 \text{ GHz} \quad (5.2)$$

$$\gamma_{O_2} = \left(\frac{4.13}{(f - 62.5)^2 + 1.1} + \frac{0.19}{(f - 118.7)^2 + 2}\right) f^2 . 10^{-3}\,, \quad (5.3)$$

$$62.5 < f < 350 \text{ GHz}\,.$$

5.2.2. Specific attenuation due to water vapour

At frequencies lower than 350GHz, water vapour shows three absorption lines, at frequencies of 22.3GHz, 183.3GHz and 323.8GHz. Furthermore, at sub-millimetre wave range and infrared frequencies there are many lines, some of them very intense. This means that the sum of the sidebands of these lines has a significant effect on the centimetre wave range and the millimetre wave range. This effect, sometimes called residual absorption, can be represented by a single term proportional to the square of the frequency.

For very low concentrations of water vapour attenuation can be treated as being proportional to concentration. However, this relation is only an approximation and leads to an underestimate of signal reduction, which becomes significant above about 10g/m^3.

Using the Van Vleck and Weisskopf line pattern and taking into account the available results and measurements the specific attenuation in dB/km due to water vapour in the lower part of the atmosphere can be represented by the following approximate expression:*

$$\gamma_{H_2O} = \left(0.067 + \frac{2.4}{(f - 22.3)^2 + 6.6} + \frac{7.33}{(f - 183.5)^2 + 5} + \right. \quad (5.4)$$

$$\left. + \frac{4.4}{(f - 323.8)^2 + 10}\right) f^2 . \rho \, 10^{-4}$$

where f is expressed in GHz and ρ in grams of water vapour per m^3 of air. For frequencies lower than about 100GHz only the first two terms need be taken into account.

Figure 5.1 shows the attenuation values obtained using Equations (5.1) to (5.4). For water vapour, we chose a value of ρ equal to 7.5g/m^3, which corresponds to 1% water vapour molecules mixed with 99% molecules of dry air. Note that this

* Since the coefficients in these equations are determined in a semi-empirical way, other sources may give slightly different values.

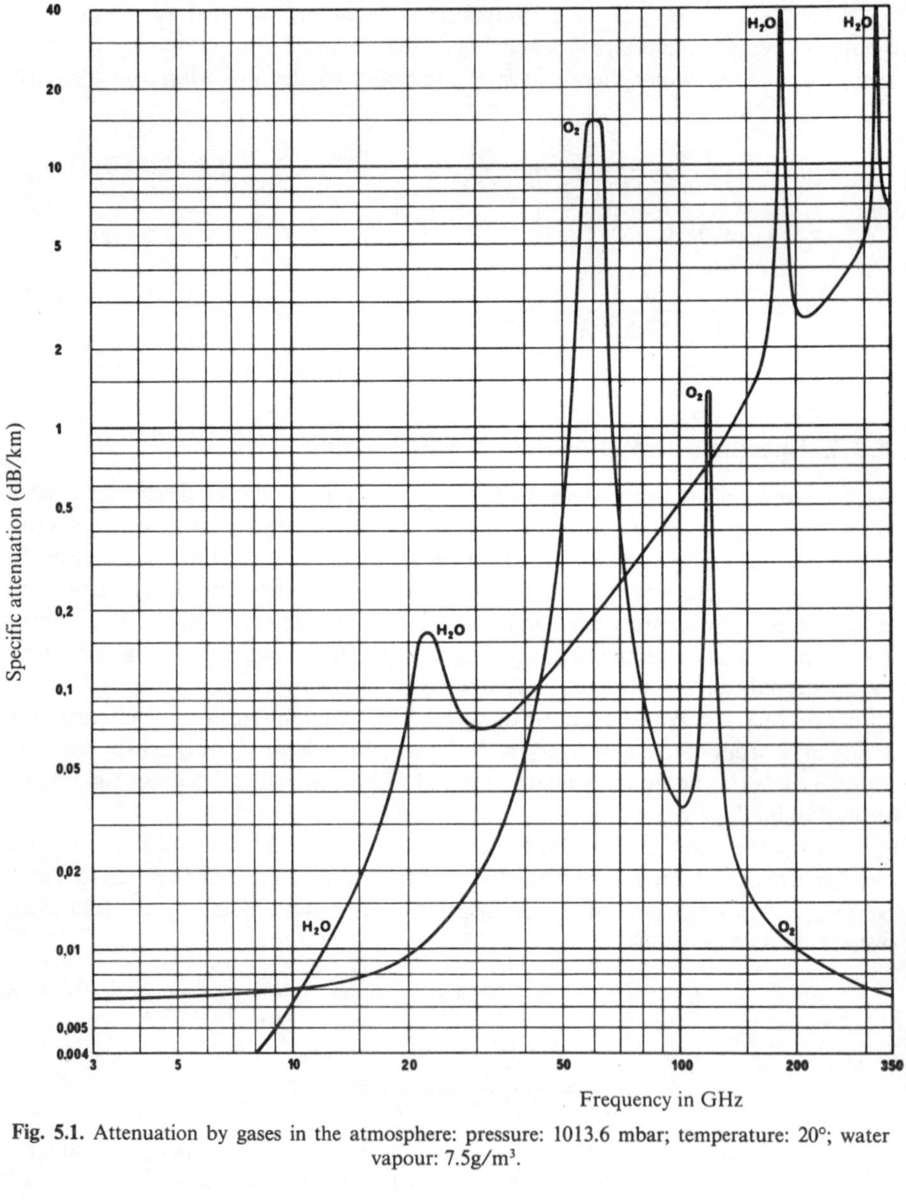

Fig. 5.1. Attenuation by gases in the atmosphere: pressure: 1013.6 mbar; temperature: 20°; water vapour: 7.5g/m^3.

value corresponds, at ground level, to 50% relative humidity if the temperature is 16.5°C or 75% at 10°C.

5.2.3. Influence of temperature

The effect of temperature on attenuation is relatively complex, since it affects both the intensity and the width of the absorption lines. Nevertheless, as the amplitude of relative variations in absolute temperature in the atmosphere is generally less than 10% and the temperature of a mass of air is not homogeneous or exactly known, only a rough idea of the effect of temperature is needed.

For many applications the effect of temperature can be completely neglected. If we want to take temperature into account, we can treat Equations (5.1) to (5.4) as valid for a temperature of 15°C and then subtract 0.7% per degree to correct for other temperatures (attenuation increases as temperature falls).

5.2.4. Attenuation along a path

Along a terrestrial path it can generally be assumed that the atmosphere is homogeneous, and total attenuation can be calculated by multiplying the length of the path by the sum of specific attenuations due to oxygen and to water vapour.

Along an Earth-satellite path attenuation occurs primarily in the lower part of the atmosphere. The effect is as if the real atmosphere had been replaced by a homogeneous atmosphere, with the same characteristics everywhere as it has near the Earth's surface, but with a height h_e called the equivalent height. Outside the absorption bands h_e is practically a constant and equal to about 6km for oxygen and 2.2km for water vapour.

Figure 5.2 gives the theoretical attenuation on a communication link between a satellite and the Earth in the direction of the zenith, either through dry atmosphere or through a temperate moderately humid atmosphere (7.5g/m^3 at ground level).

Attenuation A_φ at an angle of elevation φ can be deduced for the attenuation at the zenith A_z by taking account of the equivalent length, which depends on φ and h_e. We have the relation:

$$\frac{A_\varphi}{A_z} = \frac{l_e}{h_e} = \frac{2}{\sqrt{\sin^2 \varphi + \dfrac{2\,h_e}{Ka}} + \sin \varphi}, \tag{5.5}$$

the effective radius of the Earth, Ka, can be taken as equal to 8500km.

For angles of elevation greater than about 5° we can use the 'cosecant law':

$$\frac{A_\varphi}{A_z} \simeq \frac{1}{\sin \varphi}. \tag{5.6}$$

Note, moreover, that in the intense absorption band (57 to 62.5GHz) the calculations given above are not entirely valid as in the upper part of the atmosphere the different lines in this absorption band are clearly separated. This means that attenuation in this band can no longer be represented by a single value.

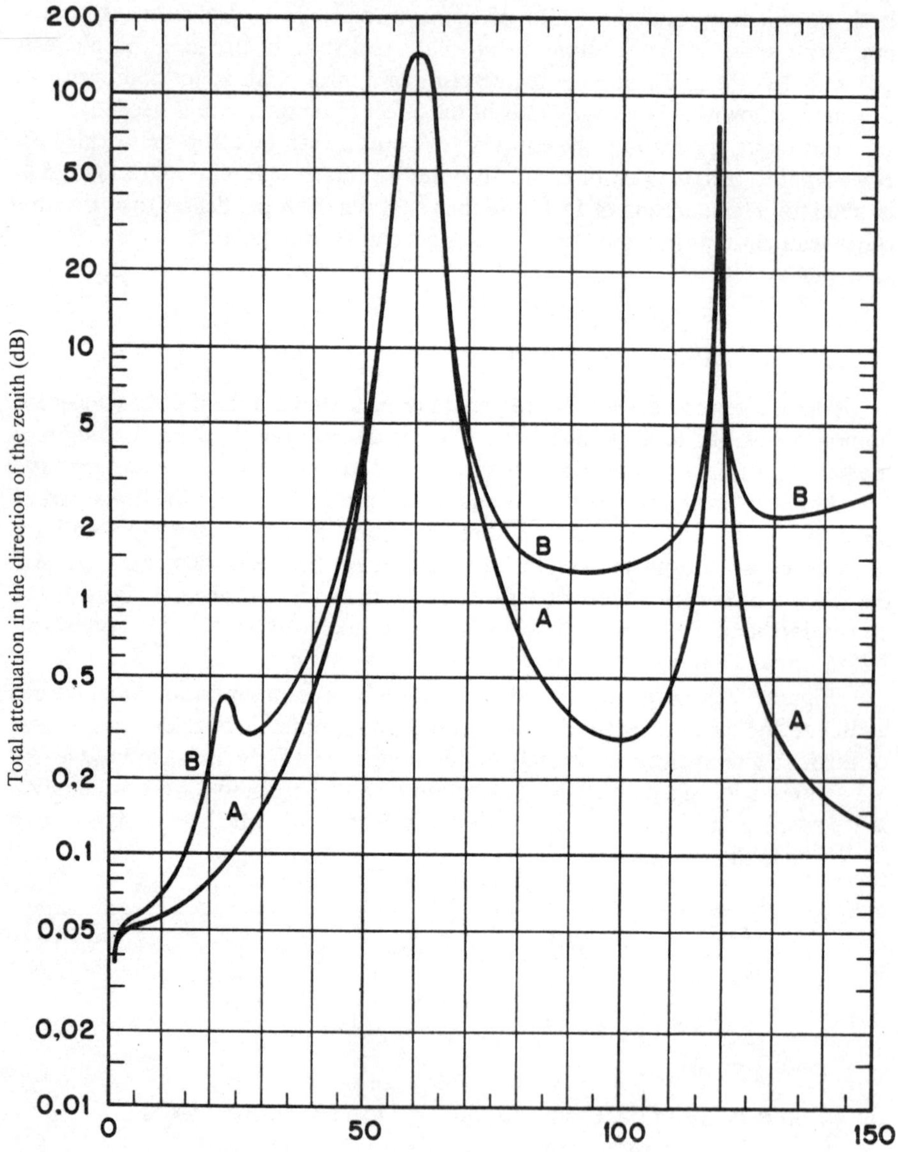

Fig. 5.2. Total attenuation due to atmospheric gases along a vertical path: A: dry atmosphere; B: humid atmosphere (7.5g/m³).

5.3. Attenuation due to hydrometeors (rain, fog, clouds, snow, hail)

Of the non-gaseous particles in the atmosphere, raindrops cause the greatest attenuation. This attenuation is due both to energy absorption from ohmic losses in the raindrops and to scattered energy by water droplets.

We start with a few remarks on the meteorological characteristics of rain, after which we consider in turn attenuation by water droplets of a given diameter, by uniform rain over the entire path and by real, random irregularly shaped raindrops.

5.3.1. Characteristics

Rain is the most important hydrometeor. We therefore briefly examine those characteristics of rainfall that are most important for wave propagation; they are not necessarily recorded with the greatest possible accuracy by weather stations.

5.3.1.1. Time characteristics of rainfall at a particular place

From the point of view of telecommunications an essential characteristic of rainfall levels at a particular place is the statistical distribution of rainfall intensities. This is measured using a rain gauge. It is made up essentially of a funnel of given diameter and a device making it possible to measure the amount of water collected over a given period of time. This device may be a calibrated bucket which tips up every time it is full and emits an electric impulse or a level detector (float or pressure sensor) in a cylindrical container, etc. Rain intensity is usually expressed in mm/h but measurements need to be taken at one minute intervals. Weather stations often provide statistics at intervals of 5 to 10 minutes, which are too long for propagation requirements since heavy rainfall intensities are underestimated. Nevertheless, observations supplied by weather stations are worth using because of the long periods they cover. Empirical correction is therefore necessary in order to deduce the distribution of intensities over one minute intervals.

We find that the distribution of rainfall intensities obtained at the same station over different years can vary considerably. The more years they cover, the more useful these statistics therefore are. Figure 5.3 gives graphs of statistical distributions measured at a certain number of stations. Although the curves were obtained from measurements made at different periods and with different equipment and analytic methods, they do provide an example of a range of rainfall levels.

As extreme values are difficult to obtain from direct measurements, it is worth setting up statistical models to represent these experimental distributions. The two most common models are the log-normal distribution and the gamma distribution. Some information on these two distributions appears in Appendix 1. Here we merely consider their mathematical forms. The probability that rainfall intensity is less than a given value R is equal to:

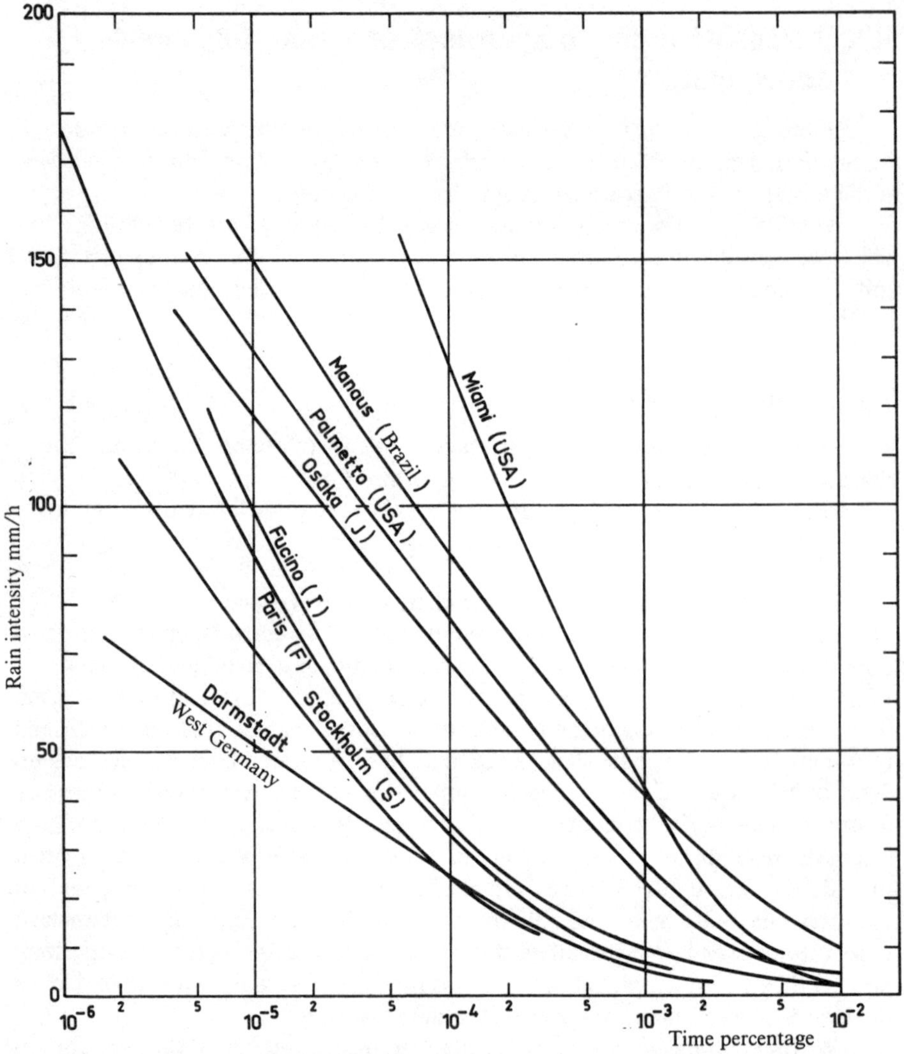

Fig. 5.3. Statistical distributions of rainfall intensities.

log-normal distribution:

$$\text{Prob}\,(r < R) = \frac{1}{\sigma\sqrt{2\,\pi}} \int_0^R \frac{1}{u} \exp\left[-\frac{1}{2}\left(\frac{\ln u - m}{\sigma}\right)^2\right] \mathrm{d}u \tag{5.7}$$

gamma distribution:

$$\text{Prob}\,(r < R) = \frac{\alpha^\nu}{\Gamma(\nu)} \int_0^R u^{\nu-1}\, \mathrm{e}^{-\alpha u}\, \mathrm{d}u\,. \tag{5.8}$$

The behaviour of these distributions is extremely different for extreme values of the variable. For very small time percentages the log-normal distribution leads to rainfall intensity values considerably greater than observed. Experience shows that the log-normal distribution is a good model for low and medium rainfall intensities, but beyond 10 to 40mm/h, depending on the climate, the gamma distribution gives a more satisfactory model.

Irrespective of the distribution selected it seems logical in order to determine the parameters of the distribution to take into consideration in the statistical analysis only those periods during which rain is non-zero (this is generally between 3 and 5% of total time). This is how the log-normal distribution is applied. For the gamma distribution, given the very small values of parameter ν (see Appendix 1), probability is approximately proportional to ν. Consequently, we can move from statistics established for rainfall times to statistics established for total time simply by multiplying the value of ν by the proportion of the time during which rainfall is non-zero. Generally, for models based on a gamma distribution we consider statistics for the total time.

Since we have to change from a log-normal distribution to a gamma distribution for rainfall intensity values with high probability, attempts have been made to find a practical model to cover all useful values with a single function. For example, if only a relatively narrow range of rainfall intensities is considered, a model can be used in which probability is proportional to a certain power of R. The index of R is generally between –2 and –3. In the same way, a model in which the property of R is given by a relation of the form:

$$\alpha \frac{e^{-\beta R}}{R^{\gamma}}$$

is an excellent approximation for rainfall intensities greater than a few mm/h, if suitable values of α, β and γ are chosen.

5.3.1.2. Spatial characteristics of rainfall

Since a radio wave path has a certain length, a knowledge of rainfall characteristics at a single point is not sufficient. We need to know the spatial structure of the rainfall zone. This structure depends considerably on the type of precipitation. In temperate regions there are generally two types of precipitation: stratiform and convective. In tropical or equatorial regions there are two further types of precipitation: cyclones and monsoon rainfall.

Stratiform precipitation covers a vast region, up to several hundred kilometres across, of low intensity rainfall, though there may be local rain intensity reaching as much as 25mm/h. Precipitation is stratified horizontally: rain to an altitude where temperatures are close to the melting point of ice (which may be below 0°C because of supercooling), and then snow to a height of around 7km, and finally ice crystals to about 9km. Stratiform precipitation is often associated with the passage of a front, i.e. a temperature discontinuity between air masses. It can therefore extend a long way in a direction perpendicular to the movement of the front.

Areas of convective precipitation are not more than 15km across horizontally, but can be very deep, reaching the level of the tropopause. This air mass is disturbed by very intense vertical movements due to its location above ground subject to considerable heat from the sun. Precipitation may be very intense, accompanied by thunder, and extends horizontally over several kilometres. It can last some tens of minutes and produces storms.

Monsoon rainfall is made up of a sequence of intense convective precipitation bands followed by intervals of stratiform precipitation. The bands may be 50km wide and several hundred kilometres in length. They produce heavy rain which can last several hours.

A cyclone is characterized by several bands of intense precipitation rolled into a spiral and covering several hundred kilometres. At the centre of the spiral there is a calm zone (the eye of the cyclone).

As convective rainfall, which is of major importance for telecommunications, is made up of relatively small rain zones, models of them have been developed in terms of rain cells, which are simple in shape (e.g. circular) and characterized by rain intensity directly related to the dimensions of the cell. Other authors, on the other hand, work in terms of considerably larger cells which contain nuclei, each of which is an area of heavy rain intensity. Some writers prefer to use a spatial rain correlation coefficient, determined experimentally. As the real structure of rain zones is always very complex, these models are often excessively simplified. Note that the spatial structure of precipitation is often studied using radar operating at a frequency of several Gigahertz and taking account of the fact that precipitation will reflect, by back scatter, a part of the energy incident on it. However, the accuracy of the precipitation intensity measurements made is always unsatisfactory: radar is more suitable for determining the size of the precipitation zone than the intensity of the precipitation within it. Rain gauges must also be used.

5.3.1.3. Drop size characteristics of rainfall

Rainfall is made up of roughly spherical water drops of diameter between about 0.1 and 6mm. Below 0.1mm the speed at which droplets fall is so low that one cannot really talk of precipitation. Small droplets may grow bigger by fusing together, which increases the rate at which they fall, but also heightens the effect of air resistance. Consequently, large drops have a flattened shape which can even show a depression on their lower face (the Pruppacher and Pitter model), leading to breaking of the drop, and consequently limiting its maximum diameter. As a result of the imperfect sphericity of large drops, we characterize their size in terms of the diameter of a sphere with the same volume.

The distribution of drop diameters in rainfall has been the subject of much research. The results depend on the type of rainfall or storm and on the rainfall levels of the region in question. The distribution may even change between the beginning and the end of a rainstorm. Most calculations lead to a nearly exponential distribution, i.e. the number of raindrops of diameter D, to within $\mathrm{d}D$, per unit volume, $N(D)\,\mathrm{d}D$ is such that:

$$N(D) = N_0 \exp\left(\frac{-D}{D_m}\right). \tag{5.9}$$

Instead of the mean diameter D_m the median diameter D_0 is often used such that half the quantity of water is contained in raindrops of diameter less than D_0. For an exponential distribution:

$$D_0 = 3{,}672\ D_m . \tag{5.10}$$

The use of different distributions depending on the type of rain (Joss) is not of much value for telecommunications, and the Laws and Parsons or Marshall and Palmer global distributions are the most frequently used, even though they are the oldest. These distributions give very similar results. For the Marshall and Palmer distribution:

$$N_0 = 8\,000\ \text{mm}^{-1}\ \text{m}^{-3} \tag{5.11}$$

$$D_0 = 0.89\ R^{0{,}21}\ \text{mm}. \tag{5.12}$$

To determine the effect of the rain, we also need to know the fall speed of the water drops. It can be assumed that all drops reach their terminal velocity, for which the values obtained by Gunn and Kinzer are generally used. Some approximate values are given in Table 5.1.

Table 5.1. Terminal velocity for raindrops

Diameter (mm)	Terminal velocity (m/s)
0.25	1
0.5	2
1	4
2	6.4
3	8
4 and over	9

If we know the distribution of water droplet diameters and their fall speed we can calculate the intensity of precipitation using the relation:

$$R = 0{,}6 . 10^{-3} \int D^3 . V(D) . N(D) . \mathrm{d}D , \tag{5.13}$$

where:

R is the intensity of the rain in mm/h,
D is the diameter of the drops in mm,
$N(D)$ as defined above is expressed in $\text{mm}^{-1}\ \text{m}^{-3}$;

the integral covers the range of all possible diameters, i.e. from about 0.1 to 9mm.

In the same way, we can calculate the water content of precipitation using the relation:

$$M = \frac{\pi}{6} 10^{-3} \int D^3 N(D) \, \mathrm{d}D \tag{5.14}$$

where M is expressed in g/m³.

Another characteristic of raindrops, which must be taken into account for certain propagation phenomena (transpolarization), is the angle of the axis of symmetry of non-spherical drops with respect to the vertical. If drops are falling through still air, or air moving with constant horizontal speed, that angle will be zero. Variations in wind speed met while falling produce an inclination in large drops. The angle of inclination may reach several tens of degrees, and in addition drops may oscillate about the mean canting angle, and there may even be oscillations in the shape of the drops themselves, particularly if they are falling through turbulent zones.

5.3.1.4. Characteristics of clouds and fog

Clouds and fog are made up of extremely small water droplets (between about 0.001 and 0.1mm) which may be mixed with particles of ice. Water contents are around 0.1g/m³. At high altitude at temperatures less than 0°C certain clouds are made up of supercooled water, and others of ice particles. These particles may be irregularly shaped, sometimes very elongated, and consequently they may, if they are electrostatically charged, align themselves with the force lines of the electric field. This phenomenon has a certain importance in propagation, and is examined in Section 5.4.2.

5.3.2. Wave absorption and scattering by particles

When a plane wave meets particles, e.g. raindrops, some of the wave's energy is absorbed by the particles, some further energy is scattered by them, and the rest passes through the zone occupied by the particles. Therefore generally:

$$P_i - P_t = P_a + P_d \tag{5.15}$$

where:

P_i is the incident power;
P_t is the transmitted power;
P_a is the absorbed power;
P_d is the scattered power;

which is sometimes more concisely expressed as:

attenuation = absorption + scatter.

Part of the scattered energy returns in the direction of the incident wave and may interfere with the transmitted wave. This phenomenon is practically negligible in cases of scatter by small particles, but becomes very important if irregularities in the transmission medium reflect energy principally forwards, as in the case of irregularities in the refractive index (see Chapter 4).

Each particle may be regarded as a minute receiving antenna. The energy absorbed corresponds to the energy received, while scattered energy is called reradiated energy.

In the following discussion we assume that the energy scattered successively by two particles is negligible (no multiple scatter). 'Equivalent areas' can be associated with these different energies, though here we tend to use the expression effective cross section, which expresses the ratio between the power considered (absorbed or scattered) to the power flux density of the incident wave:

$$\sigma_t = \sigma_a + \sigma_d \tag{5.16}$$

where:

σ_t is the total cross section (attenuation);
σ_a is the absorption cross section;
σ_d is the scatter cross section.

Let us therefore consider a medium containing N identical particles on average in a volume V. The attenuation of a plane wave of power flux density S passing through this medium over a distance dx is given by the relation:

$$\frac{\mathrm{d}S}{S} = \frac{N}{V} \sigma_t \, \mathrm{d}x \, . \tag{5.17}$$

The specific attenuation γ is then equal to:

$$\gamma = \frac{\mathrm{d}S}{S \, \mathrm{d}x} = \frac{N}{V} \sigma_t \, . \tag{5.18}$$

If the particles are not identical, and in particular if they are spheres whose radius a lies between a_0 and a_1, the product $N\sigma$ must be replaced by the expression:

$$\int_{a_0}^{a_1} N(a) \, \sigma(a) \, \mathrm{d}a \, , \tag{5.19}$$

where $N(a)$ da represents the number of particles with a radius between a and $(a + \mathrm{d}a)$:

The scattering theory was first elaborated by Lord Rayleigh in the years after 1871 for very small purely dielectric particles. The phenomenon considered was therefore one of scatter without absorption. The general theory of absorption and scatter by particles with a complex refractive index was established by G Mie in 1908 for spheres. The case of non-spherical particles was analysed later, in particular by

Atlas in 1953 and Oguchi in 1960.

Theoretical calculations make the assumption that the scattered field outside the particle and the field within it are produced by electromagnetic oscillations induced in the particle by the incident wave. These fields may be considered as the sums of multipolar fields, i.e. of electric and magnetic fields with 2 *n* poles ($n = 1$ corresponds to the dipole case).

Calculations are fairly difficult even for spherical particles, so we only give the results obtained. We find that the smaller the particle is with respect to the wavelength, the smaller is the maximum number *n* to be considered. In particular, if we limit our expansion to terms in $(D/\lambda)^6$, where *D* is the diameter of the particle, we need only consider electric and magnetic terms corresponding to $n = 1$, and the magnetic term corresponding to $n = 2$. The expression for σ_t then takes the following form:

$$\sigma_t = \frac{\lambda^2}{2\pi}\left(\frac{\pi D}{\lambda}\right)^3\left[C_1 + C_2\left(\frac{\pi D}{\lambda}\right)^2 + C_3\left(\frac{\pi D}{\lambda}\right)^3\right]. \tag{5.20}$$

Coefficients C_1, C_2 and C_3 depend on the electrical characteristics (permittivity and conductivity) of the particles. C_1 and C_2 are zero for zero conductivity, i.e. for purely dielectric particles. The first two terms are therefore principally due to absorption. Term C_3 is equal to:

$$C_3 = \frac{4}{3}\,\mathrm{Re}\left(\frac{\eta - 1}{\eta + 2}\right)^2 \tag{5.21}$$

where η is the complex refractive index of the particle:

$$\eta = \varepsilon - \mathrm{j}\,60\,\sigma\lambda\,. \tag{5.22}$$

It is therefore clear that C_3 exists even for dielectric particles, i.e. it is associated with scatter.

If the particles are spherical, attenuation does not depend on wave polarization; this is no longer true if the particles are randomly shaped.

5.3.3. Attenuation by rain drops of given diameter

If we apply these considerations to cases where the particles are raindrops, we get the following results.

a) If D/λ is extremely small, e.g. in mist, clouds or extremely fine rain, for frequencies lower than 50GHz, only coefficient C_1 needs to be taken into account; it is therefore clear that the cross section is proportional to D^3, i.e. to the volume of the drop, or to the mass of water. This property still applies even in cases where there is a distribution of various drop diameters, as long as they are all very small. Consequently, the specific attenuation due to cloud or extremely fine rain depends only on the water contents of the atmosphere, measured for example in g/m^3.

Theory shows, moreover, that coefficient C_1 is proportional for practical purposes to frequency over a very wide frequency range, particularly at temperatures that are not too low. Consequently, attenuation is approximately proportional to the square of frequencies. Figure 5.4 allows us to calculate attenuation for different temperatures, including a negative temperature (supercooled water). Note that the water content of cloud or mist is of the order of 0.1 to 1g/m³.

b) if D/λ is not very small, but is still significantly less than 1, which is the case for moderate rainfall at frequencies between 3 and 20GHz, we have to take particular account of terms C_2 and C_3. Theory shows that, up to about 20GHz, C_2 is approximately proportional to the frequency and C_3 is constant. Consequently,

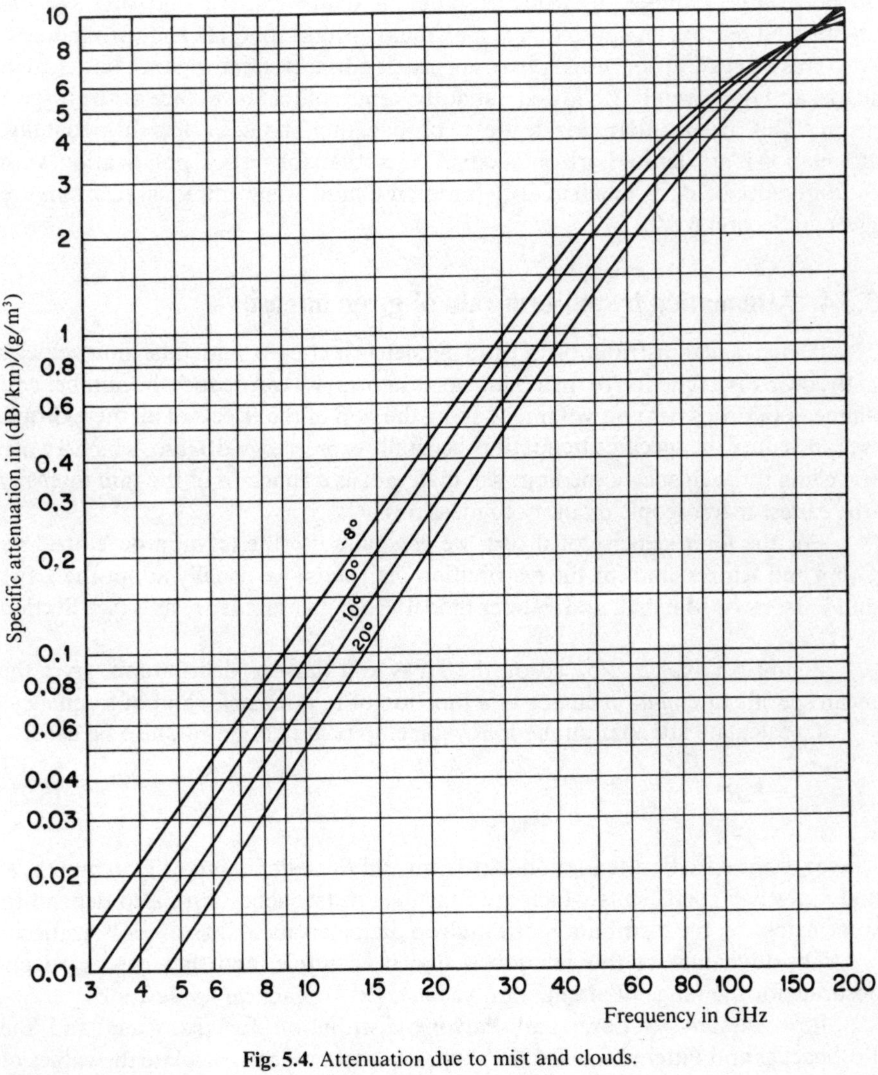

Fig. 5.4. Attenuation due to mist and clouds.

in Equation (5.20), if the last two terms are predominant, σ_t is proportional to F^4. As for the influence of drop diameter, we have a function of D^5 for the C_2 term, and a function of D^6 for the C_3 term. In any case, as there is no longer a function of D^3, we have to take drop diameter distribution into account to calculate attenuation.

c) If D/λ is much greater than one, which is the case for heavy rain at frequencies over about 20GHz, or for moderate rain at frequencies over about 50GHz, Equation (5.20) is no longer valid and a greater number of terms has to be considered. We can predict that the effective cross section of water drops stops increasing and that it tends towards the apparent geometric area. As rainfall is made up of a large number of drops of different diameters, we find that specific attenuation reaches an approximate maximum around 200GHz and then reduces.

This discussion concerns spherical particles. In fact, raindrops are flattened in shape, and may even be hollowed out at the centre of the lower face of the biggest drops. This means that, for a wave propagating more or less horizontally, attenuation is greater for horizontal polarization than for vertical polarization, with the difference being as much as 20% for intense rain. Some empirical relations are given in Section 6.6.2.

5.3.4. Attenuation by uniform rain of given intensity

If a particular distribution of drop diameters is chosen, and if the limit velocity of the drops as a function of their diameter is known, we can deduce the number and diameter of drops per unit volume. Taking the sum of the effects of all these drops, we can deduce the specific attenuation, generally expressed in dB/km, which a wave travelling through such a medium will undergo, as a function of the rain intensity (the easiest macroscopic quantity to measure).

For the limit velocity of drops, we generally use the relation developed by Gunn and Kinzer, and for the distribution diameters, we usually adopt the Laws and Parsons or Marshall and Palmer model. Both give similar results (see Section 5.3.1).

Figure 5.5, which is based on the Laws and Parsons distribution, gives the mean specific attenuation curves as a function of rain intensity and of frequency.

To calculate attenuation the following empirical relation can also be used:

$$\gamma = KR^{\alpha} \tag{5.23}$$

where γ is the specific attenuation in dB/km and R is rainfall intensity in mm/h. K and α are two coefficients which are functions of frequency, and also depend to some extent on the distribution of raindrop diameters as well as on polarization.

The advantage of this relation is that it is simple, and that it is relatively accurate for the range of attenuation values used in practical applications.

If we assume a Laws and Parsons distribution for diameters and the Pruppacher and Pitter model for the shape of drops, we can calculate the values of

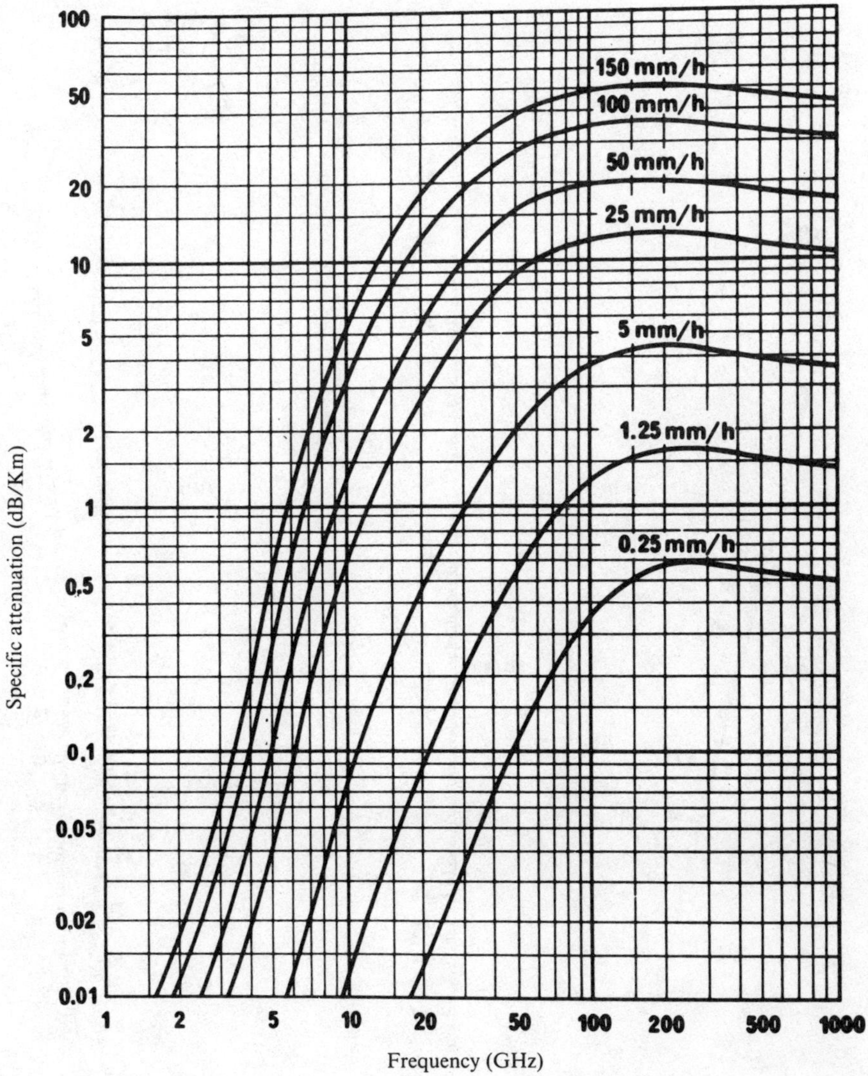

Fig. 5.5. Specific attenuation due to rain. Distribution of raindrop sizes (Laws and Parsons).

K and α as a function of frequencies. Table 5.2 gives a series of these values. It is clear that the exponent α is always very close to 1. It reaches a maximum for a frequency of the order of 7GHz. Coefficient *K*, on the other hand, increases very rapidly up to about 100GHz, and then remains virtually constant. For frequencies not given in Table 5.2 an interpolation formula has to be used, which is linear for α and logarithmic for *K*. We used Table 5.2 and Equation (5.23) to draw up the nomogram in Fig. 5.6.

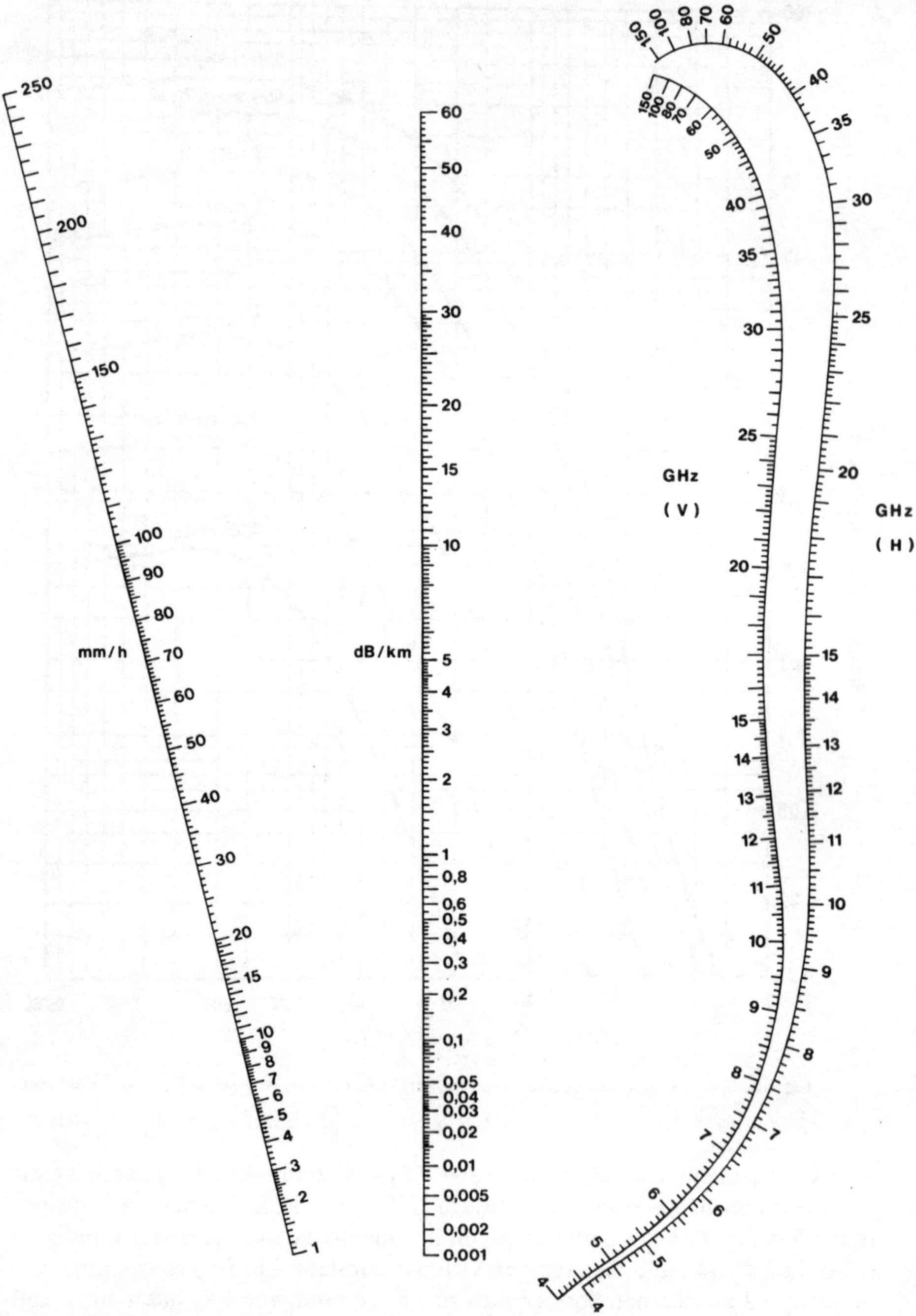

Fig. 5.6. Specific attenuation due to rain. Distribution of raindrop sizes (Laws and Parsons).

Table 5.2. Values of K and α.

Frequency (GHz)	K_H	K_V	α_H	α_V
1	0.000 038 7	0.000 035 2	0.912	0.880
2	0.000 154	0.00 138	0.963	0.923
4	0.000 650	0.000 591	1.121	1.075
6	0.001 75	0.001 55	1.308	1.265
8	0.004 54	0.003 95	1.327	1.310
10	0.010 1	0.008 87	1.276	1.264
12	0.018 8	0.016 8	1.217	1.200
15	0.036 7	0.033 5	1.154	1.128
20	0.075 1	0.069 1	1.099	1.065
25	0.124	0.113	1.061	1.030
30	0.187	0.167	1.021	1.000
35	0.263	0.233	0.979	0.963
40	0.350	0.310	0.939	0.929
45	0.442	0.393	0.903	0.897
50	0.536	0.479	0.873	0.868
60	0.707	0.642	0.826	0.824
70	0.851	0.784	0.793	0.793
80	0.975	0.906	0.769	0.769
90	1.06	0.999	0.753	0.754
100	1.12	1.06	0.743	0.744
120	1.18	1.13	0.731	0.732
150	1.31	1.27	0.710	0.711
200	1.45	1.42	0.689	0.690
300	1.36	1.35	0.688	0.689
400	1.32	1.31	0.683	0.684

5.3.5. Predicting fades due to rain on a terrestrial or space path

Any method intended to predict fades due to rain will be a compromise between the two following methods: one is a purely physical meteorological method, whereas the other is only concerned with radio wave phenomena.

In a physical meteorological method, we assume that all the physical characteristics of raindrops as well as all the spatial and time characteristics of precipitation are known, making it possible, without having to carry out radio measurements, to deduce both specific attenuation γ and fading over a path of length L by integration:

$$A = \int_0^L \gamma(x)\,\mathrm{d}x\,. \tag{5.24}$$

The main difficulty lies in the extreme complexity of the spatial structure of rainfall events, which means that the models developed are excessively simplified. In a method based entirely on direct measurement of attenuation there is no need in principle to know the rain characteristics. We need only know the curves established for the distance and frequency considered. The difficulty here is the very large number and long duration of the radio measurements necessary to establish these curves for all fequencies, distances and climates, etc., although for practical purposes, in a telecommunications project, such a method is the easiest to apply. There are still few regions where sufficiently reliable fading curves have been established, and then only for very limited frequency and distance ranges.

In most cases, we have to start with the distribution of rainfall intensities, observed in the region in question (or in a region in which the rainfall level is similar) and then use a roughly empirical method to deduce attenuation over a given path at a given frequency. We first consider terrestrial paths and then consider space paths.

5.3.5.1. Terrestrial paths

Here we only consider methods which make use of rainfall intensity distributions, measured at a particular place regarded as representative of the region in which the communication link has been established. These methods may be divided into two categories depending on whether they use a spatial model of rain or not. On the other hand, in all methods we have to use a statistical model of the rainfall at a point, to smooth measurements and extrapolate them to high rainfall intensities (Section 5.3.1).

Methods employing a spatial model of rain use either a spatial self-correlation coefficient or a 'rain cell' of the appropriate shape and size. The self-correlation coefficient is not always known, and the concept of a rain cell involves certain difficulties: we have to consider relatively large cells (e.g. 10 to 20km) containing a greater or smaller number of 'nuclei' some kilometres in diameter, which are areas of high rainfall intensity. Calculation of attenuation involves integrating effects along the path through the cells, and generally requires the use of a computer.

Methods which do not explicitly use a spatial model of rain take account of the non-homogeneity of the rain zone and its limited extent, by introducing the concept of an effective distance or a reduction coefficient, so that we can move from the real distance of the communication link to a corresponding representative distance assuming uniform rainfall. Since these methods only depend on the use of simple formulas or curves, a convenient calculation procedure is given in Section 6.6.2.

5.3.5.2. Earth-satellite paths

For an oblique communications link crossing through the entire atmosphere the vertical structure of precipitation also must be taken into account. Consequently, the problem becomes rather more complex than for horizontal links as the angle of elevation appears as a factor in our calculations, and the influence of the angle of elevation is not simple, as it depends on the type of rainfall.

For stratiform rain, which is highly extensive, but associated with low or moderate precipitation intensities, attenuation is proportional to the length of the path through the precipitation, and falls roughly in inverse proportion to sin φ, where φ is the angle of elevation, at least for value of φ greater than a few degrees. On the other hand, for convective rain corresponding to more intensive precipitation, attenuation is caused by less extensive rain cells. The length of the path through the precipitation, which is inversely proportional in this case to cos φ, increases with the angle of elevation, but the probability of interception with this type of cell diminishes as the angle of elevation increases. Effects cancel each other out in part, and fading due to very heavy rain diminishes slowly once the angle of elevation is greater than 20°.

It is possible to imagine prediction methods based on models of rain cells in three dimensions, but the uncertainty associated with the model is very large. Calculation methods proposed by the CCIR tend rather to be based on the use of an effective distance through uniform rain, this distance being determined in a partially empirical way. A convenient calculation procedure based on this method is given in Section 6.6.3.

5.3.5.3. Attenuation due to the snow and hail

Water in the form of ice causes very little attenuation. Consequently, fading due to snow or hail depends to a great extent on the proportion of water present in liquid form. In particular, there can be extensive and long lasting sleet precipitation. Such phenomena can lead to a complete breakdown in the communications link. It is only the relatively low probability of such occurrences that reduces their importance in the overall statistics.

While on this subject, it should also be noted that ice and snow have an influence on antennas. A layer of frost or dry snow, even if it is relatively thick, introduces little perturbation to the radiation pattern, except perhaps as a result of mechanical deformation due to overloading. On the other hand, attenuation becomes very important at melting point because of the layer of water that forms.

5.4. Transpolarization due to hydrometeors

As well as effects due to refraction in clear air which we examined in Section 4.3, transpolarization may also be produced by hydrometeors because of their non-spherical shape. Unlike transpolarization in clear air, which only really affects horizontal communication links, hydrometeors lead to transpolarization above all on Earth-space communications links. We consider in turn the effects due to raindrops and those due to ice crystals.

5.4.1. Effects of raindrops

We have seen that raindrops, and in particular the biggest of them, are not spherical in shape but tend rather to be flattened ellipsoids because of air resistance. They may even have a slightly hollow lower surface. They nevertheless generally maintain the shape of solid of revolution, and therefore have an axis of symmetry, but this axis is not necessarily vertical, particularly if the rain is falling through a turbulence zone.

If a radio wave crosses a medium in which there are flattened drops, in such a way that the electric field is not parallel to the axis or the equatorial plane of the drops, it gives rise, by scatter, to a wave with orthogonal polarization to that of the incident wave. The relative level of this orthogonal wave depends on many factors such as the angle of incidence of the wave, the canting angle of the drops, the intensity of the rainfall, the distribution of drop diameter, frequency, etc. There has been both theoretical and experimental research into this subject. It is particularly worth knowing the relation between the level of orthogonal polarization created by a rain zone and the attenuation due to the same zone under the same conditions. The relation can be established from some fairly simple considerations.

Consider a plane wave of clearly defined polarization which crosses a layer of homogeneous rain of length l along the direction of propagation. Let E_0 denote the field strength of the incident wave and E_c and E_x the field strengths of the copolar and crosspolar waves leaving the rain layer. At a distance z from the point of entry, the field of the copolar wave is therefore equal to:

$$E_{cz} = E_0 \, \mathrm{e}^{-\gamma z} \tag{5.25}$$

where γ is the specific attenuation.

The crosspolar component is zero at the entry point and increases gradually, as it is due to the sum of the fields radiated by all the drops. If we assume that the distribution of diameters and the orientation of drops are the same over the entire path, and if we ignore multiple scatter, radiation from all the drops will be in phase at the exit point, but will have faded over the path. The crosspolar field strength from a layer of width d_z is therefore given by:

$$\mathrm{d}E_{xz} = KE_{cz} \, \mathrm{d}z \, \mathrm{e}^{-\gamma(l-z)} \tag{5.26}$$

where K is a parameter characterizing transpolarization within the medium under the communication conditions.

Replacing E_{cz} by its value derived from Equation (5.25) we obtain:

$$\mathrm{d}E_{xz} = KE_0 \, \mathrm{e}^{-\gamma l} \, \mathrm{d}z \, . \tag{5.27}$$

Over the entire rain zone we therefore have:

$$\frac{E_c}{E_0} = \mathrm{e}^{-\gamma l} \, , \tag{5.28}$$

$$\frac{E_x}{E_c} = Kl \, . \tag{5.29}$$

The ratio of the corresponding powers is equal to the square of the expressions above. Consequently, eliminating l:

$$\frac{E_x}{E_c} = -\frac{K}{\gamma} \ln \frac{E_c}{E_0} . \tag{5.30}$$

This expression is generally written in logarithmic form as follows:

$$20 \log \frac{E_0}{E_c} = (CPA) \tag{5.31}$$

copolar attenuation in decibels:

$$20 \log \frac{E_c}{E_x} = (XPD) \tag{5.32}$$

polarization discrimination in decibels from which the general expression can be obtained:

$$(XPD) = U - V(CPA) . \tag{5.33}$$

Parameter V should equal 20 from Equation (5.30). More accurate study shows that beyond about 15GHz its value increases slightly to reach 23 at 35GHz. Parameter U covers all the characteristics of the path. The following formula has been proposed by the CCIR for the 8 to 35 GHz range:

$$U = -10 \log \frac{1}{2}(1 - 0.94 \cos 4\tau) - 40 \log(\cos\varphi) + 30 \log F \tag{5.34}$$

in which:

τ is the polarization tilt angle with respect to the horizontal plane.
φ is the angle of elevation of the path.
F is the frequency in GHz.

In the case of a wave with circular polarization, τ must be taken to equal 45°. We deduce that the degradation in polarization discrimination is at a maximum for circular polarization or for linear polarization with a polarization tilt angle of 45°. Degradation falls as polarization on arrival approaches horizontal or vertical polarization, in consequence of the symmetry of raindrops. This conclusion is also true for terrestrial paths. (It should be remembered that for vertical polarization the electric field is not necessarily vertical. It is the magnetic field which is horizontal.)

5.4.2. Effect of ice crystals

There is another cause of transpolarization which only appears in satellite communications and is due to wave diffraction from ice crystals at high altitudes. These ice crystals are found beyond the 'bright band' detected by radar which is due to a zone near the 0°C isotherm, where ice crystals are partially transformed into water. Depolarization is a consequence of these ice crystals being aligned with the electric field found at these altitudes. Transpolarization due to ice crystals differs from that due to rain because it is not associated with major fades and therefore Equation (5.33) does not apply.

The relative influence of rain and ice crystal effects depend on the climate. It should however be noted that as far as interference is concerned, transpolarization due to ice crystals can have a more damaging effect as it is accompanied by very small attenuation.

5.5. Scatter by hydrometeors

We saw in Section 5.3.2 that attenuation by raindrops is due to two simultaneous phenomena, which are relatively difficult to separate: energy dissipation in the form of heat in drops and energy scatter in all directions around the drops. Unlike dissipated energy, scattered energy remains in radio wave form and can be received, at least partially. It is therefore possible to detect storms using radar, which is of major importance both for studying storms and for avoiding them in air transport. It can also lead to interference at times when the beam to a receiving antenna crosses the beam from a transmitting antenna for another service operating at the same frequency, and if the rain is falling through the volume common to both antenna beams. This problem is examined in Chapter 10.

5.5.1. Scatter by particles

The theory underlying scatter by particles is the same as for attenuation by particles discussed in Section 5.3.2. We have to sum the effects of electric and magnetic multipoles induced in the particle by the incident wave. In the present case, we are only concerned with scattered (i.e. reradiated) power and not absorbed power. On the other hand, we are concerned with scatter in all directions taking into account the polarization of the incident wave.

Since the problem is generally extremely complicated, we often take a simpler elementary case as a reference, assuming that the particles are spheres of relatively small diameter with respect to the wavelength. (The diameter must be small not only with respect to the wavelength in air, but also with respect to the wavelength in the medium making up the particle, which can lead to very small diameters if the permittivity of the particle is very great.) If we limit ourselves, as was the case for attenuation, to terms in D/λ^6, we find that we only need take a single term in the multipolar expansion into account, which means considering only radiation from an elementary electric dipole.

The directional pattern does not depend, for all practical purposes, on the diameter of the particle (which is assumed to be small), but it does depend on the polarization of the transmitted wave. In optics, we consider natural light, which is non-polarized, and the directional pattern of the power can then be represented by the following function:

$$f(\theta) = \frac{1 + \cos^2 \theta}{2}, \tag{5.35}$$

where θ is the angle with respect to the direction of the incident wave. This pattern is therefore a surface of revolution about that direction. There is a maximum in the forwards direction, and another in the backwards direction, and both maxima are equal. Moreover, in the transverse direction, polarization is rectilinear irrespective of the polarization of the incident wave.

For radio waves, assuming rectilinear polarization, the directional pattern is a surface of revolution about the axis of the dipole induced in the particle, and this axis is orthogonal to the direction of the incident wave. However, the directivity in the forward and backward directions is independent of the polarization and equal to 3/2.

Back-scatter is of major importance in radar and leads to simplification in calculations, it is generally taken as the reference case. If we consider only the first term in D/λ, we find that the cross section for radar back-scatter is given by:

$$\sigma_d = \frac{\lambda^2}{\pi}\left(\frac{\pi D}{\lambda}\right)^6 \left|\frac{\eta - 1}{\eta + 2}\right|^2 = \pi^5 \left|\frac{\eta - 1}{\eta + 2}\right|^2 \frac{D^6}{\lambda^4}. \tag{5.36}$$

It is clear that σ_d is proportional to the square of the volume of the particles. The presence of the term in λ^4 corresponds to a frequency selectivity, which in particular explains the blue colour of the sky, due to the scattering of white light from the sun by air molecules.

If we then consider particles of increasing size, we find the following differences:

1. The forward directivity increases (the opposite would be true for infinitely conducting particles).
2. The effective scattering cross section deviates from the value given by Equation (5.36), which only gives the first term in the expansion.
3. The effect of wave polarization in transverse directions is reduced.

If we consider a medium containing a mean value of N particles in a volume V and if we ignore multiple scatter, the total effective scattering cross section σ_D per unit volume of the medium is equal to the sum of the effective cross sections of the particles:

$$\sigma_D = \frac{N}{V}\sigma_d. \tag{5.37}$$

If the particles do not all have the same effective cross sections:

$$\sigma_D = \sum \frac{N_i \sigma_i}{V}. \tag{5.38}$$

In cases of very small particles therefore, given Equation (5.36):

$$\sigma_D = \frac{\pi^5}{\lambda^4}\left|\frac{\eta - 1}{\eta + 2}\right|^2 \sum_i \frac{N_i D_i^6}{V}. \tag{5.39}$$

We generally introduce a quantity Z defined by:

$$Z = \frac{1}{V} \sum_i (N_i D_i^6) \tag{5.40}$$

which is called the radar reflectivity factor of the medium. Hence:

$$\sigma_D = \frac{\pi^5}{\lambda^4} \left| \frac{\eta - 1}{\eta + 2} \right|^2 Z, \tag{5.41}$$

where Z has the dimensions of a volume, but is generally expressed in mm^6/m^3, which introduces a coefficient of 10^{-18} in Equation (5.41).

If the particles are not extremely small, Equation (5.36) is no longer valid, and, consequently, we can no longer write Equations (5.40) and (5.41). In practice, for reasons of convenience, we use Equation (5.41), but replacing Z by a quantity Z_e called the effective radar reflectivity factor, which takes account of corrections with respect to Rayleigh's Law. This means redefining Z_e with respect to the measured equivalent reflectivity factor σ_D using relation:

$$Z_e = \frac{\lambda^4}{\pi^5} \frac{\sigma_T}{\left| \frac{\eta - 1}{\eta + 2} \right|^2} 10^{18}. \tag{5.42}$$

5.5.2. Scatter by rain: atmospheric reflectivity factor

In cases where the particles are raindrops the following properties apply:

1. The function of η which appears in Equation (5.32) is virtually independent of frequency, even though the real and imaginary parts of η vary considerably for frequencies above 10GHz. We find that between low frequencies and about 100GHz we can write to within 3% accuracy:

$$\left| \frac{\eta - 1}{\eta + 2} \right|^2 \simeq 0.93. \tag{5.43}$$

   On the other hand, for ice this quantity will not be greater than 0.2.
2. The directional pattern of this form of scatter does not obey Rayleigh's law. The difference is generally taken into account using a partially empirical correction term.
3. The difference between Z and Z_e only becomes significant at frequencies of the order of 20GHz or above for intense rainfall (150mm/h), 30GHz or above for moderate rainfall (30mm/h) and 40GHz and above for light rainfall (10mm/h).
4. The radar reflectivity z is associated with the intensity of the rain, by a relation depending on the type of rain that may generally be represented by a function:

$$Z = aR^b. \tag{5.44}$$

The CCIR has adopted the following relation which takes overall account of the effects of different types of rain:

$$Z = 400\,R^{1.4}\,. \tag{5.45}$$

It is found that for a given time percentage the reflectivity factor remains approximately constant from the Earth's surface up to a transition altitude of 3 to 4km, and then decreases regularly upwards. For example, for 10^{-4} of the time, the reflectivity factor is around 45 to 55dB up to the transition altitude, and its mean rate of decrease is of the order of 6 to 7dB/km above that altitude. The transition altitude corresponds roughly to the zone where particles of ice are partially melted, and it appears under radar observation in the form of a bright band.

For polarization it has to be noted that if circular polarization radar is used and if the water drops are small enough to be treated as spherical, the energy scattered backwards has inverse circular polarization and is therefore not received by the radar antenna, while the echo sent back by a significantly larger obstacle will be received. A circular polarization radar therefore makes it possible, to a certain extent, to eliminate echoes due to rain. On the other hand, a meteorological radar must have linear polarization.

Chapter 6

Problems due to propagation on line of sight links

6.1. Introduction

The previous chapters of this book analysed the effects of the Earth's surface and the atmosphere separately, but over a line of sight link the two act together and their effects reinforce each other. For example, over a given link, depending on the value of the vertical gradient of the refractive index, the Earth's surface may play a role either by the creation of a reflected ray (interference), or by interception of the direct ray (diffraction).

In this chapter, we systematically study the effects of propagation on the quality of the signal (loss, distortion, etc.), over a line of sight link as well as the practical measures that may be taken to reduce the influence of such anomalies.

6.2. Major phenomena

The major phenomena vary depending on the type of link. In general terms, there are three types of line of sight link.

1. Long overland links, i.e. over about 50km. This is the case for certain radio relay links and radar detection of distant low targets. It is the atmosphere that then plays the main role. Rapid fades are produced by multiple atmospheric paths, over intervals of time lasting several hours, at certain times of year and, if frequency is high, attenuation may be caused by intense rainfall. On certain paths there may be very long fades due to ducting of energy into a tropospheric duct containing neither the receiving nor the transmitting antenna. At certain times there may be obstructions in the way of the beam. Atmospheric fluctuations may also lead to distortions in the signal where broad bandwidths are being used.
2. Short overland links. This is the case of certain links in radio relay networks, particularly at frequencies over 10GHz. Fading is then due above all to reflection from the Earth, losses caused by gases, and to rain. The latter two factors are particularly important between about 15 and 20GHz.
3. Oblique links, i.e. links where the angle of elevation is greater than a few degrees. This is the case of links with satellites and many radars. Multiple atmospheric paths play a minimal role. Attenuation by rain and by gases is the main factor. However, it should not be forgotten that the antenna gain in azimuth is not negligible, and this may lead to interference with other

links, even if they are distant, in periods of intense ducted propagation.

6.3. Fading due to large scale variations in atmospheric refraction

We have seen that the curvature of trajectories is directly related to the vertical gradient of the refractive index. Here we consider only the mean gradient, measured, for example, at a height of a few hundred metres, without taking account of local heterogeneities. The mean gradient can also be expressed by the value of the factor multiplying the Earth's radius (see Chapter 4).

We have also shown that the mean gradient in the lower part of the atmosphere underwent extremely large variations with time, which can lead at certain moments to anomalies, of which the most important are the following:

1. A higher degree of subrefraction may reduce the clearance of antennas to such a point that the link is partially masked.
2. Major variations in refractive index can mean that the direct path no longer falls within the line connecting the antennas, producing an effect equivalent to antenna displacement.
3. A high degree of super-refraction may lead to ducted propagation in tropospheric ducts.

We study these three phenomena and the means to protect against their effects.

6.3.1. Path clearance on terrestrial links

For a link to be regarded as line of sight, the direct path between the transmitting and receiving antennas must never be too close to possible obstacles, and in particular to the Earth's surface. We have, however, seen in Chapter 4 that because of variations in the refractive index of the atmosphere the direct path may show varying degrees of curvature. In certain conditions of subrefraction the path may approach the Earth's surface or even reach the Earth. Losses may then increase considerably, since propagation takes place by diffraction. We therefore need to set up antennas in such a way, with particular attention to their height, that the probability of their being intercepted by irregularities in the ground is negligible.

We also saw in Chapter 4, in situations leading to high levels of subrefraction, that the structure of the atmosphere is unstable, as the denser layers are uppermost. This means that on a given path, positive or slightly negative values of the gradient cannot occur at the same time at every point along the path. Consequently, the 'mean value' of K for a path does not reach minimum values as low as those that might be deduced from measurement at a randomly selected point along the path. This is a very useful effect from the point of view of keeping the beam free of obstacles.

Studies of various line of sight links in temperate climates have led to Fig. 6.1

being drawn up to show the minimum value of K observed for a path of given length. In view of the relatively low number of links available to establish this curve, we cannot associate precise figures for time percentages with these values of K. We can only assume that the probability of finding values of K less than those shown in Fig. 6.1 can be neglected when planning new links.

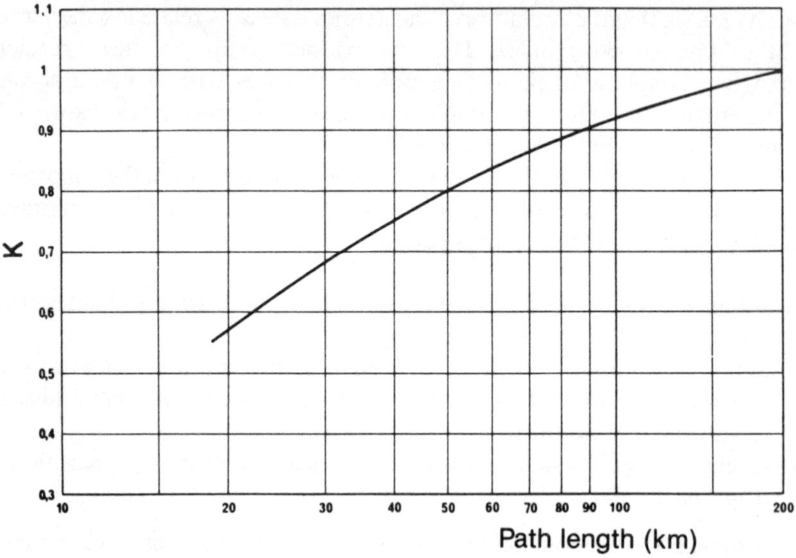

Fig. 6.1. Minimum value of K for a path, temperate continental climate.

To obtain practical rules concerning three different obstacles, we also have to take into account the size of the Fresnel ellipsoid. However, we must avoid aiming for very high levels of clearance from obstacles, so that we do not create propagation difficulties in cases of super-refraction. Experience shows that the following two rules, which have to be obeyed together, generally provide an acceptable compromise.

Rule 1: The height of the antennas must be at least such that the straight line joining the two antennas is tangential to the profile of the Earth's surface drawn with the minimum value of K as shown in the curve in Fig. 6.1.

However, for short links, where the distance between the ray joining the two antennas and the Earth's surface is subject to only small variations as a function of variations in the apparent Fresnel ellipsoid radius, part of the first would be partially hidden at all times if we took radio visibility only for the minimum radius. The same may be the case for long links because of close obstacles. For the mean received level not to be lower than the free space level except for small time percentages, the following rule must also be applied.

Rule 2: The height of the antennas must be at least the height necessary for the first Fresnel ellipsoid to be free from obstacles on a profile drawn with K=4/3.

It should be borne in mind that these rules give minimum heights, and other considerations can lead to the selection of higher values.

6.3.2. Errors in the angle of elevation of antenna beams

In Earth-satellite communication links, large diameter antennas at Earth stations always include an automatic tracking system taking into account variations in atmospheric refraction. For small diameter antennas, variations in refraction have a negligible effect if the angle of elevation is greater than 10° (see Chapter 4).

In terrestrial links, antennas are fixed and the effect of refraction is important. With very high gain antennas it is possible that, as a result of refraction, the direct path is no longer in the axis of the antenna at certain moments. We attempt to evaluate this effect.

It can be shown that, if we write G_0 for the gain along the axis of a highly directional antenna, gain G in a direction within the main lobe and making an angle φ with the axis, is given by the following approximate expression in which φ is expressed in degrees and G and G_0 as power ratios.

$$10 \log G = 10 \log G_0 - 1.28\, G_0\, \varphi^2 . \qquad (6.1)$$

Let us consider a link of length d and the antennas are placed at heights h_1 and h_2. The sighting angle of antenna 1 with respect to the horizontal is giving by:

$$\theta = \frac{h_2 - h_1}{d} - \frac{d}{2\,Ka} . \qquad (6.2)$$

For the second antenna h_1 and h_2 are simply reversed.

If factor K varies, angle θ also varies and its variations do not depend on the heights of the antennas. As a value of K passes from 1 (homogeneous atmosphere) to infinity (appearance of ducts), the variation in θ is equal to:

$$\Delta\theta = \frac{d}{2\,a} . \qquad (6.3)$$

For example, for a 50km link θ takes a value of around 4mrad, 1 for the transition from a homogeneous atmosphere to a normal atmosphere and 3 for the transition from a normal atmosphere to the appearance of ducts.

If we assume that the antenna is correctly oriented for a normal atmosphere, the reduction in gain (expressed in decibels) in the direction of the link as K becomes infinite is equal to:

$1.2\ 10^{-5}\ G_0$

which, when we consider both antennas, gives:

2.4dB for antennas with 50dB gain;
7.6dB for antennas with 55dB gain.

We can therefore see that the effect becomes significant once the antenna gain reaches 50dB. Note too that these calculations assume correctly oriented antennas for normal refractive conditions. If this is not the case, the effect could be far greater. For example, an error in orientation leading to a loss of 0.5dB in normal atmosphere for an antenna assumed to have a gain of 50dB could lead to a loss for that antenna of as much as 3.3dB as K becomes infinite (as opposed to 1.5dB with no orientation error).

6.3.3. Tropospheric ducting on terrestrial links

We have seen that the existence of a tropospheric duct produces considerable field heterogeneity, as certain zones may be reached by many different trajectories, while others are reached by none. When conditions leading to the formation of ducts appear, we generally observe periods of high and fluctuating field intensity alternating with long deep fading.

As the number of modes possible within a duct is generally very high for decimetric or shorter waves, we can use the trajectory (see Chapter 4) to derive some interesting conclusions. In particular, for surface ducts the following conditions apply.

1. If the Earth's surface is reflecting, the probability that there is at least one path between the antennas is at a maximum if they are at the same height as each other. If, furthermore, this height is less than 3/4 of the thickness of the duct, there is at least one trajectory between them. In this case, the maximum distance reached is limited only by the length of the duct.
2. In all other cases, there is a maximum distance beyond which certain zones cannot be reached, in particular if the antennas are above the duct. In the latter case, the maximum distance reached increases with the height of the antennas.
3. If the Earth's surface is not reflecting, the maximum distance reached is limited by the first 'hop' in the trajectory, and that distance is inversely proportional to the height of the antennas. For example, in a duct 200m high with a gradient of 500N/km, the maximum distance reached without reflection from the Earth is around 60km if the antennas are 50m high, and only 25km if they are 175m high.

All these properties taken together suggest the following rules as ways to provide at least partial protection against the effects of tropospheric ducts for paths on which they are likely to appear, i.e. particularly on paths above flat and damp ground.

1. Keep links reasonably short. This is obviously not always possible, in particular if the link is between two islands. Moreover, this increases the number of links and makes communication less reliable.
2. Make sure that transmitter and receiver heights are as close as possible at the same level, unless the link is very short and sufficiently different antenna

heights can be used for the angle between the direct path and the horizontal to be greater than about 1°.

3. Use an auxiliary link where the antennas are at a far lower height than those of the main link, so that they are always in the lower part of the duct. This link obviously undergoes very great losses due to spherical diffraction under normal propagation conditions but in periods of ducted propagation it replaces the faded normal link.
4. If local radio and meteorological conditions are such that the thickness of the duct is always low, set up the antennas as high as possible.

It has to be said however, that these rules are only intended to avoid radio holes leading to prolonged fades. Even under nearly normal propagation conditions, there may still be rapid fades due to multi-paths. To protect against them, space or frequency diversity is used.

6.4. Fades and distortion due to tropospheric multi-paths

Even if ducting conditions are not reached over the entire path, the occurrence of super-refraction leads to the appearance of tropospheric multi-paths. If one of these paths stands out particularly from the others we simply get scintillation around the received mean level. On the other hand, if energy is distributed between a number of paths of comparable importance, we may have rapid deep fades. It is this phenomenon which is generally called multi-path propagation.

6.4.1. Statistical distribution of received levels

We saw in Chapter 4 (Section 4.2.4.1) that the distribution of received levels on a terrestrial link undergoing multi-path effects can be modelled in general using the Nakagami-Rice distribution, with parameters associated with the characteristics of the link. In particular, Fig. 4.13 gives a distribution for a 4GHz link above moderately hilly terrain in a temperate climate. Figure 4.13 can be used for other frequencies and other climates as long as we replace the real length of the link with an effective length. We find that for a link at frequency f in the same region, we only need to replace the real length d of the link by an effective length d_{ef} given by:

$$d(F/4)^{0.25} \qquad (6.4)$$

For example, a 40km link at 11GHz has the same distribution of received levels as a 51.5km link at 4GHz.

When it comes to climatic effects the coefficient multiplying the distance must be determined from measurements. For example, results have been published suggesting that a 42km link in the United States has the same statistical distribution as a 65km link in Western Europe.

6.4.1.1. Asymptotic distribution

If we only consider that part of Fig. 4.13 in which curves are virtually parallel straight lines, i.e. if we consider only relatively long links and very low received levels (major attenuation), the model becomes much simpler. In particular, the following approximate equation is often used because of its simplicity:

$$p = k \frac{W}{W_0} F^{\alpha} d^{\beta}, \tag{6.5}$$

where:

p is the probability of the received power being less than W;
W is the received power under free space conditions;
k is a coefficient depending on climate and terrain;
F is the frequency;
d is the distance.

For the worst month, in North West Europe, and for moderately hilly terrain, the CCIR gives the following numerical values:

$\alpha=1 \quad \beta=3.5 \quad k=1.4\times10^{-8}$

where F is expressed in Gigahertz and d in kilometres.

Using a logarithmic scale and writing A for the attenuation in decibels with respect to free space, the formula can be rewritten:

$$A = 10 \log F + 35 \log d - 10 \log p - 78.5 . \tag{6.6}$$

The coefficient of 10 multiplying log F is not always accepted and a smaller value seems more plausible at least above 1GHz.

In other parts of the world, similar formulas to Equation (6.5) are used, but the coefficients are different. However, the exponents of F and d always have values close to those shown above. A multiplying factor is sometimes introduced to take into account the 'roughness' of the ground. This coefficient increases as the terrain becomes smoother. However, roughness is not easy to determine for a given path, and the use of such a coefficient leads to considerable uncertainty.

The accuracy of Equation (6.6) fails very rapidly if we consider path lengths lower than 35km. For very short distances, we need rather to use a gaussian distribution for modelling, as emerges from the properties of the Nakagami-Rice distribution for low attenuation (see Appendix 3).

In certain cases, we need to know losses corresponding to a time percentage differing considerably from the extreme percentages. For example, the CCIR uses the 80% figure to define transmission qualities for radio relay links; this is the so-called 20% clause. The following approximate formula can be used:

$$A_{80} = 3\ 10^{-2}\ d \sqrt{F} \tag{6.7}$$

which gives the loss, in decibels, due to tropospheric multi-paths not exceeded for 80% of the time over a link of length d (km) at frequency F (GHz).

6.4.2. Number and duration of fades due to tropospheric multi-paths

It is essential to know the global figure of total time at which the level is less than a certain value but it is not always enough to define quality. Other important quantities are the number and duration of individual fades. Although not all observers are in complete agreement over the statistical data concerning these quantities, we can, however, give the following empirical rules for frequencies greater than about 3GHz.

For a given length, greater than around 40km, the duration of fades due to multi-paths greater than a given depth obey a log-normal distribution, at least for fades greater than 10db. The standard deviation of this distribution shows little frequency dependency and falls off slightly as we consider lower levels. The mean duration is proportional, for practical purposes, to the square root of the received power, and inversely proportional to the frequency:

$$\bar{T} = k_1 \frac{\sqrt{W}}{F}. \tag{6.8}$$

To give an idea of the orders of magnitude, it has been observed on a 58km link at 11.7GHz that fades greater than 30dB last on average 5s.

The number of multi-path fading occurrences greater than a given depth is broadly proportional to the square root of the received power, and grows more rapidly than frequency, following an approximate relation of the form:

$$N = k_2 \sqrt{W} F^q \tag{6.9}$$

where q is of the order of 1.4.

An interesting conclusion can be drawn from these two relations. The product $N\bar{T}$ is the total duration of fades exceeding a given depth:

$$N\bar{T} = k_3 W F^{q-1}. \tag{6.10}$$

It is clear that this duration is proportional to the received power and depends little on frequency. This proportionality to received power shows that the distribution of received levels tends asymptotically towards a Rayleigh distribution.

6.4.3. Effect of multi-paths on the transmissible bandwidth

Though the troposphere is not a dispersive medium (refractive index does not depend on frequency), a frequency selectivity effect can nonetheless appear because of the differences between radio multi-path lengths which cause phase interference. In the case of line of sight links we can generally assume that there is a direct path and several secondary paths, each of them characterized by the amplitude and relative phase of the wave associated with it.

At any instant, we can sum all waves, and consider a two-ray model covering only two paths. This method makes it possible to describe the phenomenon of attenuation observed on a fixed frequency. In the case of a non-zero bandwidth we have to introduce additional considerations, and a two-ray model does not make it possible to explain all the phenomena observed. It does however give a good approximation and we briefly study it here. We write r and φ for the amplitude and relative phase of the wave with respect to the direct wave. The resultant amplitude is given by:

$$a = 1 + r\,e^{j\varphi}. \tag{6.11}$$

If τ is the difference in propagation times along the two paths, and ω is the angular frequency considered:

$$a = 1 + r\,e^{j\omega\tau}. \tag{6.12}$$

This function of ω may be regarded as the transfer function of the atmosphere in the two-ray model. If A and Φ are the modulus and argument of this function:

$$A^2 = 1 + 2\,r\cos\omega\tau + r^2, \tag{6.13}$$

$$\tan\Phi = -\frac{r\sin\omega\tau}{1 + r\cos\omega\tau}, \tag{6.14}$$

$$\frac{1}{\tau}\frac{d\Phi}{d\omega} = \frac{-\,r(r + \cos\omega\tau)}{1 + 2\,r\cos\omega\tau + r^2}. \tag{6.15}$$

It is clear that the atmosphere behaves as a filter with an infinite number of frequency pass-bands centred on frequencies defined by:

$$\cos\omega\tau = 1 \tag{6.16}$$

and an infinite number of attenuated frequency bands centred on frequencies defined by:

$$\cos\omega\tau = -1. \tag{6.17}$$

When $\omega\tau$ is close to 1, the selectivity of the filter is very strong in the attenuated frequency bands.

In fact, the only useful part of the above functions is in general that which is near the first frequency with maximum attenuation:

$$F_0 = \frac{1}{2\tau}, \tag{6.18}$$

$$A_0 = 1 - r, \tag{6.19}$$

$$\Phi_0 = 0\,, \tag{6.20}$$

$$\frac{1}{\tau}\left(\frac{\mathrm{d}\Phi}{\mathrm{d}\omega}\right)_0 = \frac{r}{1-r}. \tag{6.21}$$

In Fig. 6.2 we have shown variations of A and $\Phi/\mathrm{d}\omega$ near this frequency for various values of r as a function of the product $\tau.\ \Delta F$ where ΔF is the deviation with respect to frequency F_0.

In order to determine quantitatively the effect of multi-paths on a given transmission, we must therefore know the statistical distribution of r and τ for the link in question. Despite many studies, our knowledge in this field is still far from good. We can however obtain order of magnitude figures.

The relative amplitude r of the wave resulting from the spurious paths can take all possible values between 0 and 1, and these values are related to those of the maximum fade depth.

Delay τ depends on the height of the atmospheric layers likely to reflect part of the incident wave, given the angle of incidence on the layers. Experience shows that, in most cases, the delay is no greater than a nanosecond. In certain models hypothetical paths with much longer delays may be considered. To give an idea of the magnitudes involved, on a 50km link, such a delay can be obtained with a path containing a point of reflection at around the middle of the link and 87m from the axis of the beam. As distance to the axis increases proportionally to the square root of the delay time, it is clear that only fluctuations in the refractive index at most a few hundred metres from the axis play any significant role.

This distance could be reduced by using extremely directional antennas. If G is the gain of an antenna, expressed as a ratio of powers, the half-width at 3dB of the main lobe is given in degrees by the following approximate relation:

$$\alpha = \frac{1.5}{\sqrt{G}}. \tag{6.22}$$

For a 50km link and a reflecting element near the middle of the path, the element is in the central part of the antenna beams if it is at most 120m from the axis for antennas with 50dB gain. We can consequently assume that beyond 50dB the antenna reduces the delay time of the spurious wave. The use of very large antennas would therefore make it possible to increase the transmission bandwidth, but for long hops this would lead to excessively big antennas and to risks of incorrect pointing (Section 6.3.2).

6.4.3.1. Effects on signal quality

The effect of multi-paths on the quality of the signal depends not only on the bandwidth occupied by the signal but also on the type of modulation used (e.g. analog or digital) and on the sensitivity of the signal to perturbations (e.g. a colour television signal is more sensitive than a black and white television signal). It is therefore difficult to give general rules.

We know that the performance of radio equipment can be characterized, for

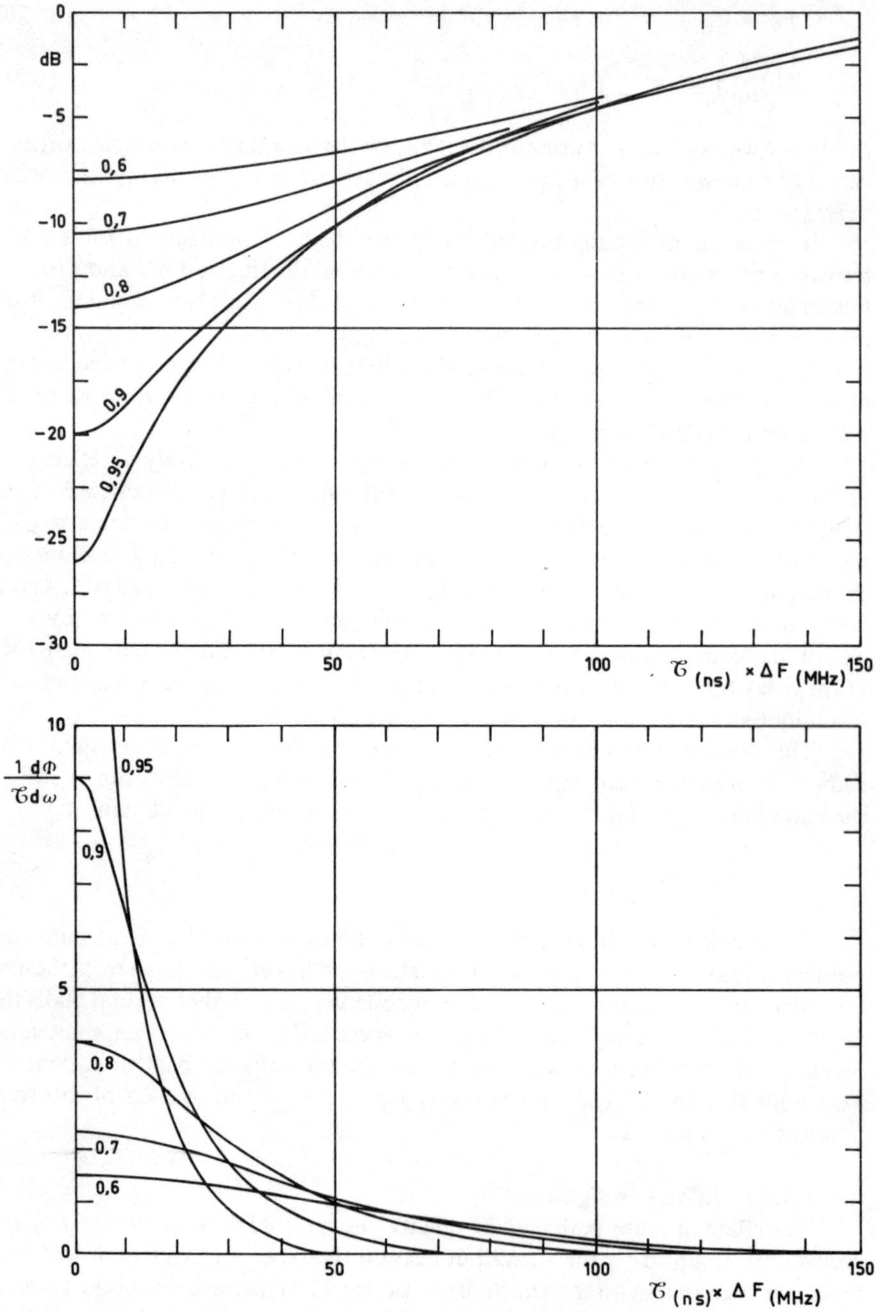

Fig. 6.2. Selectivity of fades due to multi-paths.

a given link, by the maximum depth of fade, making it possible to obtain a specified noise level or error rate. This maximum fade depth, expressed in decibels, gives the equipment margin. It is generally defined and measured on the assumption that fading is non-selective: it is called the flat fade margin.

If fades are selective in frequency, as is the case, for example, in multi-path propagation, we find that the minimum quality specified is obtained for a fade depth of which is less than the flat fade margin, i.e. it is as though the equipment margin was reduced in the presence of multi-paths. This allows us to define the net margin, which is the fade depth exceeded for the same time percentage as the specified noise level or error rate.

For analog transmission the difference between the net margin and the flat fade margin is relatively small up to 1800 telephone channels. On the other hand, for digital transmission the difference becomes significant for much smaller numbers of channels. Figure 6.3 shows, by way of example, the relation between the net margin and the flat fade margin for links between 40 and 60km long operating at around 11 to 13GHz, with 8PSK modulations. This relation is obtained using a model based on measurements. It is clear that, above a certain value, increases in flat fade margin (obtained, for example, by increase in the transmission power) only lead to a very small increase in net margin essentially for high bit rates. In this case, improvements in the net margin can only be obtained using diversity reception and adaptive equalization systems.

The curves in Fig. 6.3 can be represented by the following empirical relations:

$$M_n = a + \frac{M_f - a}{\left[1 + \left(\frac{M_f - a}{30}\right)^4\right]^{1/4}} \tag{6.23}$$

with:

$$M_l = 73.5 - 20 \log s$$

and:

$$a = M_l - 30$$

where:

M_f is the flat fade margin in decibels;
M_n is the net margin in decibels;
M_l is the limit margin in decibels;
s is the bit rate in Megabits/s.

The way in which the net margin depends on the length of the path, the frequency and the modulation is not clearly understood. We can nonetheless assume that the net margin bears little relation to the length of the path, that it increases with frequency and that it diminishes as the number of phase states increases.

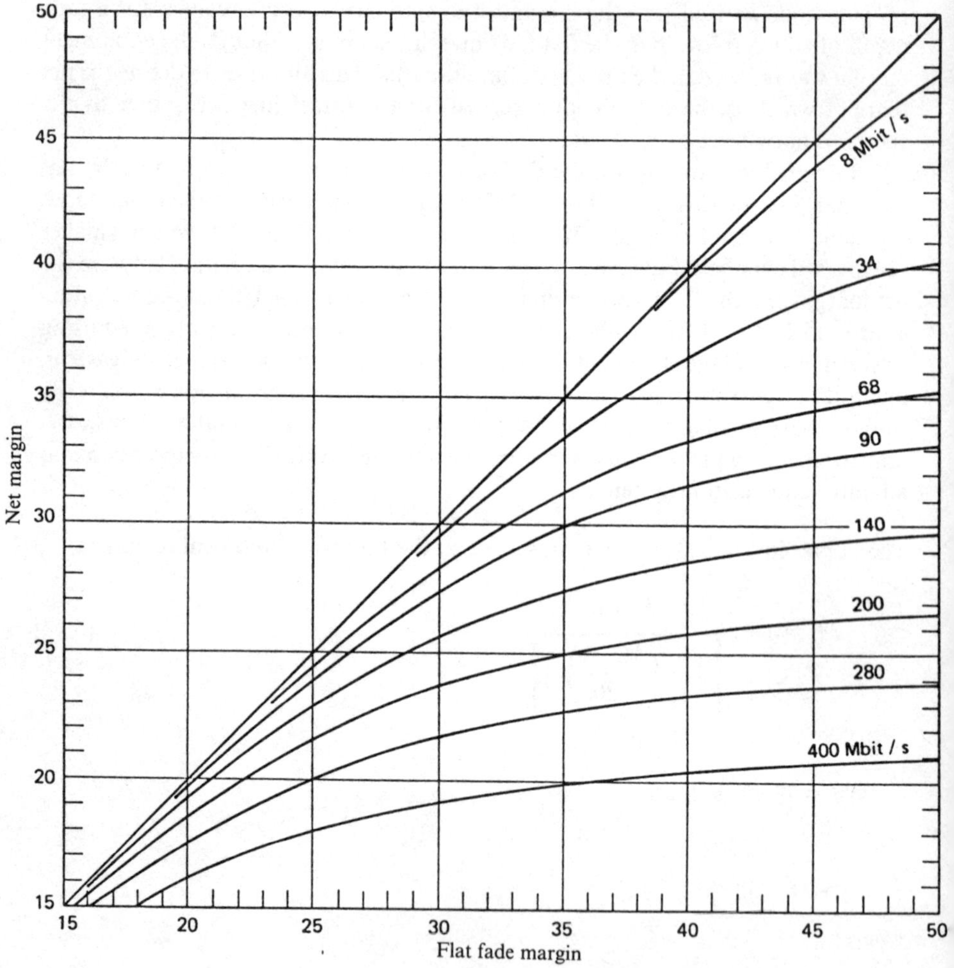

Fig. 6.3. Net margin as a function of flat fade margin.

6.4.4. Scintillation

If one of the paths between transmitter and receiver is clearly the most important, the others can only create roughly rapid fluctuations in the received level, the amplitude of these fluctuations being of the order of a few decibels. We then talk of scintillation rather than multi-paths, but the basic phenomenon is the same and it obeys the same rules. In particular, in Chapter 4 we showed that the amplitude variations showed a gaussian distribution about the free space level.

Scintillation does not generally exert any major influence on transmission quality for terrestrial links because of the large operating margin of the equipment used, but it must sometimes be taken into account for satellite links. Fluctuations in level diminish as the angle of elevation increases, since this reduces the length of the path in the atmosphere, and the reflection coefficient from the tropospheric layers (see Section 2.4.1.4).

It is however important to note that on a path between two points on the Earth's surface the amplitude of fluctuations can be very much amplified by the existence of a stable reflected ray on the ground, as it may at certain moments considerably reduce the mean level received. The relative importance of fluctuations then increases. This phenomenon is discussed in Section 6.5.

6.4.5. Protection against fades due to tropospheric multi-paths: diversity reception

As the probability of fading due to multi-paths increases rapidly with the length of the path, it is worth reducing that length, e.g. by bringing the radio relay network towers closer together. This method, however, has the disadvantage of increasing transmission costs as well as the probability of interruptions due to equipment failure. It should be said that the improved reliability of equipment, obtained by using semiconductors, and reduced operating costs thanks to remote monitoring, have led to the reduction in the length of radio relay links.

The most effective method to guard against the effects of these fades is the use of diversity reception. One can either use vertical space diversity by installing two antennas a certain distance apart on the same tower, or frequency diversity by transmitting the same signals on two sufficiently widely spaced frequency channels. In either case, we use two receivers and then combine their signals on output. Combination can take place by simple automatic switching to the signal with the higher level, but there are more efficient methods.

We generally measure the effect of diversity over a given link by using the concept of an improvement factor, which is the time percentages during which a given fade depth fading is exceeded, with or without diversity, respectively (see Appendix 6.1). Various formulas to predict the improvement factor have already been published, but none of them seems sufficiently general, probably because of the small number of measurements available. The following formulas give the best available compromise.

For space diversity (6.25)

$$I_s = 1 + 0.18\left(\frac{s}{d}\right)^{1.5}\frac{w_0}{w}$$

where I_s is the improvement factor, s is the vertical space between the antennas in metres, d is the length of the path expressed in kilometres, w is the received power and w_0 is the power received under free space conditions. The ratio w_0/w, expressed in decibels, shows the fade depth in question.

It emerges that the main parameter is the s/d ratio, i.e. the angle subtending the diversity antennas at the other end of the path. Frequency does not seem to be a major factor. In practice, however, for a given time percentage, it influences ratio w_0/w.

For frequency diversity:

$$I_f = 1 + 3.7\,\frac{1}{F}\left(\frac{\Delta F}{F}\right)^{1.2}\frac{w_0}{w} \qquad (6.26)$$

where f is the frequency of the link in GHz and $\Delta F/F$ is the relative frequency difference assumed to be less than around 6%, i.e. $\Delta F/F < 0.06$.

It is clear that the length of the path is not involved in the formula. It would appear, in fact, that the length of the path has a negligible influence at least up to 70 or 80km. Beyond that level, a few results of measurements over greater distances, though they are hard to reconcile with each other, seem to suggest that the diversity improvement diminishes as distance increases. This is probably due to the fact that very long paths require that the station altitudes are high, and consequently the atmosphere does not have the same characteristics over the entire path.

6.5. Fading due to reflection from the Earth's surface

A line of sight link set up above a reflecting surface, and in particular above the sea, can be perturbed by the existence of a roughly stable reflected ray.

If reflection takes place on a maritime path free of obstacles, we can apply ray theory to it, and obtain the amplitude and relative phase of the reflected ray, as well as the resultant field intensity, from geometric considerations.

If reflection takes place on a terrestrial path, the problem is generally more complicated since the reflecting area is often relatively small and is not necessarily located near the point of geometric reflection. In this case, diffraction phenomena must be taken into account. All these problems are discussed in Chapter 3 (Sections 3.6 and 3.7).

6.5.1. Combined effects of reflection from the Earth's surface and tropospheric scintillation

To assess the effect of reflection, we need to take account of the fact that the link is not in a vacuum but in the atmosphere, and that at certain moments scintillation may occur. In the absence of reflection from the Earth's surface, scintillation is nearly always negligible (see Section 6.4). On the other hand, if reflection from the ground takes place, scintillation may become more damaging, as can be seen in Fig. 6.4, which uses a rotating vector representation.

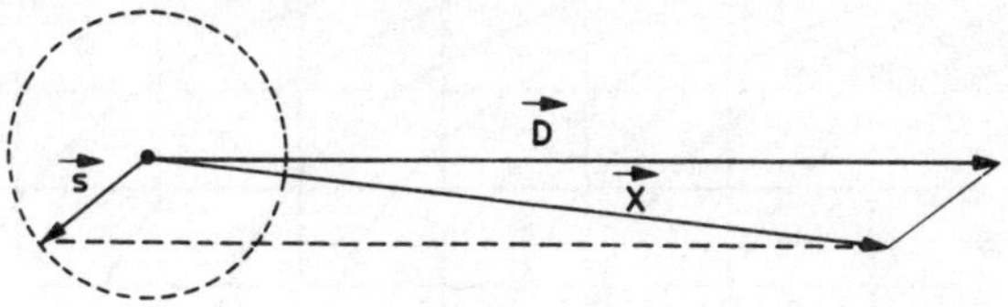

a) No reflected ray
X = D + s

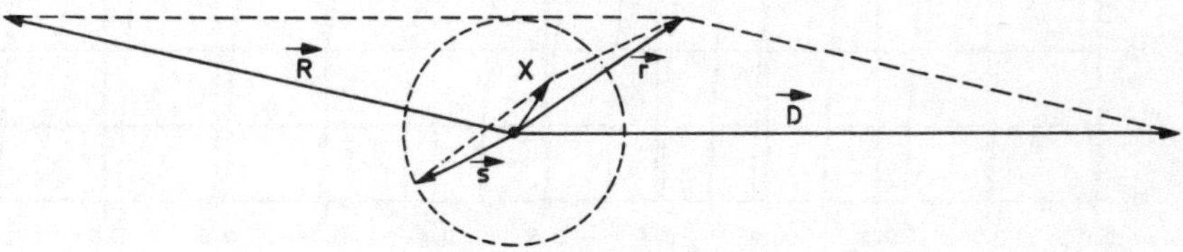

b) A reflected ray
X = D + R + s
or x = r + s with r = D + R

Fig. 6.4. Combination of scintillation (vector s) and reflection from the ground (vector **R**) with a direct ray (vector **D**).

Without reflection (Fig. 6.4a) the received level is the sum of the direct vector and a small random vector. Variations in the level are therefore small. If any significant degree of reflection from the ground takes place, the resultant direct reflected wave leads to fading which is stable or varies slowly, which would be acceptable for transmission if there was no scintillation. On the other hand, with even a little scintillation, the resultant vector $\vec{x}$ can be very small at certain times (Fig. 6.4b).

Figure 6.5 gives the minimum level as a function of the ground reflection coefficient and the amplitude (in decibels) of the scintillation due to the atmosphere. It is clear, for example, that a reflection coefficient of 0.7 gives a maximum fade level of 10.5dB, in the absence of scintillation. If the amplitude of the scintillation is 2dB, the fade depth reaches 20.5dB. If the reflection coefficient increases to 0.8, this can lead to a complete breakdown in transmission.

6.5.2. Protection against fades due to ground reflection

The ideal method of protection, if it is possible, is to choose a path for which there is little ground reflection. In particular, this method should always be considered when determining a path for a radio relay network. Obviously, in certain

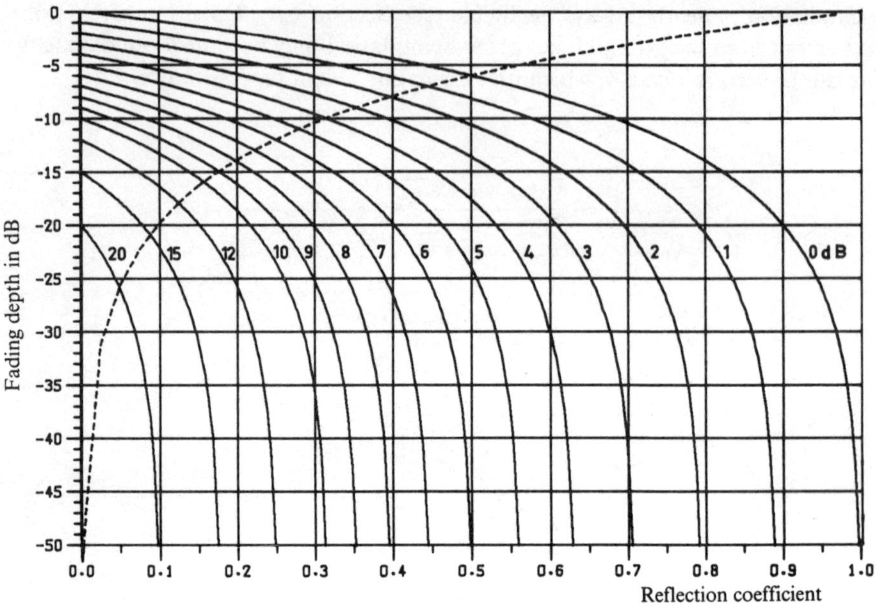

Fig. 6.5. Fading created by a reflected ray in the presence of atmospheric scintillation. Parameter: peak amplitude of scintillation.
----------------- : relative amplitude of reflected ray.

cases this is not possible, for example if the link has to cross an expanse of water or a very flat humid region. But then certain steps can be taken to reduce the effect of reflection.

The main difficulty is caused by the major variations in refraction in the atmosphere, and, in the case of reflection from the sea, by variations in height due to tides, leading to variations in the path difference between the direct and reflected rays. This means that certain compensating factors that might be useful to protect against reflection under given refractive conditions will increase the problem under other conditions.

6.5.2.1. Influence of antenna gain on fade depth

The presence of a ray reflected from the ground leads to the presence, above the location of the receiver, of a system of interference fringes with regularly spaced maxima and minima. If the refraction conditions in the atmosphere remain constant with time, we will only need to place the receiving antenna at a point where the field is at a maximum. But because of variations in atmospheric refraction, the maximum is likely at certain times to be replaced by the minimum, particularly if the link is fairly long.

In general, the antenna's size is not negligible with respect to the distance between a maximum and the next minimum. The received level is therefore a result of a kind of mean field strength over the antenna aperture, and this reduces very considerably the effect of refraction variations. The effect can be determined as follows.

Consider a link of length d and antennas are placed at heights h'_1 and h'_2 above the plane tangent to the ground at the point of reflection. Let d_1 and d_2 be the distances from the point of reflection to the ends of the links. We then have the following relations (Fig. 6.6):

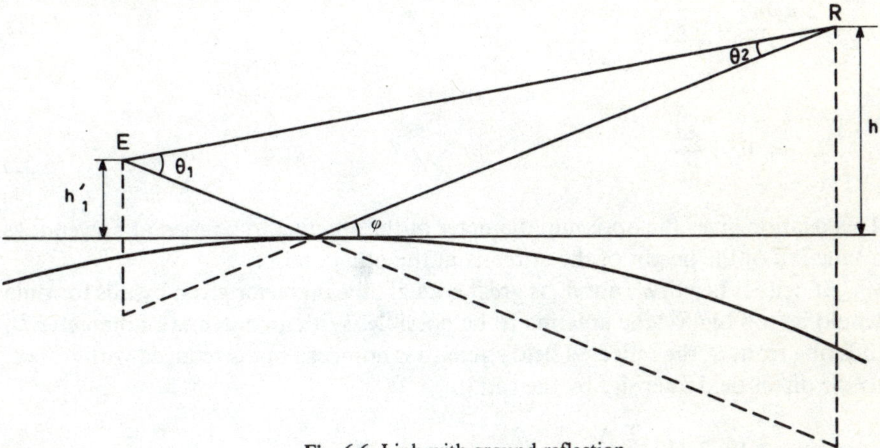

Fig. 6.6. Link with ground reflection.

$$h'_1 = h_1 - \frac{d_1^2}{2\,Ka} \qquad h'_2 = h_2 - \frac{d_2^2}{2\,Ka} \tag{6.27}$$

and also:

$$\theta_1 = \frac{2\,h'_2}{d} \qquad \theta_2 = \frac{2\,h'_1}{d}. \tag{6.28}$$

It is further assumed that the antenna is a paraboloid of diameter D with uniform illumination. It is known that its radiation pattern is then given by:

$$\frac{E(\theta)}{E(0)} = \frac{2\,J_1\left(\dfrac{\pi D}{\lambda}\sin\theta\right)}{\dfrac{\pi D}{\lambda}\sin\theta}. \tag{6.29}$$

As angles θ_1 and θ_2 are always small, and writing R for the ground reflection coefficient (which may include the divergence factor), the ratio between the direct wave and the reflected wave is given by:

$$\frac{E_r}{E_d} = \frac{J_1\left(\frac{2\,\pi D h'_1}{\lambda d}\right)}{\frac{\pi D h'_1}{\lambda d}} \; \frac{J_1\left(\frac{2\,\pi D h'_2}{\lambda d}\right)}{\frac{\pi D h'_2}{\lambda d}} R\,. \tag{6.30}$$

The reflected field strength is zero if the refected ray is in the direction of a zero on the radiation pattern. The first zero of function j_1 is obtained for:

$$\frac{2\,\pi D h'}{\lambda d} = 3.832 \tag{6.31}$$

i.e.

$$D_1 = 0.61\,\frac{\lambda d}{h'_2}\,. \tag{6.32}$$

This relation gives the optimum diameter of the antenna to be used at one end, as a function of the height of the antenna at the other end.

If neither height h'_1 nor h'_2 is great enough, the diameter given by this formula would be too big for the antenna to be possible. With an antenna of diameter D_1 differing from D, the reflected field strength is non-zero but is reduced with respect to the direct field intensity by the ratio:

$$z = \frac{2\,J_1\left(3.832\,\frac{D}{D_1}\right)}{3.832\,\frac{D}{D_1}}\,. \tag{6.33}$$

Writing R for the ground reflection coefficient the field strength lies between:

$1 + zR$ and $1 - zR$;

Figure 6.7 gives the value as a function of the ratio D/D_1.

Example: Consider a 20km link at 7.5GHz (λ=4cm) and assume that at one of the ends the antenna is 100m above the plane tangent to the Earth's surface. We can calculate that at the other end the optimum diameter of the antenna should be:

$$D_1 = \frac{0.61 \times 0.04 \times 20\,000}{100} = 4.88\text{m}$$

If we use an antenna 2m in diameter we have:

$$\frac{D}{D_1} = 0.41$$

On a reflecting terrain the maximum amount of fading possible due to ground reflection is 11dB. With 3m antennas the figure would be 5.3dB.

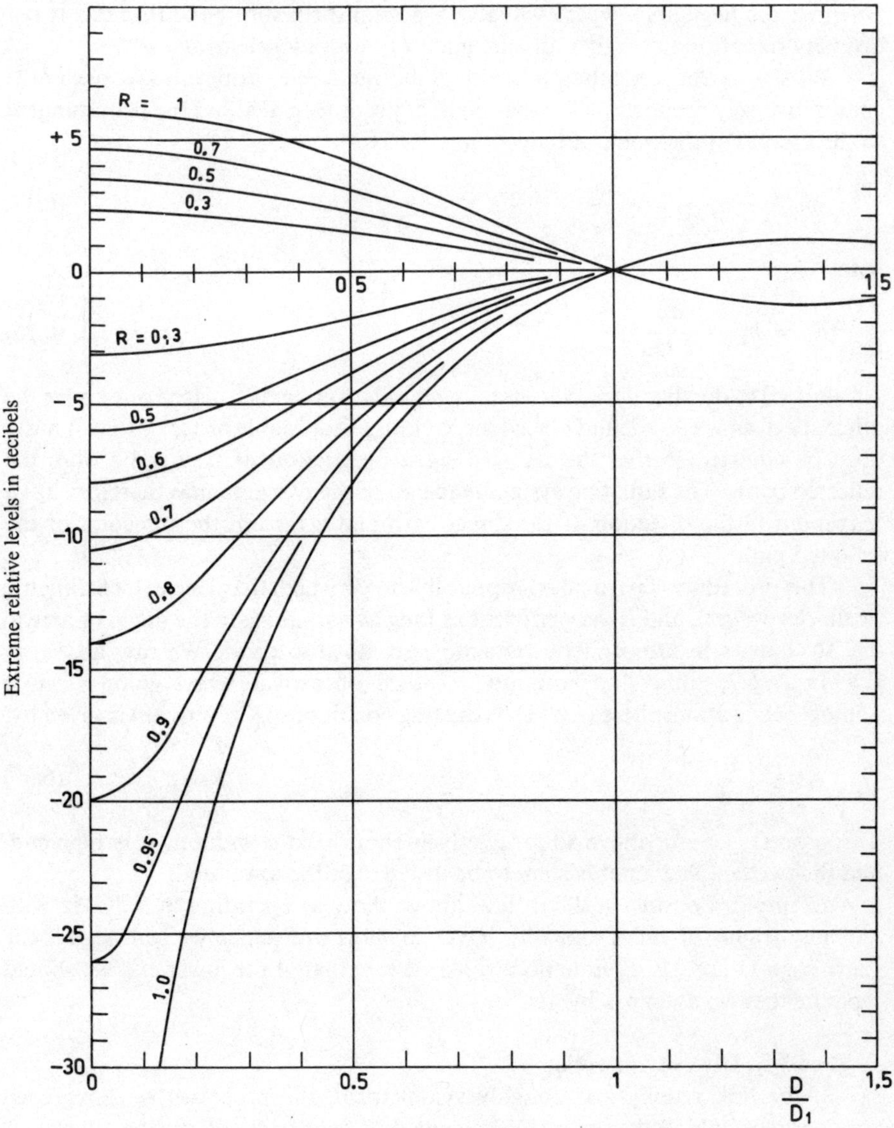

Fig. 6.7. Link with ground reflection. Influence of the antenna diameter.

D_1 = optimum diameter
R = ground reflection coefficient

6.5.2.2. Use of two coupled antennas

If the calculation above leads to an excessive antenna diameter, we can replace the single antenna by a system of two coupled antennas, separated vertically by a suitable distance. To calculate that distance, we assume that the sizes of the antennas are negligible, which will give a pessimistic result. Note that this is not diversity reception but rather an antenna array with two elements.

We saw in Chapter 4 that the width of the interference fringes above one of the ends is inversely proportional to the height of the other end above the plane tangent to the ground at the point of reflection:

$$\delta h_2 = \frac{\lambda d}{2\, d_1\, \varphi} = \frac{\lambda d}{2\, h'_1} \tag{6.34}$$

with:

$$h'_1 = h_1 - \frac{d_1^2}{2\, Ka}. \tag{6.35}$$

All we need to do, therefore, is to install a set of two antennas, placed one above the other at a distance $\delta h_2/2$, and to feed them through adjustable phase shifters. Phases must be adjusted so that the received signals cancel out if they come from the reflected paths. The radiating system made up of the two antennas therefore has a maximum in the direction of the direct path and a zero in the direction of the reflected path.

This procedure is particularly applicable to very high frequencies (centrimetric or shorter waves), and is very efficient as long as variations in the angle of arrival due to changes in atmospheric refraction are not too great. We saw above, in Equation (6.2), that the variation in angle of arrival when going from a homogeneous atmosphere ($K = 1$) to ducting conditions (K is infinite) is given by:

$$\Delta\theta = \frac{d}{2\, a}. \tag{6.36}$$

This process is useful above all for relatively short lengths with one very high end, and the two coupled antennas are to be installed at the lower end.

Example: Consider a 20km link above the sea, operating at 7.5GHz with antenna heights of 100m and 20m. Over so short a distance we can neglect the Earth's curvature. Using Equation (6.34) we find that at the lower end we should separate the two antennas by 4m.

6.4.2.3. Diversity reception

If the link is long and roughly symmetrical, the processes we have been considering would lead to excessive antenna diameters or to too great a separation between antennas. A small variation in the vertical gradiant of the refractive index would then be enough to replace a maximum by a minimum in the radiation pattern. The only possible solution would be to use reception with vertical space diversity. The arrangement of the antennas is the same as before but each of them is associated with a receiver, and signals are either combined or added after re-phasing and demodulation.

The diversity distance has to be carefully selected for maximum efficiency, but unlike the previous procedure, diversity cannot produce undesirable effects. Given that when K is small, the divergence effect on a convex Earth (see Chapter 4) reduces the amplitude of the reflected ray, it is better to optimize the separation between antennas for very large K. A simple practical rule is to choose a separation which is three-quarters of that, corresponding to half the inter-fringe distance for infinite K, which leads to

$$\delta h_2 = \frac{3}{8}\frac{\lambda d}{h_1} \tag{6.37}$$

or:

$$\delta h_1 = \frac{3}{8}\frac{\lambda d}{h_2}\,. \tag{6.38}$$

Example: Consider a 50km link operating at a frequency 7.5GHz and antenna heights are 30 and 50m:

$\delta h_2 = 25\text{m} \quad \delta h_1 = 15\text{m}$

6.6. Attenuation due to hydrometeors

6.6.1. General considerations on prediction methods

In Chapter 5 we analysed the mechanism leading to the attenuation of waves by raindrops. Here we make a few comments on the way in which the attenuation over a link during showers or storms may be predicted statistically using rain gauge data in the region in question. Two points need to be taken into account when establishing a practical prediction method.

1. For a given link, the range of significant attenuation values is limited by the operating margin of the equipment used. For example, for radio relay links the equipment margin is generally less than 45dB, and in addition we can generally ignore attenuation less than a few decibels. We only therefore need a prediction method which gives accurate results between about 4 and 45dB. For satellite links, the range of significant values is between about 1 and 15dB.
2. We know that the slope of the rain intensity distribution varies from one type of rain to another. For example, for the same value of rainfall intensity exceeced for 0.1% of the time, the value exceeded during 0.001% of the time may differ considerably from region to region. However, these great variations do not appear in statistics of attenuation. Regions in which there is a high slope to the distribution of rainfall intensities are generally those in which there occur intense but not very extensive storms. On the other hand, those regions where the slope is low are regions where the rainfall has a lower maximum intensity but covers a far greater area. These two effects to a considerable extent cancel out, from the point of view of determining attenuation to such a point that in certain cases, between 0.1 and 0.001% of the time, the distribution of attenuation can be entirely determined from the

value obtained for 0.01% of the time, at least to a first degree of approximation.

After studying many types of method intended to predict attenuation due to rainfall, the CCIR now only considers methods using effective distances or distance reduction coefficients, for the purposes of practical calculation. It is only these methods, therefore, that we consider here.

6.6.2. Terrestrial paths

To predict attenuation due to rainfall, we can use various procedures, depending on the meteorological or radioelectrical information at our disposal.

First case: We know the statistical distribution of rainfall intensities in the region containing the path. We may then take account of all the quantities involved: distance d, rainfall intensity R, time percentage p, frequency f, on the assumption that attenuation A can take the form of the product of two terms:

$$A(d, R, p, F) = \gamma(F, R).d_e(p, d)\,. \qquad (6.39)$$

$\gamma(F,R)$ is the specific attenuation in (dB/km) given by: $\gamma = KR^{\alpha}$ in which K and α are functions of frequency. For practical calculations, the nomogram in Fig. 5.6 or Table 5.1 may be used. For frequencies not mentioned in Table 5.1 interpolation has to be carried out using a logarithmic scale for K and a linear one for α (taking into account that α reaches a maximum for frequencies near 7GHz).

$d_e(p,d)$ is the effective distance which may be calculated using the following empirical relation:

$$d_e = \frac{1.55 + 0.1 \log p}{1 - 0.08 \log p \cdot d^{0.75}} \qquad (6.40)$$

This relation seems to be applicable to a very broad field, covering all cases met in practice. We can see, in particular, that for very short paths, the effective distance may be greater than the real distance.

This prediction method is an equi-probability method which calculates the attenuation corresponding to a given time percentage, starting from the point at which the rainfall intensity corresponds to the same time percentage. However, for time percentages of the order of 0.1% or greater, attenuation is due in part to hydrometeors other than rain. Consequently, it may be preferable, for these time percentages, not to use the value corresponding to the rainfall intensity but to base oneself on the second case described below.

Second case: We do not have sufficiently accurate information about rainfall intensities for time percentages close to 0.01%. This is a fairly common case, as for very small time percentages, statistics are based on too few samples and for large time percentages, statistics are not always available. We therefore calculate, as above, attenuation for 0.01% of the time, and then deduce attenuation for other time percentages using an empirical formula for p less than about 0.2:

$$\frac{A_f}{A0.01} = 0.12p^{-(0.546 + 0.043 \log f)} \tag{6.41}$$

Third case: We do not know the statistical distribution of the rainfall. On the other hand, we do know the distribution of attenuation but at frequencies different from the considered frequency. This is often the case when we have to use several frequencies over a path, and measurements have only been carried out for one of them.

We find that, for a given link, attenuation exceeded for the same time percentage at two frequencies F_1 and F_2 are practically proportional, and that the coefficient of the proportionality depends, to a first approximation, only on the frequencies F_1 and F_2 by a relation of the form:

$$\frac{A_2}{A_1} = \frac{\Phi(F_2)}{\Phi(F_1)}. \tag{6.42}$$

Several empirical expressions have been obtained for ΦF in particular the following where F is in Gigahertz, which seems to apply at least from 7 to 60GHz:

$$\Phi(F) = \frac{F^{1,72}}{1+3 \times 10^{-7}(F^{1.72})^2}. \tag{6.43}$$

Simpler relations may be used, which apply to smaller frequency ranges:

$$\Phi(F) = F - 6 \quad \text{for} \quad 9 < F < 20 \text{ GHz} \tag{6.44}$$

$$\Phi(F) = F - 10 \quad \text{for} \quad 20 < F < 40 \text{ GHz} \tag{6.45}$$

$$\Phi(F) = F^{1.72} \quad \text{for} \quad 7 < F < 30 \text{ GHz}. \tag{6.46}$$

Fourth case: We know the statistical distribution of attenuation for vertical polarization and want to deduce the distribution for horizontal polarization, or vice versa. Experience shows that, except for very low attenuation, we can establish an approximate relation between the attenuation for the two types of polarization that does not involve the rainfall intensity distribution. For attenuation greater than about 5dB we can use the following empirical relation:

$$A_V = \frac{300\ A_H}{335 + A_H} \quad \text{or} \quad A_H = \frac{335\ A_V}{300 - A_V}, \tag{6.47}$$

A_H and A_V are attenuation expressed in decibels for horizontal and vertical polarizations, respectively.

Example: Assume we wish to establish a 30km link at 15GHz in a region where rainfall intensities for 0.1%, 0.01% and 0.001% of the time are 7mm/h, 32mm/h and 85mm/h, respectively. From Equation (6.40) the effective distances are 26.2km, 13.7km and 9.3km, respectively.

For specific attenuation we can use Equation (5.23) or the nomogram in Fig. 5.6. This gives the values in Table 6.1.

Table 6.1

P(%)	R(m/h)	d_e(km)	γ_H	γ_V	A_H(dB)	A_V(dB)
0.1	7	21.5	0.34	0.31	7.3	6.7
0.01	32	13.3	2.00	1.75	26.6	23.3
0.001	85	9.2	6.30	5.10	58.0	46.9

Calculating attenuation for 20GHz first, and then applying Equation (6.46), we find:

horizontal polarization: 8.3dB; 28.8dB; 56.7dB
vertical polarization: 7.2dB; 22.7dB; 45.5dB

6.6.3. Earth-space paths

As in the case of terrestrial links, we can think in terms of using an effective distance through rain, but there the effective distance depends on the latitude, the angle of elevation, the altitude of the station and the time percentage.

The following method is based on measurements carried out on Earth-satellite paths as well as on terrestrial radio measurements.

Equation (6.40), which for terrestrial links gives the effective distance d_e corresponding to a geographic distance d, can also be used for Earth-satellite links on condition that we regard d as the length of the path between the ground and the top of the rain zone. The top of that zone is at an altitude which is roughly that of the zero degree isotherm, but can differ significantly from it. In particular, at lower altitudes the height of rain to be taken into account, even for heavy rainfall intensities due to storms, is always less on average than that of the zero degree isotherm in summer.

The height of the rain h_R to take into account in calculations of attenuation is above all a function of latitude Φ, so empirical formulas have been proposed, in particular the following one in which h_R is in kilometres and Φ is in degrees. Slightly different formulas but equivalent values given the available data have been proposed by various observers.

$$h_R = 4.1 \cos\left[1.3\,(\Phi - 20)\right]. \tag{6.48}$$

Another expression has been mentioned by the CCIR:

$h_R = 4.0$ for $0 < \Phi < 36°$
$h_R = 4.0 - 0.075\,(\Phi - 36)$ for $36° < \Phi < 90°$

The distance d through the rainfall can be deduced from the geometric relation:

$$d = \frac{2(h_R - h_0)}{\sqrt{\sin^2 \theta + 2\,\dfrac{h_R - h_0}{Ka}} + \sin \theta} \tag{6.49}$$

where h_0 is the altitude of the ground. $K.a$ is the modified radius of the Earth which we can take as equal to 8500km. θ is the angle of elevation. If θ is greater than 5^0, as is generally the case, we need only write:

$$d = \frac{h_R - h_0}{\sin \theta}. \tag{6.50}$$

We can obtain the attenuation due to rain by multiplying the effective distance d_e obtained in this way by the specific attenuation calculated as for terrestrial links. We should however note that for Earth-space paths, polarization is generally neither vertical nor horizontal, and attenuation is therefore intermediate between those corresponding to each of the two polarizations.

Example: Let us suppose that we want to establish a link with a satellite at 12GHz, at an angle of 30° in a region where rainfall intensities for 0.1%, 0.01% and 0.001% of the time are 7mm/h, 32mm/h and 85mm/h, respectively. The latitude is 45°.

Using Fig. 6.8 and the nomogram in Fig. 5.6 we obtain the values shown in Table 6.2 (taking the mean loss for the two polarizations).

Table 6.2

P(%)	R(mm/h)	d_e(km)	γ(dB/km)	A(dB)
0.1	7	7.5	0.19	1.4
0.01	32	5.5	1.20	6.5
0.001	85	4.3	3.90	16.8

6.6.4. Protection against rainfall effects: path or site diversity

The only way to guard against rainfall effects is to have two paths sufficiently remote from each other for the probability of intense rainfall on both paths simultaneously is very small. This principle is obviously different in application to terrestrial links and space links.

For terrestrial links we need two entirely separate paths. It seems that a separation of 10km between paths is generally enough to produce a considerable improvement, but few measurements have yet been carried out on this subject.

For Earth-satellite paths, we need two interconnected Earth stations. Improvement due to this site diversity can be calculated from the general formula given in Appendix 6.1.

$$p_2 = \frac{p_1{}^2(1 + \beta^2)}{p^1 + \beta^2} \tag{6.51}$$

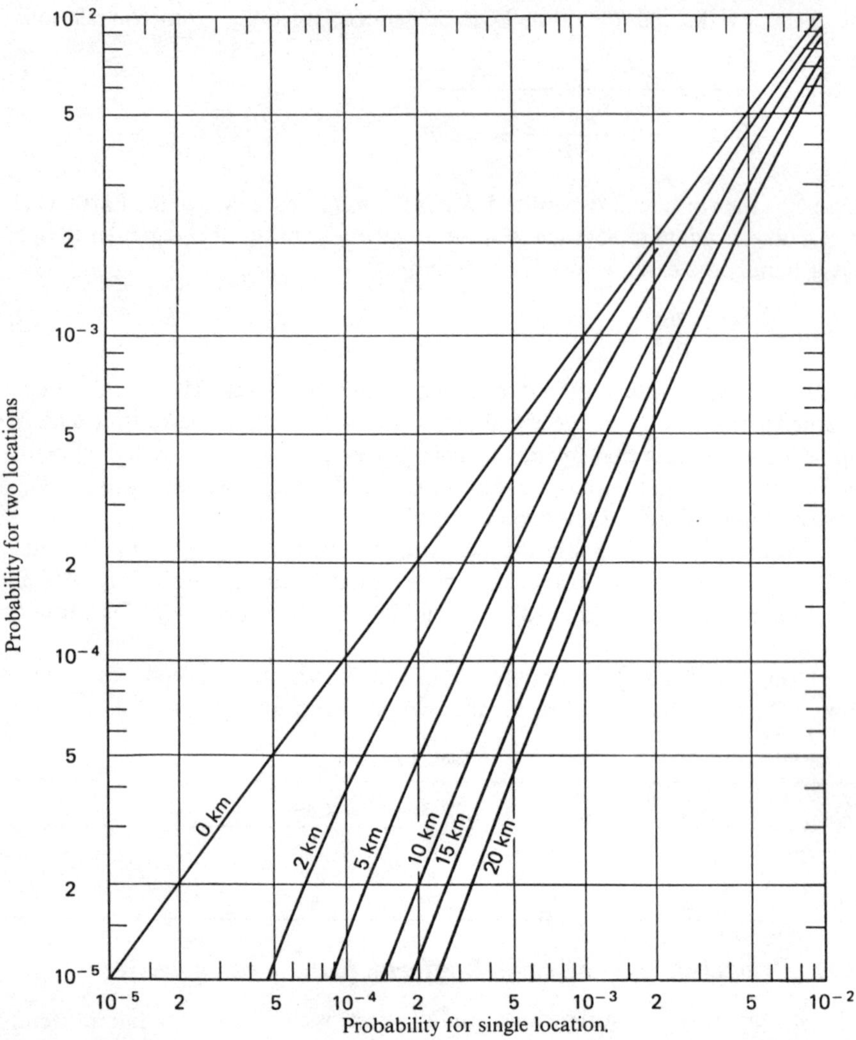

Fig. 6.8. Attenuation due to rain. Effect of site diversity. Parameter: distance between stations.

where p_1 and p_2 are the probabilities corresponding to the same attenuation value with or without diversity, respectively, and β^2 is a parameter depending on the characteristics of the link.

Available information suggests that β^2 depends essentially on the distance between the two Earth stations according to the following empirical relation:

$$\beta^2 = 10^{-4} d^{1.33} \tag{6.52}$$

where d is expressed in kilometres (Fig. 6.8). This relation is only a first approximation as it is probable that β^2 also depends a little on the angle of elevation of the path, and the orientation of the straight line joining the two stations with respect to the general direction of rain fronts in the region in question.

Appendix 6.1

Diversity reception

A.6.1. Introduction

Diversity reception involves taking advantage of decorrelation, which may only be partial, between two paths from the same transmitter and carrying the same signal, but differing from each other in a certain characteristic (frequency, reception location, etc.). The improvement that may be obtained increases as correlation between the two transmission channels falls. Clearly we can reduce correlation by increasing the difference between the values of the chosen characteristic (frequency difference, antenna separation) but this reduction cannot always make it possible to reach zero correlation, and in certain cases the residual correlation is still very large.

A6.2. Measuring the effect of diversity

There are several ways of measuring the effect of diversity. They compare statistical distributions of a given characteristic of the signal (level, error rate, etc.), obtained with or without diversity on the same path.

Fig. 1 shows the statistical distributions of received levels without diversity (curve 1) and with diversity (curve 2). For each point in curve 1 we can define two points on curve 2, point B corresponding to the same probability and point C corresponding to the same received level. The difference AB, called the diversity gain is given in decibels. The difference AC, which is called the diversity improvement is a ratio of probabilities. Diversity gain and diversity improvement may be expressed as a function of either the probability or the received level corresponding to point A.

The concept of diversity gain may be of practical value to determine a system in which quality is specified for a single value of probability. However, the concept of diversity improvement is more generally used as there are relations between probabilities p_1 without diversity and p_2 with diversity. In particular, whatever the value of p_1:

$$p_1^2 < p_2 < p_1 \qquad p_2 = 1 \text{ if } p_1 = 1 \tag{1}$$

The exact relation between p_1 and p_2 depends on the correlation coefficient between the two channels and the statistical distribution of the received level. However, we can adopt a simple relation corresponding to a plausible correlation model. In particular, the following relation covers almost all practical cases, and satisfies relations (1)

$$p_2 = \frac{p_1^2}{k^2 p_1 + 1 - k^2} \tag{2}$$

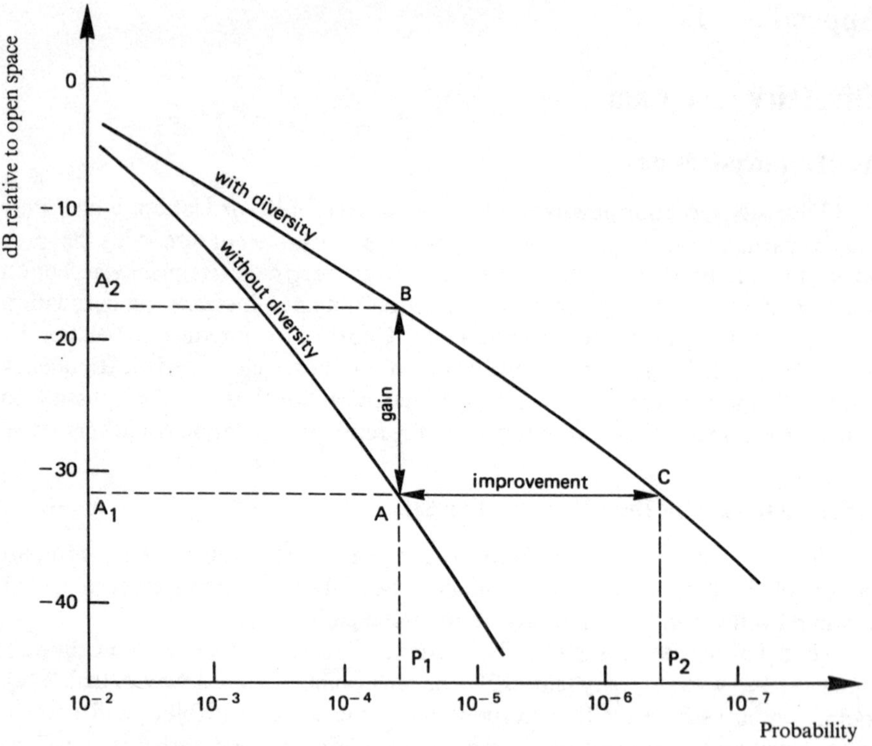

Fig. 1. Definition of diversity gain and divesity improvement.

where k is the correlation coefficient between the two channels.

For practical applications it is best to replace k by a parameter varying between 0 for a correlation equal to 1, and infinity for a zero correlation, as this parameter is zero for a zero separation in frequency or distance and increases with the separation. We need only write:

$$\frac{1}{k^2} = 1 + \beta^2 \tag{3}$$

and therefore:

$$p_2 = \frac{p_1{}^2(1 + \beta^2)}{p_1 + \beta^2} \tag{4}$$

the improvement factor I due to diversity is therefore given by:

$$I = \frac{p_1}{p_2} = \frac{1}{1 + \beta^2}\left(1 + \frac{\beta^2}{p_1}\right) \tag{5}$$

In most practical cases β^2 has a very low value and:

$$I \simeq 1 + \frac{\beta^2}{p_1} \quad (6)$$

Figure 2 shows the relation between p_1 and p_2 for various values of β^2.

We can see from Equation (5) that the improvement factor is roughly inversely proportional to the probability that attenuation of a given value is exceeded. This factor therefore increases with attenuation according to a function depending on the statistical distribution of attenuation.

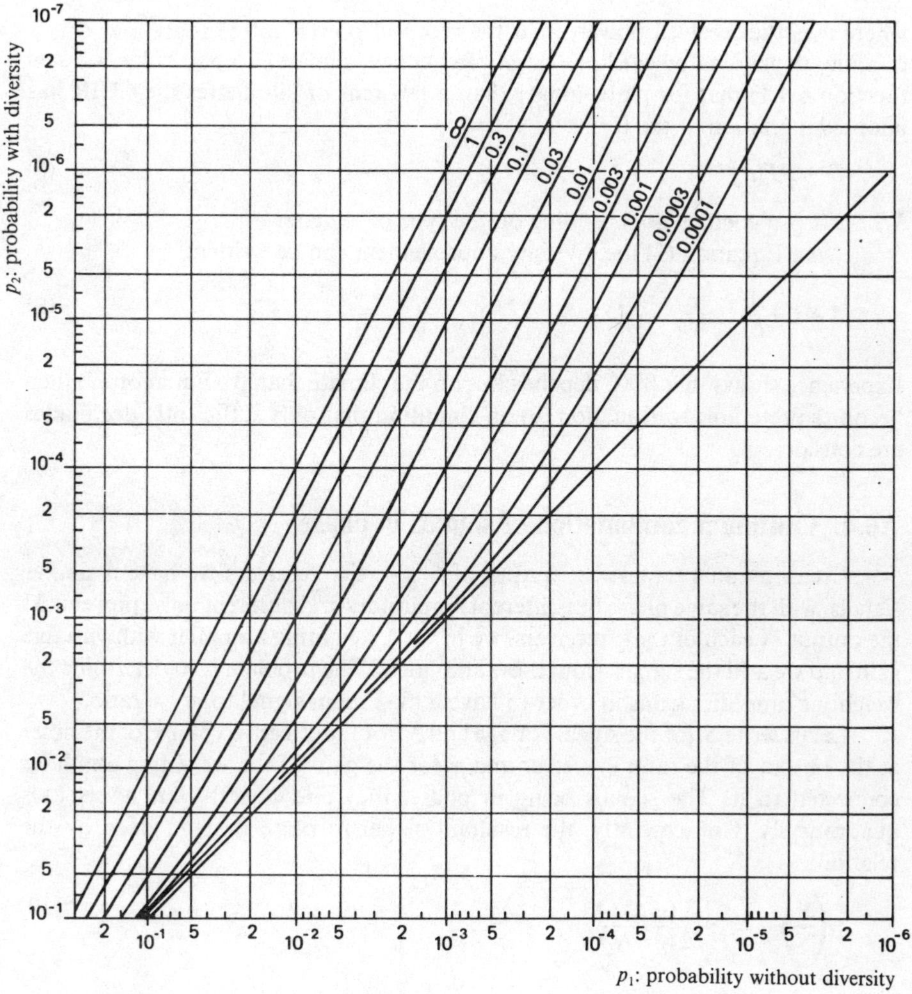

Fig. 2. General curves for double diversity, parameter β^2 $\quad p_2 = \frac{p_1\ (1 + \beta^2)}{p_1 + \beta_2}$

A6.3. Particular case: Rayleigh distribution and distributions with the same asymptotic behaviour

In a very common case where the probability of a low received power is asymptotically proportional to this power (e.g. Rayleigh distribution, Nakagami-Rice distribution, etc.) we have the asymptotic relation:

$$p_1 \frac{w_0}{w} = C \qquad (7)$$

where w is the received power, w_0 is the received power in free space and C is a quantity depending on the link (length, frequency, climate). In particular, we saw (Section 6.4.1) that for paths longer than a few tens of kilometres the CCIR has adopted a relation of the form:

$$C = AF(d)^{3.5} \qquad (8)$$

Where A is a coefficient depending on the type of climate.

Using Equation (7) the diversity improvement can be written:

$$I \doteq 1 + \frac{\beta^2}{C} \quad \frac{w_0}{\text{w}} \qquad (9)$$

Experience shows that β^2/C depends less on the climate than β^2. Equation (9) then becomes preferable to Equation (6) on condition that only sufficiently deep fades are considered.

A6.4. Optimum combination of signals in phase

Let us assume that at the output of n diversity receivers we have n similar signals, with the same phase but different amplitudes and different noise powers. At the output of each of these receivers, we insert a weighting amplifier with variable gain and we add the signals from these amplifiers. The problem is to determine the weighting amplifier gains in order to have a maximum signal to noise ratio.

Let us write S_i for the signal voltage and N_i for the effective voltage of the noise at the output of the rank i receiver and g_i for the gain of the weighting amplifier connected to it. The signals being in phase, they add linearly but noise adds quadratically. Consequently, the resultant signal to noise ratio is given by the relation:

$$\left(\frac{S}{N}\right)^2 = \frac{[\Sigma\,(g_i\,S_i)\,]^2}{\Sigma\,(g^2{}_i\,N_i)} \qquad (10)$$

The condition we require can be obtained by making the derivative with respect to g_i equal to zero:

$$\Sigma\,(g_i^2 N_i^2).\Sigma\,(g_i S_i).2S_i - [\,\Sigma(g_i S_i)\,]^2 2g_i N_i^2 = 0 \qquad (11)$$

which can be written as:

$$\frac{g_i N_i^2}{S_i^2} = \frac{\Sigma\,(g_i^2 N_i^2)}{\Sigma\,(g_i S_i)} \tag{12}$$

Since the right-hand side of this equation is independent of the rank *i* of the receiver considered:

$$\frac{g_i N_i^2}{S_i} = K \tag{13}$$

Where K is a constant depending on the characteristics of the link. Consequently:

$$g_i = K\,\frac{S_i}{N_i^2} \tag{14}$$

The gain of the rank *i* weighting amplifier must at any instant be proportional to the signal voltage and inversely proportional to the noise power at the output by the corresponding receiver. Under these conditions:

$$\left(\frac{S}{N}\right)^2 = \Sigma\left(\frac{S_i}{N_i}\right)^2 \tag{15}$$

The resulting optimum signal to noise ratio is the sum of the signal to noise ratios over all the channels.

Chapter 7

Propagation by diffraction

7.1. Introduction

We have already seen that if there is an obstacle near the path linking the transmitter with the receiver, we have to take account of reradiation from that obstacle. If the obstacle lies directly between the transmitter and the receiver, received power is only provided by reradiation from the obstacle. The study of diffraction is therefore essentially a 'boundary condition' problem, and it is this that makes it difficult.

In the case of wave propagation near the Earth's surface, the major obstacle is the Earth itself, but the assumptions we have to make governing the Earth's shape depend very much on the wavelength with which we are concerned. For example, for myriametric waves the Earth can be treated as a smooth sphere, whereas for centimetric waves even slight variations in terrain height can present a major obstacle.

In view of the difficulties involved in studying this kind of problem for obstacles which do not have a geometrically simple shape, only two theoretical cases have been studied in detail: diffraction by a large radius sphere and diffraction by a ridge of low or negligible thickness. Practical problems are reduced to combinations of these two theoretical cases.

7.2. Spherical diffraction

The diffraction of waves around the Earth's sphere (Fig. 7.1) is one of the oldest propagation problems. Following work by Poincaré, Sommerfeld, Van der Pol, Bremmer and Norton, the general expression giving the field was expressed as an expansion of the following series, sometimes called the residue series because of the calculation method used:

$$\frac{E}{2\,E_0} = (2\,\pi\zeta)^{1/2} \left| \sum_{n=1}^{\infty} \frac{e^{-j\tau_n\zeta}}{\delta + 2\,\tau_n} \cdot g_n(h_e) \cdot g_n(h_r) \right| , \tag{7.1}$$

ζ and δ are dimensionless quantities, defined by:

$$\zeta = \left(\frac{2\,\pi}{\lambda R^2}\right)^{1/3} d\,, \tag{7.2}$$

$$\delta = \left(\frac{2\,\pi R}{\lambda}\right)^{2/3} C_0\,. \tag{7.3}$$

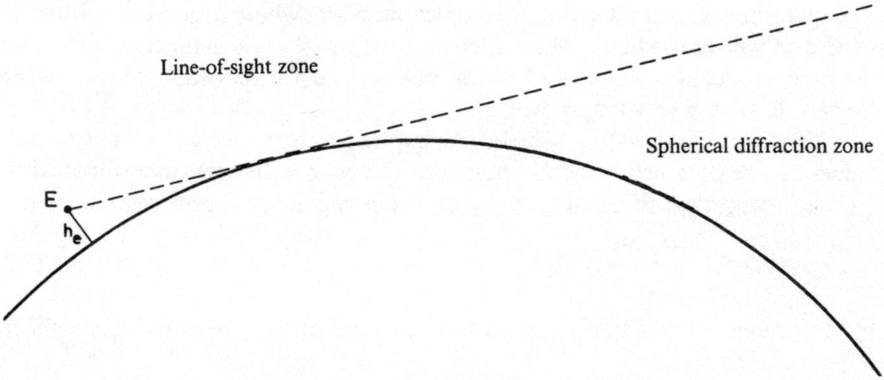

Fig. 7.1. Spherically diffracted path.

Slightly different quantities are sometimes used, see Appendix 7.1.

It is also worth introducing the quantity:

$$q = \zeta \mid \delta \mid = \frac{2\,\pi}{\lambda} \mid C_0 \mid d\,. \tag{7.4}$$

In these expressions the symbols used are:

E is the modulus of the field at a point (phase will not be considered);
E_0 is the field in free space at the same distance;
h_e and h_r are the heights of the antennas above the Earth's surface;
λ is the wavelength;
R is the Earth's radius which can be replaced by an effective radius (see Chapter 4) to take account of atmospheric refraction; note however that for very long waves, as the height of the atmosphere is not great with respect to the wavelength, the effective radius is close to the real radius;
d is the distance measured along a great circle;
C_0 is the value for $\varphi = 0$ of parameter C introduced in Chapter 3 which depends on the nature of the terrain and the polarization of the wave. If, as before, we use η for the complex permittivity of the terrain:

$$\eta = \varepsilon - \mathrm{j}\,60\,\sigma\lambda\,. \tag{7.5}$$

C_0 is given by the following expressions:
– in vertical polarization:

$$C_{0v} = \frac{\eta - 1}{\eta^2} \tag{7.6}$$

– in horizontal polarization:

$$C_{0H} = \eta - 1\,. \tag{7.7}$$

The quantities τ_n form a series of complex numbers whose argument is equal to $-\pi/3$ and whose modulus, which increases with rank n, is a function of δ. For example, modulus τ_1 goes from 1.85 for infinite δ, i.e. for very short waves, to 0.88 for zero δ, i.e. for very long waves.

The functions $g_n(h)$ form a series of complex functions, for which approximate values can be obtained in certain ranges of values of h. In particular, for small h, $g_n(h)$ is independent of n and given by the following linear approximation:

$$g_n(h) = 1 + j\frac{2\pi}{\lambda}\sqrt{C_0}\,h\,; \tag{7.8}$$

$g_n(h)$ is therefore practically constant and equal to 1 up to a characteristic height H_1 approximately equal to:

$$H_1 = \frac{\lambda}{2\pi}\left|\frac{1}{\sqrt{C_0}}\right|. \tag{7.9}$$

Beyond H_1, $g_n(h)$ is practically proportional to h up to a second characteristic height H_2 given by:

$$H_2 = \frac{1}{2}\left(\frac{R\lambda^2}{\pi^2}\right)^{1/3} \tag{7.10}$$

and above H_2, $g_n(h)$ is approximately exponential.

It is therefore clear that H_1 depends on the nature of the terrain and H_2 on the Earth's curvature. For various types of terrain and for vertical and horizontal polarization, Fig. 7.2 gives the values of H_1 and H_2 as a function of the wavelength. It is found that for horizontal polarization, height H_1 is less than the wavelength, while for vertical polarization H_1 can take some very high values, particularly above sea. We see the consequences of this later.

Theory gives the following general results:

1. For given frequency and ground characteristics if h_e and h_r are both greater than the value of H_1 corresponding to vertical polarization, the field strength is practically the same for both types of polarization. In particular, if h_e and h_r are greater than the value of H_1, corresponding to vertical polarization above sea, field strength no longer depends on the type of terrain, nor on polarization.
2. If we start with values of h_e and h_r greater than H_1, then as they decrease so will the field strength, but the decrease in field strength stops at a greater height for vertical polarization than for horizontal polarization and therefore a greater field strength is maintained down to ground level. This effect is particularly significant for low frequencies, to the extent that near the ground such frequencies (hectometric or longer wavelengths) can only be used in vertical polarization.

The successive terms in Equation (7.1) decrease rapidly, and at great distances we can consider the first term alone, though for short distances several terms have to

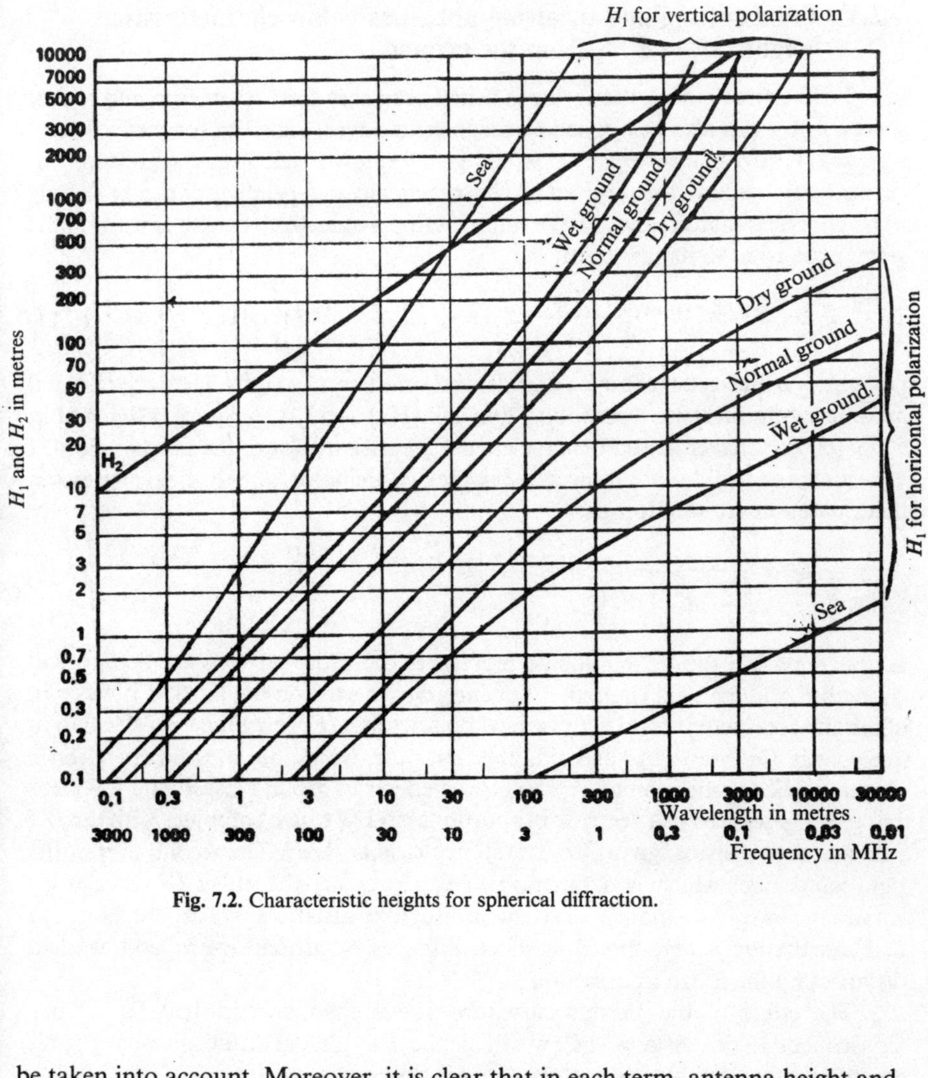

Fig. 7.2. Characteristic heights for spherical diffraction.

be taken into account. Moreover, it is clear that in each term, antenna height and distance are independent of each other, but this characteristic vanishes as the sum of several terms is taken. For this reason, it is often preferable to carry out an approximate calculation with the first term alone, rather than a more rigorous calculation with several terms. Moreover, given the rapidity of convergence of the series, the error does not become large until we come close to line of sight conditions and a correction factor can be introduced to reduce it (see Section 7.2.1).

Even if we limit the series to its first term, the calculation of the field strength depends on too large a number of parameters (ε, σ, λ, h_e, h_r, d, R) to give a simple expression for the general case. We are therefore obliged to study particular cases for which pracitcal calculation methods can be established.

7.2.1. Transmitting and receiving antennas below characteristic heights H_1 and H_2 from the ground

For vertical polarization, this is usually the case for hectometric and longer waves, but it can also be applied to decametric waves, and even for metric waves over sea. Horizontal polarization would give too low a field near the ground. For vertical polarization, the received level is practically independent of height as long as height is less than H_1 and H_2. We can therefore assume zero height. If we limit the series expansion to its first term:

$$\frac{E}{2\,E_0} \simeq (2\,\pi\zeta)^{1/2} \left| \frac{e^{-j\tau_1\zeta}}{\delta + 2\,\tau_1} \right| . \tag{7.11}$$

Quantities δ and τ_1 are complex and depend on the wavelength. They are of equal modulus for a wavelength of around 20m (50MHz) over sea, and greater than about 500m (600kHz) over land. For wavelengths significantly lower than these values, 2 τ_1 becomes negligible with respect to δ in the denominator, and the field expression takes on an interesting form:

$$\frac{E}{2\,E_0} \simeq \frac{1}{\frac{2\,\pi}{\lambda}\,|\,C_0\,|\,d}\,(2\,\pi)^{1/2}\,\zeta^{3/2}\,|\,e^{-j\tau_1\zeta}|. \tag{7.12}$$

In this equation the left-hand side represents the ratio between the actual field strength E and the field strength 2 E_0 which would exist on the Earth if it was flat, of infinite conductivity and with vertical polarization (E_0 being the field strength in free space). On the right-hand side the first part, which depends on the electric characteristics of the ground, represents the effects of ground losses, and it is clear that these losses lead to a decrease proportional to $1/d$ which combines with the $1/d$ decrease in free space, giving an overall decrease of $1/d^2$. The second part of the right-hand side, which is a function of ζ, represents the effect of the Earth's curvature, and gives exponential attenuation at long distances. We can therefore say that attenuation is determined at short distances by ground losses, and at great distances by the Earth's curvature.

The effect of the Earth's curvature is less than around 1dB for $\zeta<0.4$. Consequently, the distance below which the Earth's curvature gives negligible attenuation is given by the approximate relation:

$$d_e \simeq 10\,\lambda^{1/3} \tag{7.13}$$

where d is expressed in kilometres and λ in metres. This distance is sometimes called the flat Earth approximation limit.

For very long wavelengths, we can no longer ignore the term in 2 τ_1 with respect to δ, and the decomposition we saw before no longer takes place. This means that attenuation due to ground losses becomes relatively small with respect to attenuation due to the Earth's curvature. Furthermore, the latter form of attenuation begins to appear at distances which increase with the wavelength itself.

In the limiting case, i.e. assuming that δ is negligible with respect to 2 τ_1, the field becomes independent of the nature of the terrain. This is in fact the case for myriametric and longer wavelengths.

It is possible for these approximate calculations to avoid expanding the series if we use an expression which maintains the same general form as Equation (7.12) but which contains a correction factor determined in a partly empirical way. It can in fact be shown that we can consider the first term alone on condition we multiply quantity ζ by a factor β which is a function of δ. Then:

$$\frac{E}{2\,E_0} = U(q).V(\beta\zeta) \tag{7.14}$$

where:

$$U(q) = \frac{4 + 0.3\,q}{4 + q + 0.3\,q^2}, \tag{7.15}$$

$$V(\beta\zeta) = [1 + 2\,\beta\zeta + 2\,\pi(\beta\zeta)^3]^{1/2}\ \mathrm{e}^{-1.607\beta\zeta} \tag{7.16}$$

q is defined by relation Equation (7.4). ζ is defined by Equation (7.2) and β is given by the following approximate relation:

$$\beta \simeq \frac{|\,\delta\,|^2 + 1.6\,|\,\delta\,| + 0.75}{|\,\delta\,|^2 + 4.5\,|\,\delta\,| + 1.35} \tag{7.17}$$

where δ is defined by Equation (7.3). For horizontal polarization β is equal to 1, since δ is always very great. For vertical polarization β lies between 0.47 and 1. Figure 7.3 shows the value of β as a function of frequency for various types of ground.

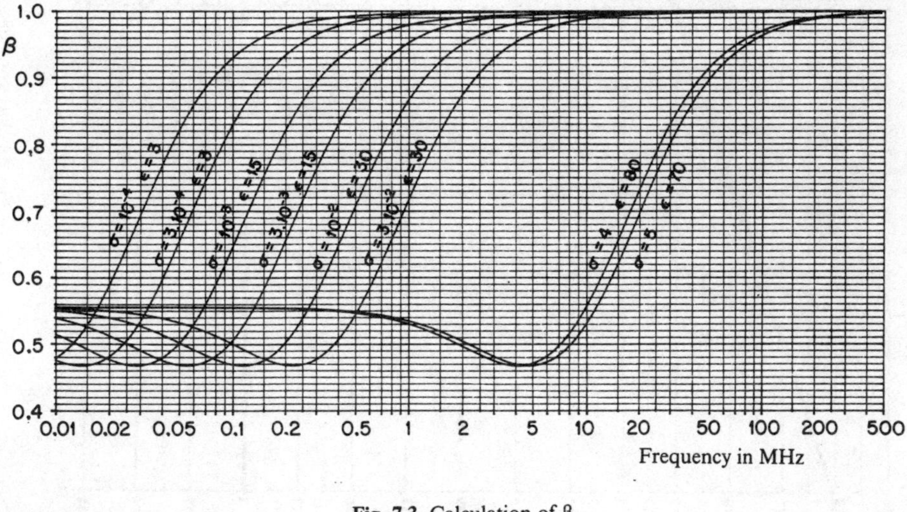

Fig. 7.3. Calculation of β.

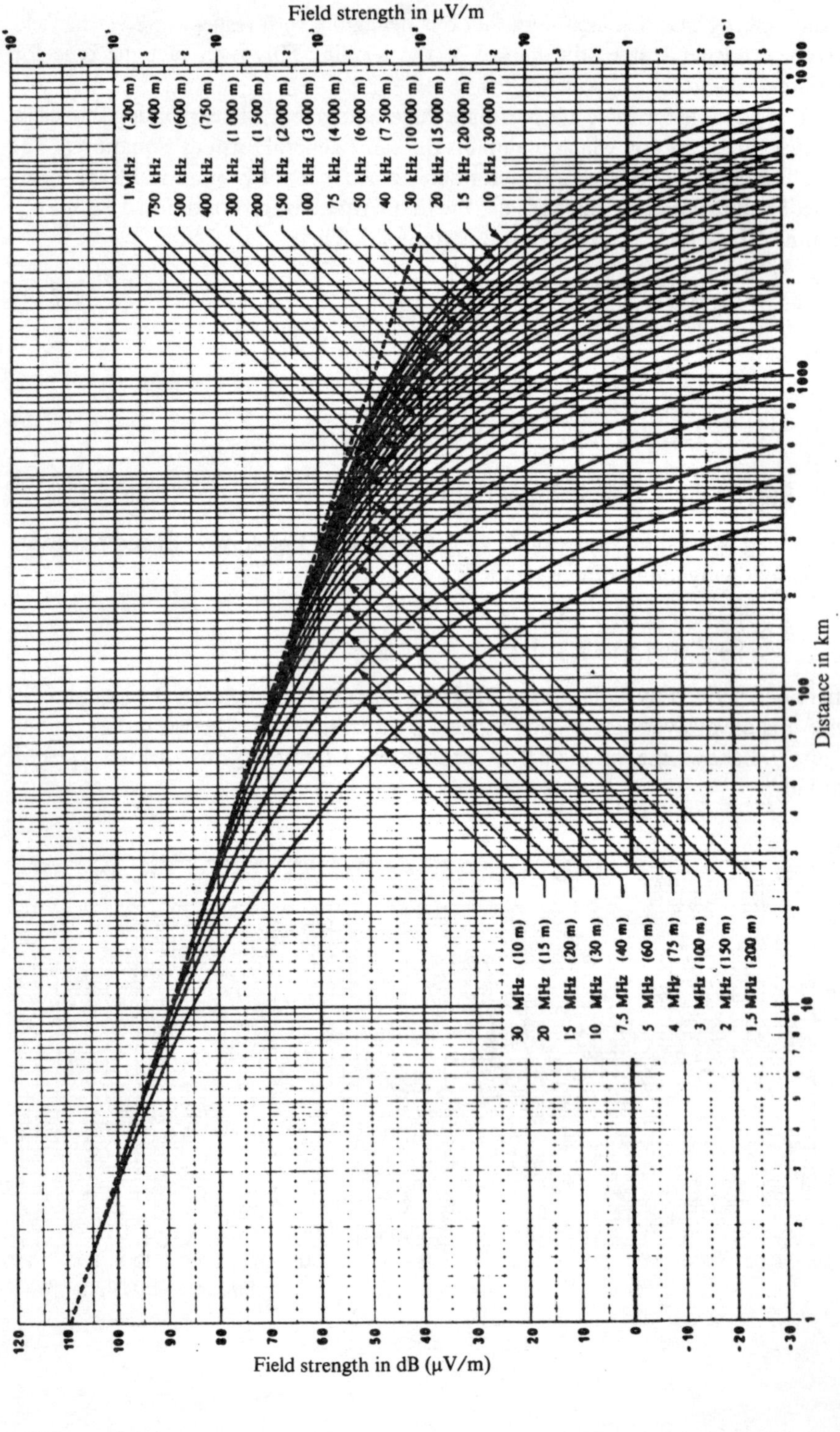

Fig. 7.4 Wave propagation over sea, σ =5S/m ε = 80. ---- reciprocal of distance.

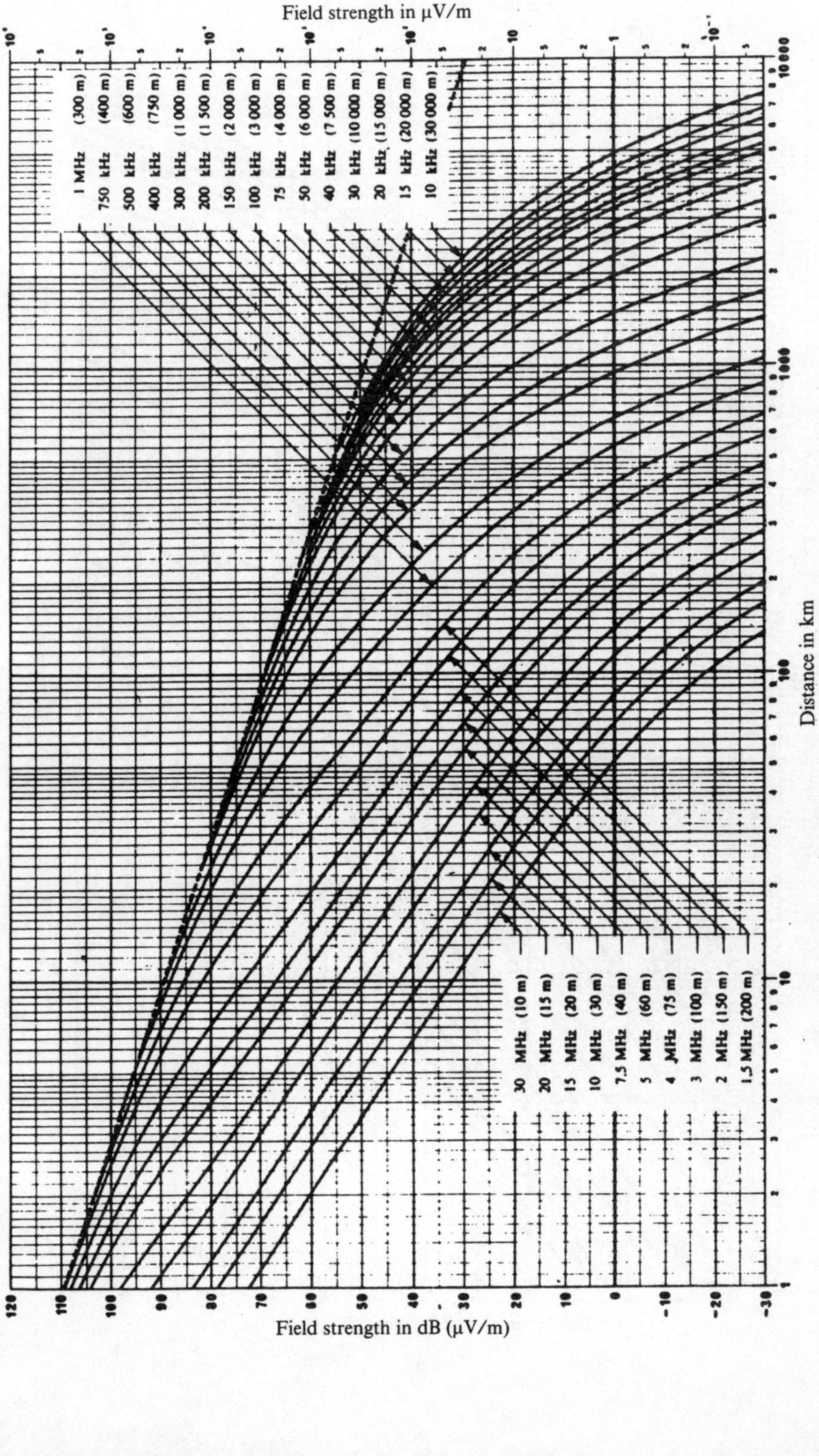

Fig. 7.5 Wave propagation over land,; $\sigma = 10^{-2}$ S/m, $\varepsilon = 4$. --- reciprocal of distance.

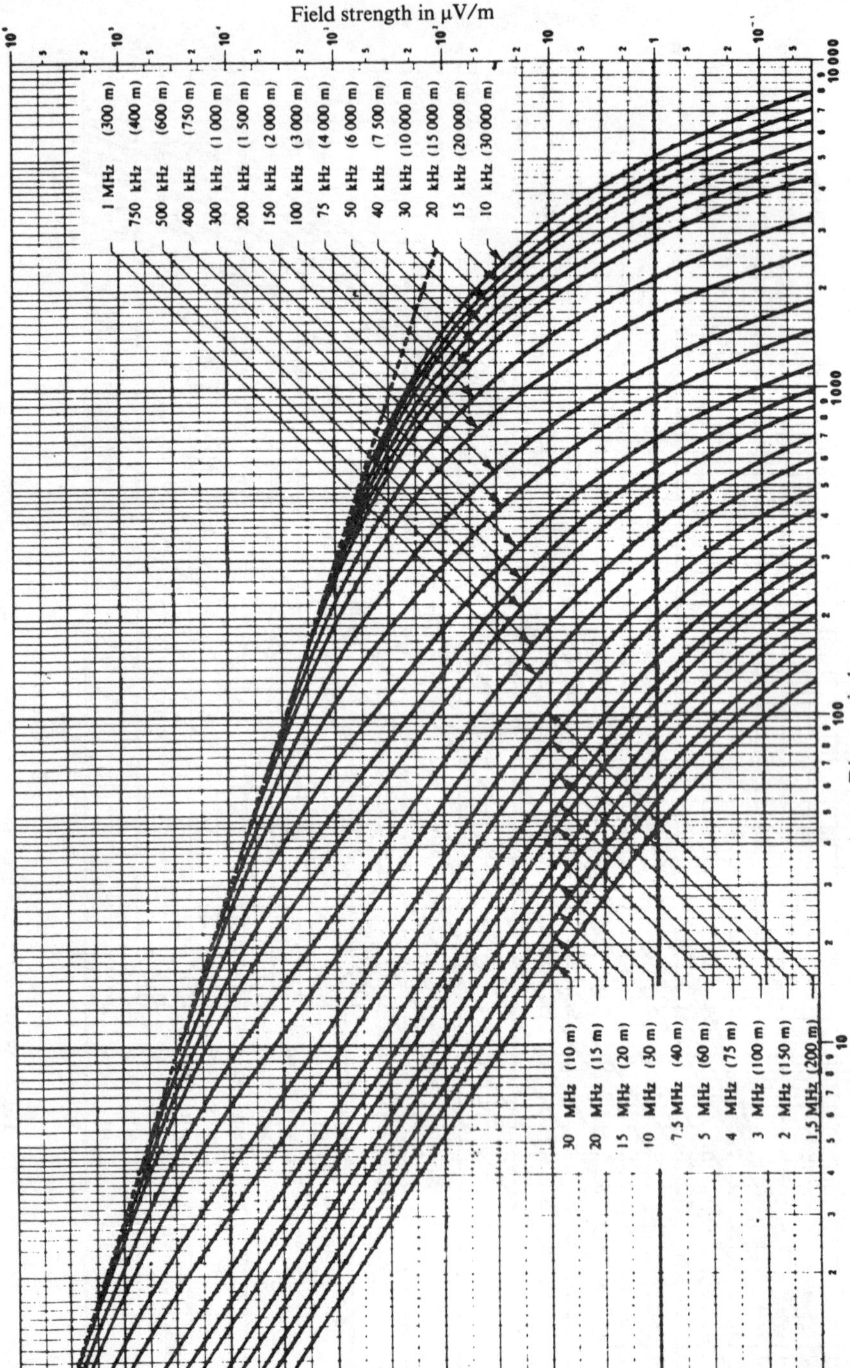

Fig. 7.6 Wave propagation over land, $\sigma = 10^{-3}$ S/M, $\varepsilon = 4$. ---- reciprocal of distance.

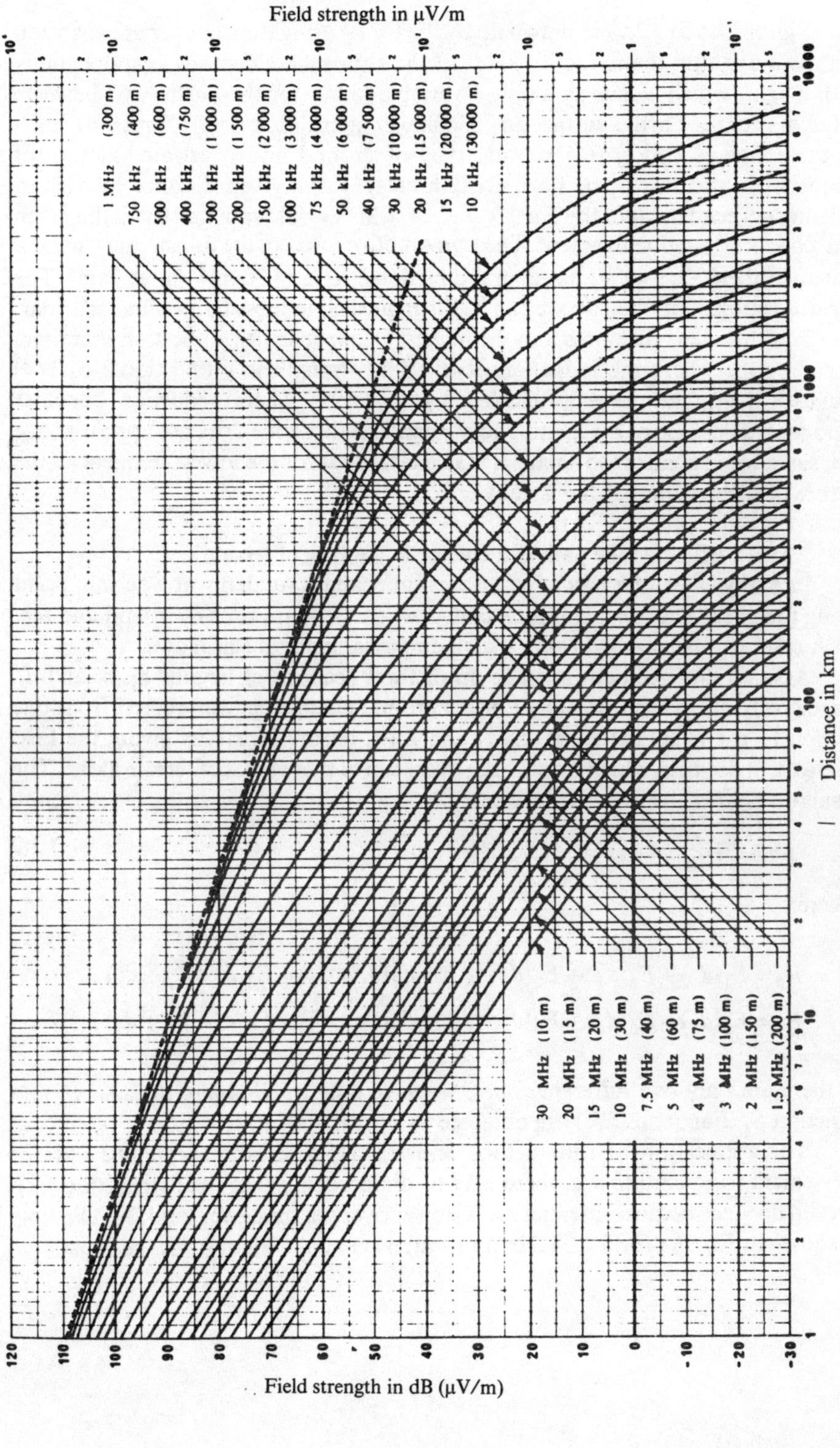

Fig. 7.7 Wave propagation over land, $\sigma = 10^{-4}$ S/m, $\varepsilon = 4$. ---- reciprocal of distance.

Figures 7.4 to 7.7 were drawn up by the CCIR using the same series expansion but assuming that the refractive index of the atmosphere decreases exponentially with height, which means assuming a variable radius for the Earth lying between 8500km near the Earth's surface and 6360km at high altitudes (cf. Chapter 4). Each of these figures concerns a specific type of ground and a certain number of frequencies, and shows the field strength created at a given distance by a short antenna perpendicular to the Earth's surface, which would radiate 1kW if the Earth was a perfectly conducting plane. The 'cymomotive force' of the transmitter-antenna system is then 300V in all horizontal directions (field of 300mV/m at 1km). The various zones mentioned above can be distinguished in these curves. In particular, over land for short distances and the highest frequencies, it can be seen that there exists a zone where the reduction of the field strength is proportional to $1/d^2$ because of ground losses, followed by the zone where reduction becomes exponential because of the Earth's curvature. The first zone does not appear above sea, since sea losses are too small. It is also clear that for the lowest frequencies the nature of the ground has hardly any effect.

7.2.1.1. Effect on the path of a change in the type of terrain

In many cases, the nature of the terrain changes suddenly at a certain point along the path. The most common case is where the path runs over land and sea. Such cases can be handled using a semi-empirical method as follows.

Assume that the path is made up of three sections of lengths d_1, d_2 and d_3, respectively, for which the nature of the ground changes as we proceed from one section to the next. Let $E_1(d)$, $E_2(d)$ and $E_3(d)$ be the functions giving the field strength, in decibels, which would exist over each of these types of terrain alone. The resultant field is given by the relation:

$$E = \frac{1}{2}(E_R + E_T), \tag{7.18}$$

where:

$$E_R = E_1(d_1) - E_2(d_1) + E_2(d_1 + d_2) - E_3(d_1 + d_2) + E_3(d_1 + d_2 + d_3), \tag{7.19}$$

$$E_T = E_3(d_3) - E_2(d_3) + E_2(d_2 + d_3) - E_1(d_2 + d_3) + E_1(d_1 + d_2 + d_3). \tag{7.20}$$

If there are only two different zones, we have only to eliminate distance d_3 and function E_3. Functions *E(d)* are obtained as shown above.

The application of this method shows, in particular, that in the case of movement from land to sea, there may be an increase in the field strength over a certain distance because attenuation is slower over sea than over land. An example is shown in Fig. 7.8 for a 1.5MHz transmitter on land 100km from the coast.

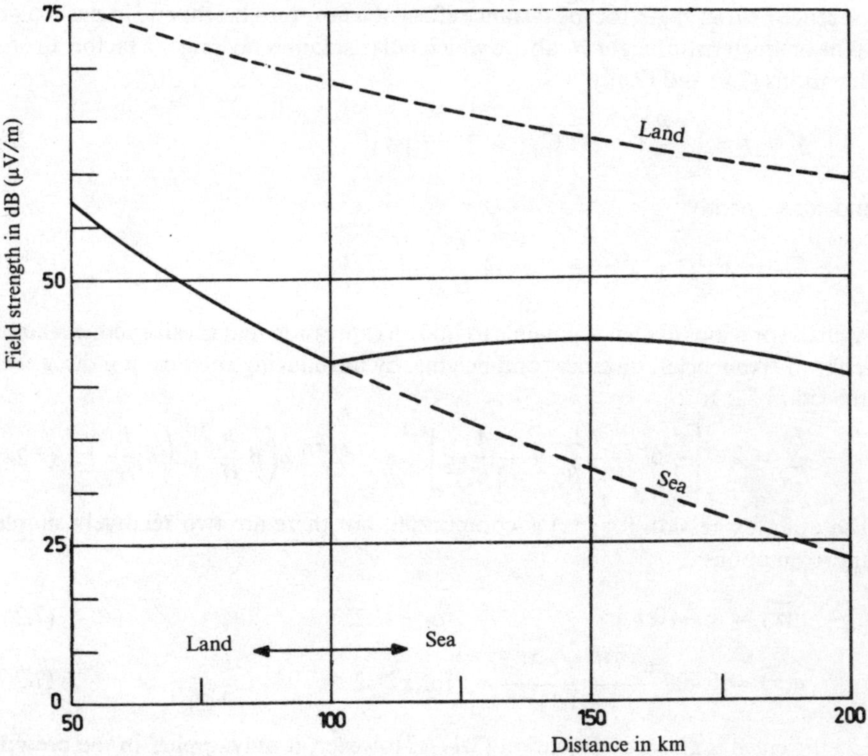

Fig. 7.8. Field strength over a mixed terrain, 1kW transmitter at 1.5MHz.

7.2.2. Transmitting and receiving antennas are at heights above the ground greater than the characteristic height H_1

This is the usual case:

1. for decimetric waves or shorter, whatever the polarization or the nature of the terrain;
2. for metric waves, except for vertical polarization above sea;
3. for decametric waves, for horizontal polarization.

To calculate the field strength, if we take only the first term of the series expansion, we obtain a good approximation except near the horizon. Moreover, for the present case, $2\tau_1$ can be regarded as negligible with respect to δ, except perhaps for vertical polarization above sea below about 30MHz. Taking the free space field strength E_0 as a reference value:

$$\frac{E}{E_0} = 2(2\,\pi\zeta)^{1/2}\,\frac{e^{-j\tau_1\zeta}}{\delta}\,g(h_e).g(h_r)\,. \tag{7.21}$$

As we are dealing with considerable heights in this case, it is better to take the reference field strength for the distance effect at a non-zero height and in particular at the characteristic height H_2 above which polarization is no longer a factor. From Equations (7.8) and (7.10):

$$g(H_2) \simeq \left(\frac{\pi R}{\lambda}\right)^{1/3} \sqrt{|C_0|} = 2^{-1/3} |\delta|^{1/2}, \tag{7.22}$$

and consequently:

$$\frac{E}{E_0} \simeq 2^{1/3}(2\pi\zeta)^{1/2} | e^{-j\tau_1\zeta} | g\left(\frac{h_e}{H_2}\right) g\left(\frac{h_r}{H_2}\right). \tag{7.23}$$

As in the previous case, it is possible to find an expression that is valid across a large range of frequencies, distances and heights, by introducing coefficient β discussed in Section 7.2.1:

$$\frac{E}{E_0} = 2^{4/3}\left[\frac{\pi}{2}\beta\zeta + \frac{1}{2\beta\zeta} + \frac{1}{(2\beta\zeta)^2}\right]^{1/2} e^{-1.607\beta\zeta} g\left(\beta\frac{h_e}{H_2}\right) g\left(\beta\frac{h_r}{H_2}\right). \tag{7.24}$$

The exact expression for $g(x)$ is complicated but there are two relatively simple approximations:

$$g(x) = x + 0.1\,x^3 \qquad \text{for } x < 2, \tag{7.25}$$

$$g(x) = 0.402\,\frac{e^{2.025(x-1.1)^{1/2}}}{(x-1.1)^{1/4}} \qquad \text{for } x > 2. \tag{7.26}$$

Coefficient β is given by Equation (7.17). However, it only applies in the present case to paths above sea. It is then possible to use an approximate expression, for example:

$$\beta \simeq 1 - 20\frac{\sigma}{K^{2/3}F^{5/3}} + 420\left(\frac{\sigma}{K^{2/3}F^{5/3}}\right)^2 \tag{7.27}$$

which is valid above sea for frequencies greater than about 30MHz.

Moreover, in order to apply Equations (7.25) and (7.26), we must remember that if either height h_e or h_r is less than H_1, it should be replaced by H_1 in the calculations.

It also becomes possible to carry out the calculation using nomograms. In practice, for frequencies greater than around 30MHz we need only consider two distinct cases, one for horizontal polarization above any type of ground or for vertical polarization above land, and the other for vertical polarization above sea. The nomograms in Figs 7.9 and 7.10 apply to the first case, and those in Figs 7.11 and 7.12 to the second. Figures 7.11 and 7.12 differ from Figs 7.9 and 7.10 only at the low end of the frequency range. Moreover, to calculate the height gain for vertical polarization above sea, if the height is less than H_1 (expressed in Fig. 7.12 by the fact that the straight line for alignments cuts across segment AB), we do not take the actual height into account but put the straight line through point A.

These figures have been drawn only for $K = 1$ and $K = 4/3$. For other values

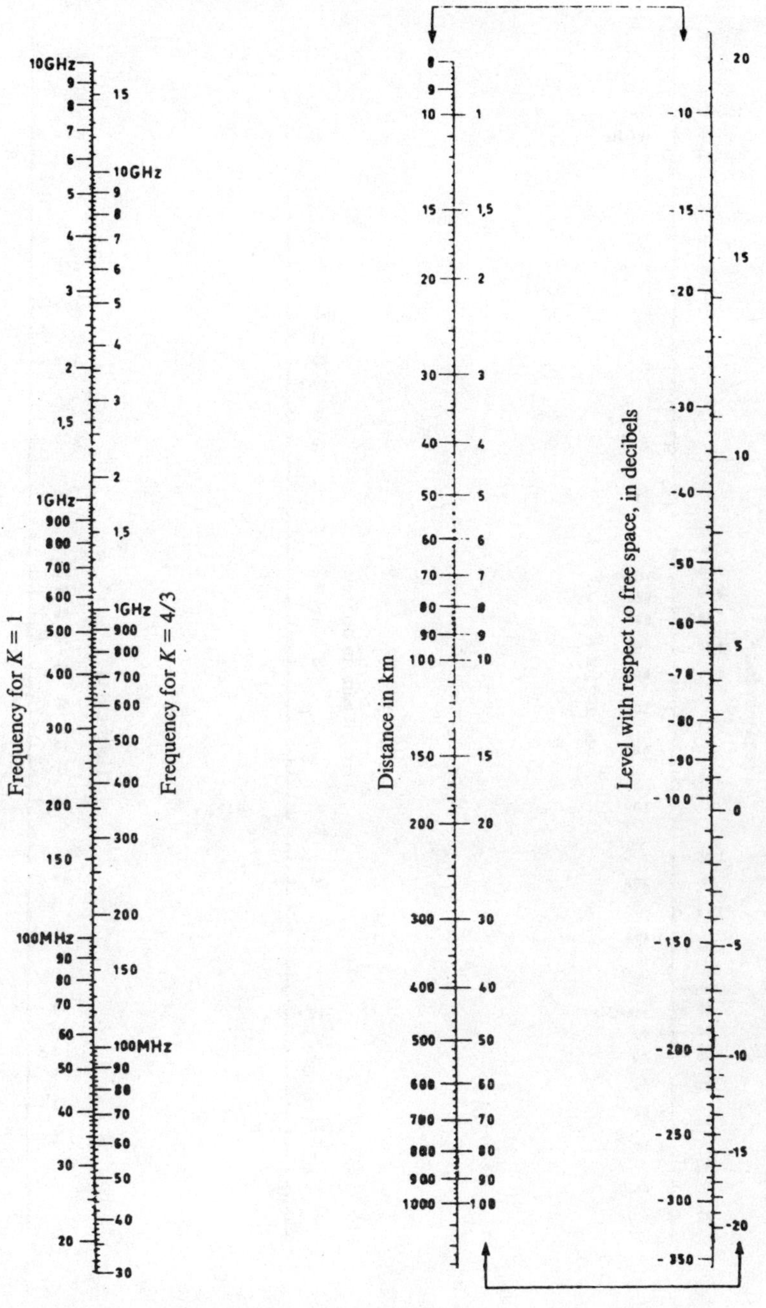

Fig. 7.9. Spherical diffraction. Effect of distance. Horizontal polarization over land or sea. Vertical polarization over land. (The scales linked by arrows should be used together.)

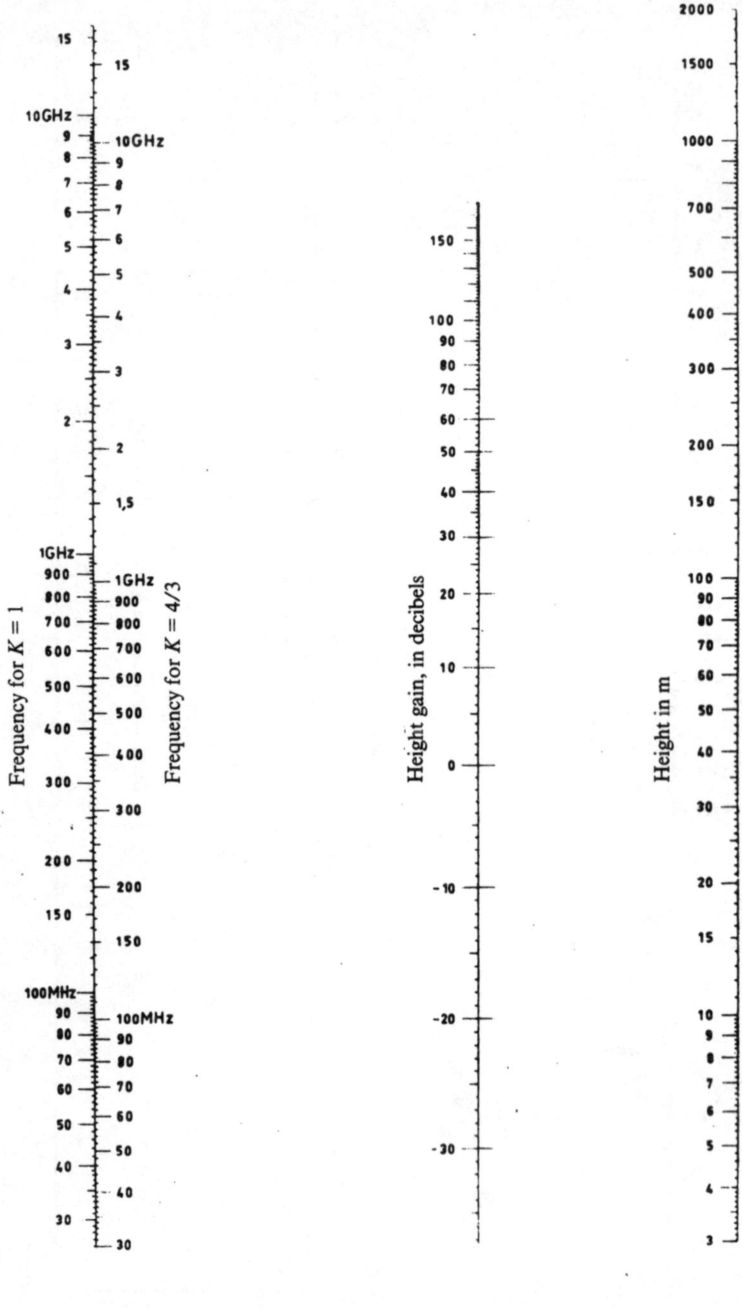

Fig. 7.10. Spherical diffraction. Height gain. Horizontal polarization over land or sea. Vertical polarization over land.

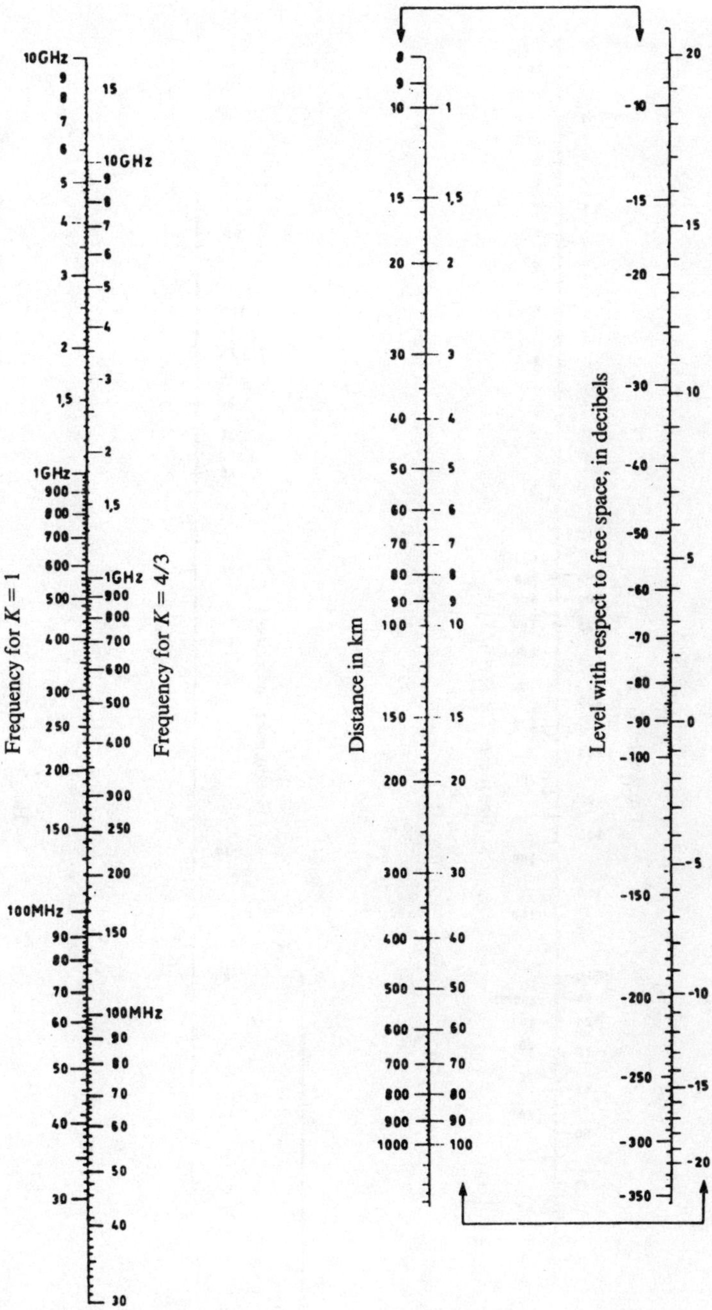

Fig. 7.11. Spherical diffraction. Effect of distance. Vertical polarization over sea. (The scales linked by arrows should be used together.)

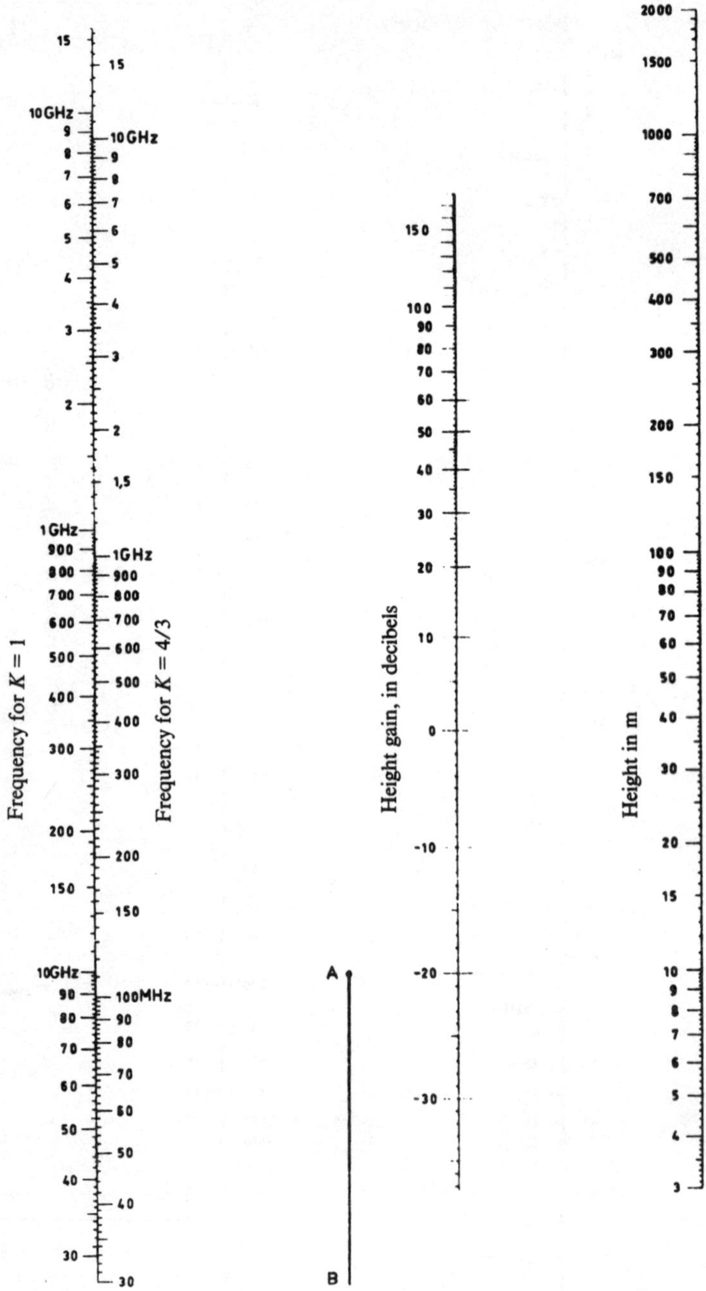

Fig. 7.12. Spherical diffraction. Height gain. Vertical polarization over sea. (If the alignment line goes through segment AB, put it through point A with the same value of the frequency.)

of K, nomograms in Figs 7.9 to 7.12 may be used if the real frequency is replaced by a hypothetical frequency. In the expression for H_2 [Equation (7.10)] R and λ appear in the product $R\lambda^2$, and in the expression for ζ [Equation (7.2)] they appear in the product $R^2\lambda$. Consequently, to calculate the attenuation corresponding to a frequency F and a value K, we only use the frequency scale for $K = 1$, but have to replace F by

F/K^2 for nomograms in Figs 7.9 and 7.11

$F/\sqrt{K}$ for nomograms in Figs 7.10 and 7.12.

7.3. Edge diffraction

In certain cases, there is no line of sight between transmitter and receiver because a mountain or hill stands between them. It is relatively difficult to work out the fading due to such an obstacle, where its shape is often irregular and its conductivity unknown. The problem is consequently simplified by treating the obstacle as a ridge of negligible thickness, as a simple cylindrical surface or as a succession of such obstacles.

7.3.1. Knife edge diffraction

There is a rigorous solution to the problem of diffraction from a semi-plane (Sommerfeld). However, the tendency is rather to use the purely scalar approximate solution from optics, as it gives results that are very close to the rigorous solution and is considerably simpler, since it does not take polarization into account.

Consider, therefore, a ridge of zero thickness at the summit (Fig. 7.13) which is at a height h (positive or negative) with respect to the straight line between the transmitter and the receiver. Let d be the total distance, d_1 and d_2 be the distances from the ridge to the transmitter and to the receiver, respectively, θ be the angle of diffraction which is assumed to be small and λ be the wavelength. If we then consider the variable:

$$v = h\sqrt{\frac{2}{\lambda}\left(\frac{1}{d_1} + \frac{1}{d_2}\right)} = h\sqrt{\frac{2}{\lambda}\,\frac{d}{d_1\,d_2}} = \theta\sqrt{\frac{2}{\lambda}\,\frac{d_1\,d_2}{d}}\,, \tag{7.28}$$

The amplitude and phase of the field strength at the location of the receiver are given by the following expression:

$$\frac{E}{E_0} = \frac{1}{1+\mathrm{j}}\int_v^{\infty} \mathrm{e}^{\mathrm{j}\pi\frac{t^2}{2}}\,\mathrm{d}t\,, \tag{7.29}$$

where E_o is the field strength without a ridge. The ratio of corresponding powers can then be written as:

$$\frac{P}{P_0} = \frac{1}{2}\left[\left(\frac{1}{2} - \xi(v)\right)^2 + \left(\frac{1}{2} - \eta(v)\right)^2\right] \tag{7.30}$$

where we write $\zeta(v)$ and $\eta(v)$ for the Fresnel integrals:

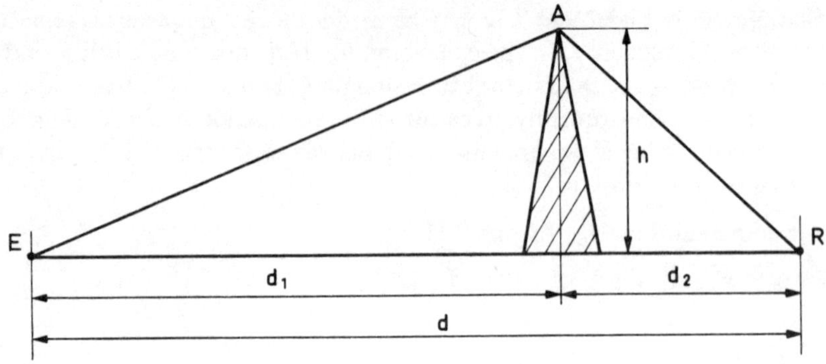

Fig. 7.13. 'Knife-edge' diffracting ridge.

$$\xi(v) = \int_0^v \cos\frac{\pi t^2}{2}\,\mathrm{d}t \qquad \eta(v) = \int_0^v \sin\frac{\pi t^2}{2}\,\mathrm{d}t\,. \tag{7.31}$$

The function P/P_0, expressed in decibels, is shown in Fig. 7.14. It is clear that for negative values of v, i.e. if the top of the ridge is below the straight line linking the transmitter to the receiver, P/P_0 oscillates about the open space level and tends towards it, while for positive values of v, P/P_0 decreases regularly as the extent of obstruction by the ridge increases. When $v = 0$ the transmitter and receiver are aligned with the top of the ridge and fading is 6dB, i.e. the received power is a quarter of the open space level (and not a half as one might have expected).

In the case of an obstruction with ($v > 0$) approximate expressions for the P/P_0 can be obtained. The most common is an asymptotic expression for $v > 1.5$:

$$\frac{P}{P_0} = \frac{1}{2\,\pi^2\,v^2} \simeq \left(\frac{0.225}{v}\right)^2 \tag{7.32}$$

or in decibels:

$$10\log\frac{P}{P_0} = -\,13 - 20\log v\,. \tag{7.33}$$

An approximation which is less simple but valid for $v > -0.7$ is:

$$10\log\frac{P}{P_0} = -\,6.9 + 20\log\left[\sqrt{(v-0.1)^2+1} - v + 0.1\right]. \tag{7.34}$$

The expression can be used, in particular, around $v = 0$.

7.3.2. Rounded top ridge diffraction

Ridges with rounded tops have been studied in terms of models based on simple cylindrical surfaces, e.g. a parabolic cylinder or a horizontal circular cylinder associated with two semi-planes. Results are reasonably similar, since the major parameter is the radius of curvature of the ridge near the top. A spherical surface

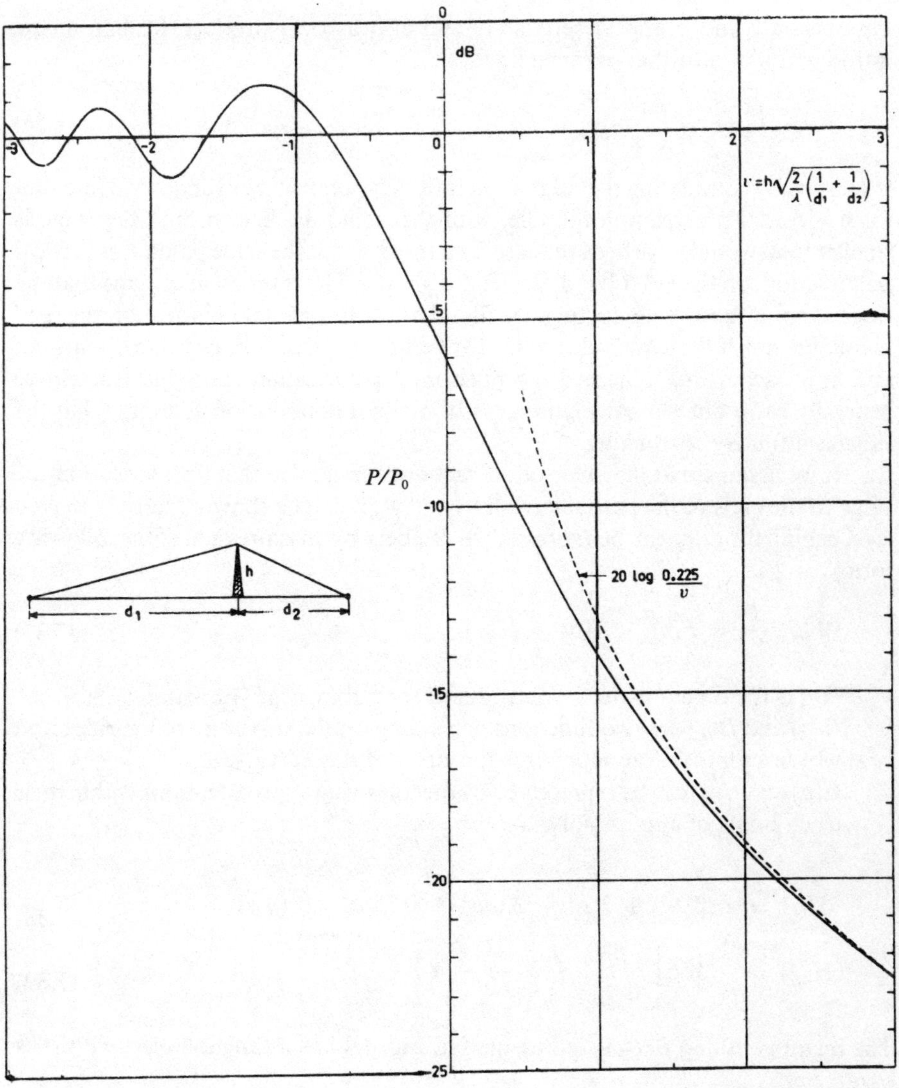

Fig. 7.14. Diffraction by a sharp ridge.

can also be used as a model, making it possible to use results from studies on spherical diffraction.

Theoretical considerations lead us to introduce the radius of curvature R at the top of the ridge using the following parameter:

$$\rho = \left(\frac{1}{d_1} + \frac{1}{d_2}\right)^{1/2} \left(\frac{\lambda R^2}{\pi}\right)^{1/6}. \tag{7.35}$$

Parameters v and ρ appear separately and also as their product, leading to the introduction of a further parameter:

$$\chi = \sqrt{\frac{\pi}{2}}\, v\rho = \left(\frac{\pi R}{\lambda}\right)^{1/3} \theta \,. \tag{7.36}$$

If we start by considering the ridge as an infinitely conducting cylinder, we find that for horizontal polarization (parallel with the ridge) the field behind the ridge is smaller than would have been the case for a thin ridge at the same point. For vertical polarization, on the other hand, the field is greater. However, if we assume that the ridge is not infinitely conducting, we find that for horizontal polarization the field maintains much the same value while for vertical polarization its value diminishes and approaches that obtained for horizontal polarization. For this reason, we generally carry out our calculations for horizontal polarization assuming that the ridge is infinitely conducting.

If we assume that the angle of diffraction is small and that distances from the ridge to the ends of the path are relatively great, it can be shown that the received level behind the ridge can be expressed, in decibels, by an expression of the following form:

$$10 \log \frac{P}{P_0} = F(v) + G(\rho) + H(\chi) \,. \tag{7.37}$$

$F(v)$ is the function already introduced for a thin ridge [Equation (7.30)];
$G(\rho)$ and $H(\chi)$ are two functions depending on the curvature of the ridge and whose sum therefore represents the effect of this curvature.
$G(\rho)$ and $H(\chi)$ can be represented by the following approximations within their usual limits of applicability:

$$G(\rho) = -\,7.2\,\rho + 2\,\rho^2 - 3.6\,\rho^3 + 0.75\,\rho^4 \qquad (\rho < 3) \tag{7.38}$$

$$H(\chi) = -\,166\left[\sqrt{1 + \frac{\chi}{8} + \frac{\chi^2}{80}} - 1\right] . \tag{7.39}$$

The quantity 10 log P/P_0 is represented in Fig. 7.15 as a function of v, with ρ as parameter.

When the angle θ is zero and consequently v and χ are zero, the function $F(v)$ is equal to –6dB. The received level is given by:

$$10 \log \frac{P}{P_0} = -\,6 + G(\rho) = -\,6 - 7.2\,\rho + 2\,\rho^2 - 3.6\,\rho^3 + 0.75\,\rho^4 \,. \tag{7.40}$$

This relation is applicable, in particular, to the case of spherical diffraction, making it possible to calculate the received level when we are exactly in the direction of the horizon for both stations. We know that under these conditions we can apply neither direct line of sight calculations, nor approximate diffraction calculations. Note that for spherical diffraction by the Earth, it is preferable to calculate ρ from the heights h_e and h_r of the antennas above the ground, which for the grazing incidence gives:

$$\rho = \left(\sqrt{\frac{R}{2\,h_e}} + \sqrt{\frac{R}{2\,h_r}}\right)^{1/2} \left(\frac{\lambda}{\pi R}\right)^{1/6} \tag{7.41}$$

where R is the Earth's radius.

7.3.3. Multiple edge diffraction

In many cases, the ground is hilly. The problem then rapidly becomes more complicated because of the growing number of geometrical parameters to be taken into account (distance, heights of ridges, radii of curvature), and generally only empirical solution have been proposed.

However, in the case of two thin ridges there is a theoretical solution presented by Millington, which is a generalization to two dimensions of Fresnel integrals used for a single ridge. Even though, from the purely mathematical point of view, the use of these integrals leads to convergence problems, the solution obtained is acceptable experimentally and is generally used as the standard to check empirical solutions, which are easier to use.

Among empirical solutions, one method devised by Bullington (1947) was to replace all ridges by a single ridge seen from the ends of the path at the same angle of elevation as the outermost real ridges. This method can introduce major errors and is no longer used.

The method devised by Epstein and Peterson (1953) assumes that each ridge is illuminated by the transmitter or by the diffracted energy from the top of the preceding ridge. We then sum the diffraction fading from each of these ridges using the above calculations. The geometrical elements involved are shown in Fig. 7.16. The first diffraction path is defined by distances a and b and by height h_1, the second path by distances b and c and height h_2.

A method devised by Deygout (1966) is based on the observation that in nearly all cases one of the ridges is predominant and is responsible for a large part of the loss, and the other ridges take only a subsidiary role. We first of all calculate the loss as though the predominant ridge was the only ridge present, and then add the losses due to the others. In Fig. 7.16 the principal diffraction path is defined by distances a and $b + c$ and by height h_1, while the subsidiary diffraction path is defined by distances b and c and by height h_2.

These two methods give results which do not differ by very much in practice. Generally, Epstein's method gives a higher field value than Deygout's. It should therefore be used for interference problems, whereas Deygout's method is safer calculating communication effects. In cases of grazing links, i.e. if h_1 and h_2 are zero, the two methods give a relative field of –12dB, while Millington's calculation gives a value between –6 and – 12dB, depending on the distance between the two ridges.

Obviously, when applying these methods, we can if necessary take into account a certain curvature at the top of the ridge, since the calculations only apply to the results obtained for a single ridge.

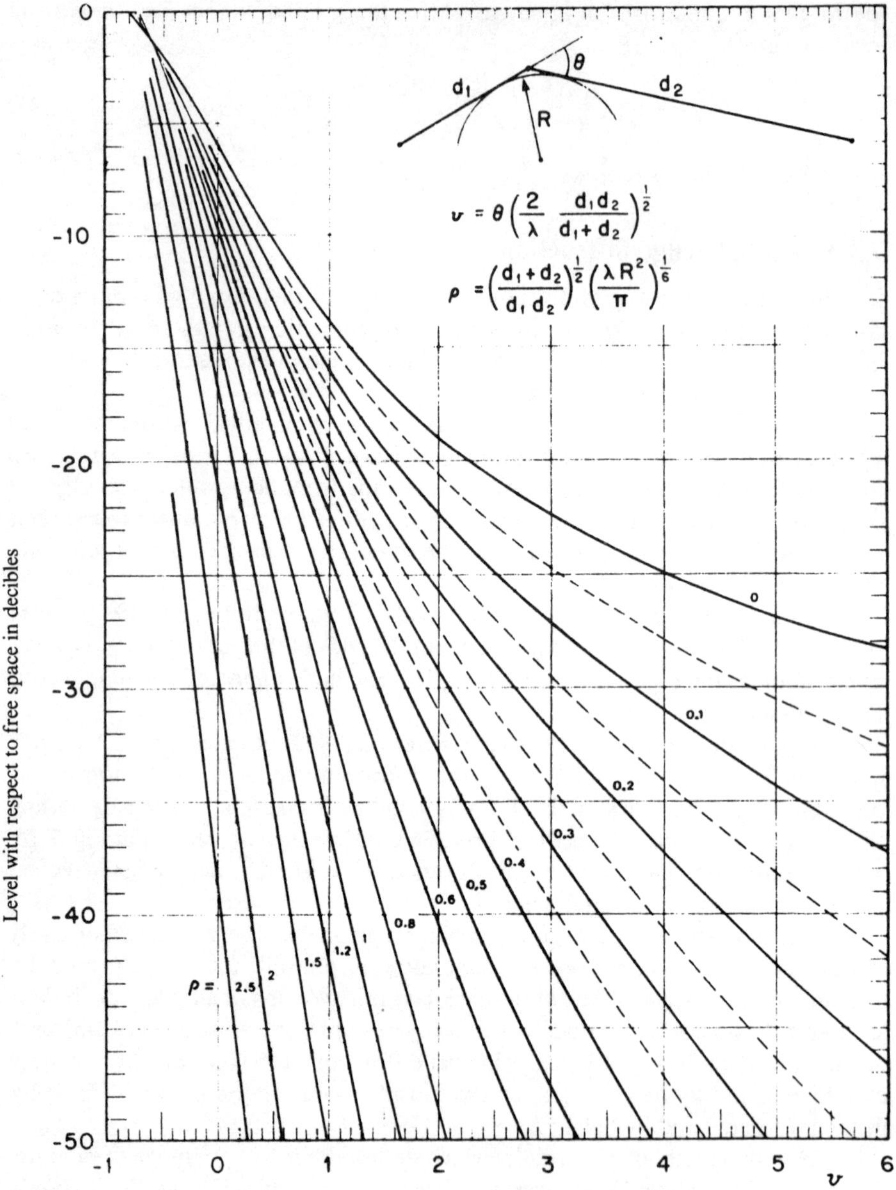

Fig. 7.15. Diffraction by a ridge with a rounded top.

7.3.4. Comparison between diffraction from a ridge and diffraction by the curvature of the Earth

The diffraction field behind a sharp ridge decreases, when we move into the shadow of the ridge, following a function which aproximates to the reciprocal of the distance to the geometric edge of the shadow, while for diffraction from a sphere the function is exponential, giving a considerably faster decrease. This mathematical difference can be explained as follows. The field strength due to diffraction is caused by radiation of currents induced by the incident field. In the case of a sharp ridge, we can regard the top of the ridge as being in line of sight both to the transmitter and to the receiver, while, for a rounded top or sphere, currents are induced from

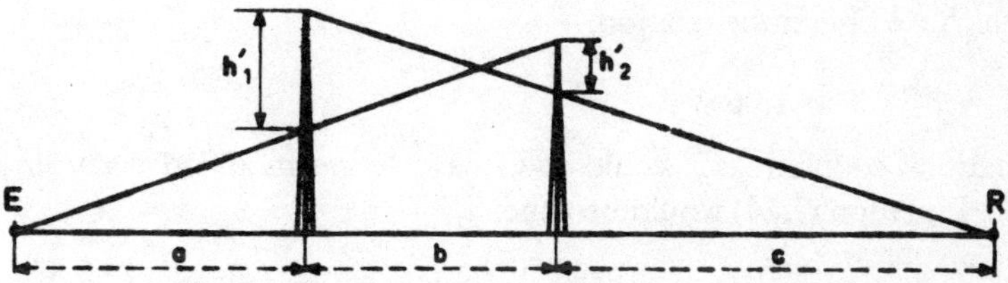

1. Epstein and Peterson's method

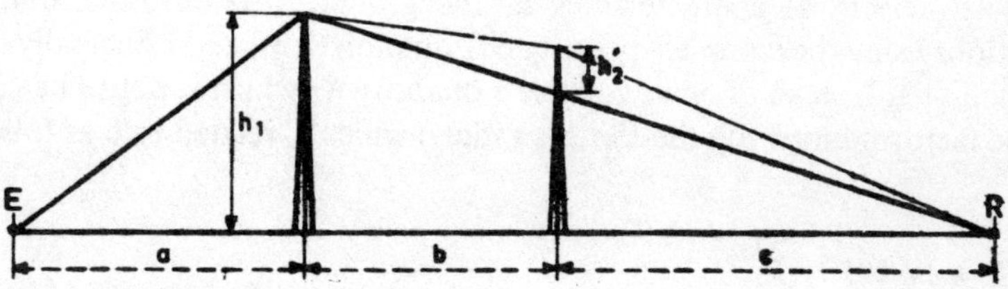

2. Deygout's method

Fig. 7.16. Diffraction by two ridges. Approximate methods.

element to element of that part of the ridge which is not in line-of-sight with either end of the path. These currents therefore decrease exponentially and radiate a lower field towards the receiver.

This phenomenon may be highlighted on certain communications links. For example, it has been observed that the field behind an island is often very much greater than the field at the same distance on a path without an island. As the peaks of the island are illuminated by a direct field very much greater than the field at the surface of the sea, they diffract much more energy despite the increase in the angle of diffraction. This phenomenon, called obstacle gain, hardly appears as soon as there is more than one ridge on the path.

Appendix 7.1

A.7.1. Spherical diffraction: variants

When studying spherical diffraction, somewhat different relations than those presented in this chapter are sometimes used because the different choices can be made for the normalized quantities. To make comparisons with other discussions of the problem easier, we mention the most common variants.

1. The normalized distance ζ [Equation (7.2)] may sometimes be replaced by a quantity X defined by:

$$X = \frac{d}{L} \quad \text{with} \quad L = \left(\frac{\lambda R^2}{\pi}\right)^{1/3} . \tag{1}$$

from which we obtain the relation:

$$\zeta = 2^{1/3}\, X \simeq 1{,}26\, X . \tag{2}$$

All equations containing ζ would then have to be modified accordingly. For example, Equation (7.24) would become:

$$\frac{E}{E_0} = 2\left[\pi\beta X + \frac{0.63}{\beta X} + \frac{1}{(2\,\beta X)^2}\right]^{1/2} \mathrm{e}^{-2.025\beta X} \cdot g\left(\frac{\beta h_e}{H_2}\right) \cdot g\left(\frac{\beta h_r}{H_2}\right) . \tag{3}$$

2. The electrical characteristics of the ground were introduced into our calculations using the complex quantity δ [Equation (7.3)] and principally through its modulus |δ|. Instead of δ, we can use a quantity K (which is not to be confused with the factor multiplying the Earth's radius) which is related to δ as follows:

$$K = |\,\delta\,|^{-1/2} \tag{4}$$

K is therefore very small at all frequencies for horizontal polarization and at high frequencies for vertical polarization. Note, incidentally, that the relations deduced from Equations (7.9) and (7.10) are:

$$\frac{H_2}{H_1} = 2^{-1/3}\,|\,\delta\,|^{1/2} = \frac{2^{-1/3}}{K} . \tag{5}$$

Chapter 8

Propagation by scatter

The model of a perfectly spherical atmosphere used in the previous chapters allowed us to explain most of the phenomena described. However, we have already mentioned that line of sight propagation is sometimes affected by interference due on small heterogeneities in the refractive index of air. According to certain theories, certain times and over certain parts of the path.

In this chapter we study propagation mechanisms entirely due to the existence of heterogeneities. These mechanisms are classed together under the general heading of scatter, and include in particular:

tropospheric scatter;
ionospheric scatter;
scatter from precipitation.

Only the first two phenomena occur all the time and can therefore be used to establish communication links, while the third, which we studied in Section 5.5, leads to one type of interference discussed in Chapter 10.

8.1. Tropospheric scatter (or 'troposcatter')

We have had to consider the structure of the atmosphere at a very much smaller scale than for meteorological studies, to explain propagation over large distances using very short wavelengths. Such propagation attracted attention relatively late, as the corresponding attenuation, though it is very much less than in diffraction by the Earth's curvature, is nonetheless considerable.

The various explanations proposed to account for such propagation are based on small heterogeneities in the refractive index of air. According to certain theories, the atmosphere, when examined at the appropriate scale, behaves like a disturbed medium which scatters, in all directions, part of the energy moving through it (theory of isotropic scatter). Other theories suggest that heterogeneities in the atmosphere are not isotropic, but are smaller in size in the vertical than the horizontal direction. These heterogeneities therefore behave more like mirrors, reflecting part of the incident rays (partial reflection theory). This model seems relatively plausible, as the scale of air movements in the atmosphere is clearly different in the vertical and horizontal directions, and the same must be true of the heterogeneities. In fact, none of these theories has led to practical results of immediate value for predicting propagation characteristics, as the atmospheric quantities used (scale of turbulence, fluctuation spectrum, etc.) are not systematically measured by weather stations, and indeed are more difficult to determine than the characteristics of propagation themselves.

Without having to choose between these theories, we can assume that energy transmission takes place according to the following mechanism: heterogeneities in the refractive index of the volume of air common to the beams from both antennas, will, when receiving energy from the transmitting antenna, retransmit a small part of it in every direction, particularly in the direction of the receiving antenna (Fig. 8.1). As these heterogeneities fluctuate with time, the received level undergoes the same fluctuations. Consequently, a study of the phenomenon can be divided into two stages, first a study of the mean level received, and then a study of fluctuations about that mean level.

8.1.1. Mean level and its long term variations

8.1.1.1. General characteristics

The mean level, while it is sometime called the hourly mean level, is determined in practice by introducing a time constant of several minutes in the received level. From the physical point of view, the mean level received depends on the number and characteristics of the heterogeneities within the common volume, but, as these parameters are not generally known, this observation is of no help in drawing up predictions. It is preferable to find a relation between the mean level received and macroscopic characteristics of the atmosphere which can be obtained from meteorological measurements.

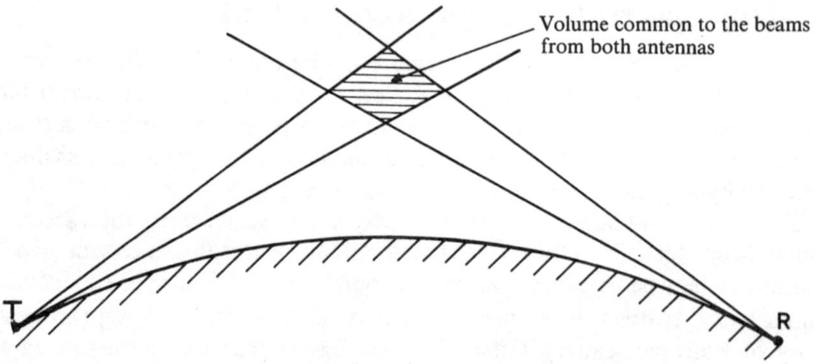

Fig. 8.1. Tropospheric scatter propagation path.

It has been experimentally determined that the mean level received is a function principally of the mean value of the vertical gradient of the refractive index within the common volume and the following empirical relation has been obtained, based on a comparison of a large number of measurements in different climates:

$$A = 30 \log F + 30 \log d + 1.5\,G + 102 \tag{8.1}$$

where:

- A is the attenuation between isotropic antennas or 'basic transmission loss' in decibels;
- F is the frequency in Megahertz;
- d is the distance in km;
- G is the refractive index gradient in the common volume, in N/km (G is generally negative). The statistical distribution of the gradient G is generally near-normal in its central part, and the same is therefore true for A (log-normal distribution for the received power).

As the values of G are negative, attenuation is lower in regions where the absolute value of G is large, e.g. in hot humid regions, while attenuation is greater in regions where the gradient of the index is low, e.g. in desert regions. However, this rule must be applied with care. In fact in humid regions with a monsoon climate the atmosphere may, at certain times, have virtually uniform temperature and humidity over a large range of heights, leading to a low value of the refractive index gradient and attenuation may consequently be relatively great though the refractive index at ground level is very high. This is, for example, the case in the Dakar region between June and November.

This observation explains why the use of the refractive index near the ground, as a means of predicting the mean level received, can lead in certain cases to major errors. The refractive index at ground level should only be used for an 'exponential' atmosphere, i.e. if the refractivity N can be represented by an exponential function of height, as in this case the index gradient is proportional to the index. This is often approximately true in temperate climates, but not in other climates at certain times of the year, especially in the lower part of the atmosphere.

In Equation (8.1) the length of the link appears in the 30 log d term and may also appear in the 1.5 G term. If the refractive index gradient depends very little on altitude, as is, for example, the case during the summer in desert climates, the refractive index gradient in the common volume is independent of its height and, therefore, of the length of the link. In this case attenuation follows a 30 log d law (instead of 20 log d for propagation in free space). On the other hand, if the refractive index gradient decreases with altitude, attenuation expressed as a function of distance grows more quickly than 30 log d, and all the more quickly as the quicker the vertical decrease in refractive index gradient itself.

It is also found that long term basic transmission loss is for all practical purposes proportional to the cube of frequency (rather than the square as in free space). This relation, which is valid between about 200MHz and 5GHz, allows us to calculate attenuation at any frequency once attenuation at a particular frequency, e.g. 1GHz, is known.

8.1.1.2. The effect of climate: experimental curves

As we do not always know the refractive index gradient at any altitude, Equation (8.1) cannot be used and it is useful to prepare sets of curves, each of which

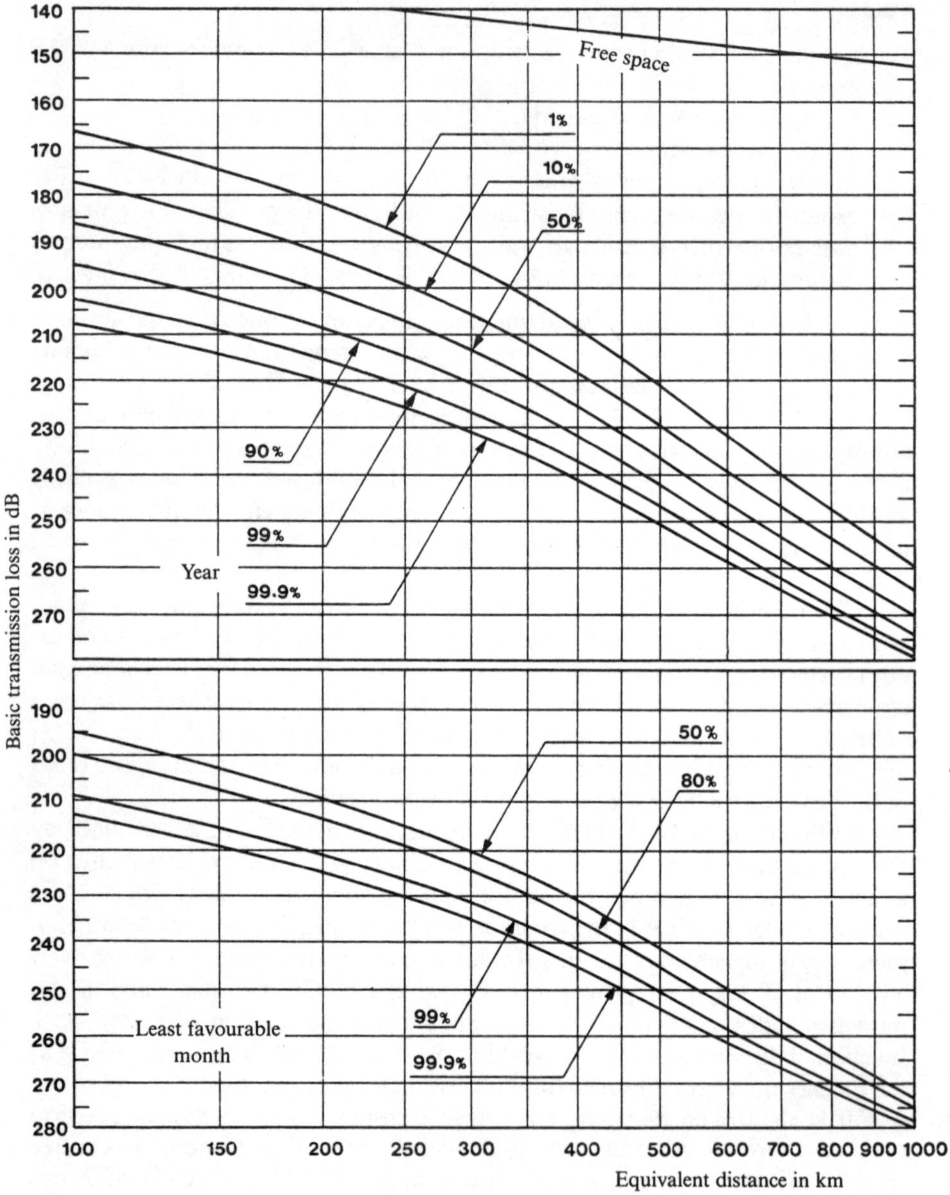

Fig. 8.2a. Mean attenuation in tropospheric scatter. 1GHz. Paths over land. Temperate climate.

is valid in a given climate. Figures 8.2a, b, c and d are curves for four climates, for the 1GHz frequency, and for various time percentages taken into consideration when planning links. For a given distance, we find that received levels obey a nearly normal distribution.

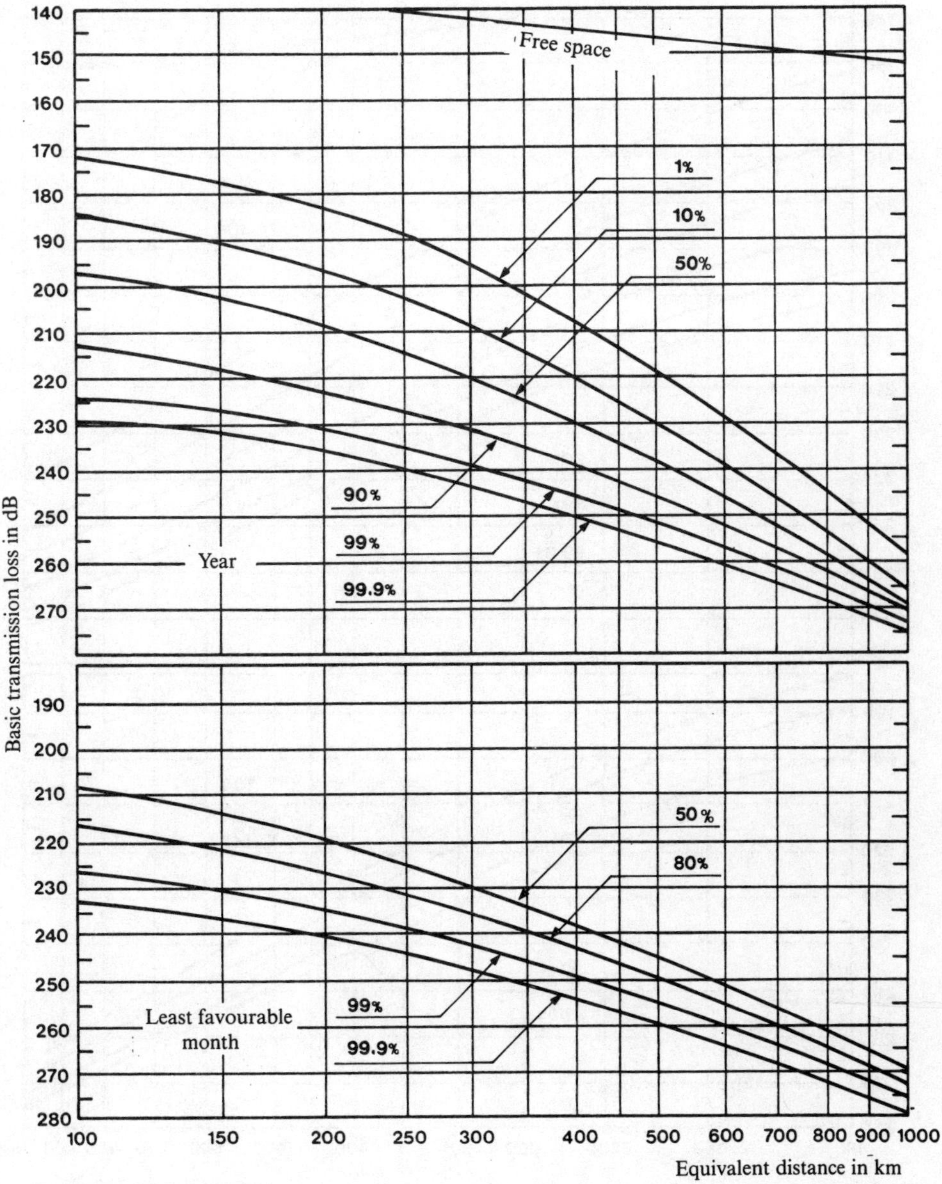

Fig. 8.2b. Mean attenuation in tropospheric scatter. 1GHz. Paths over land. Desert climate.

It needs also to be said that these curves assume that antennas are clear of nearby obstacles and that the beam from each antenna is horizontal. If the latter condition is not met, a correction has to be made, but this only means replacing the real length of the link by an effective distance, in such a way that the angle between

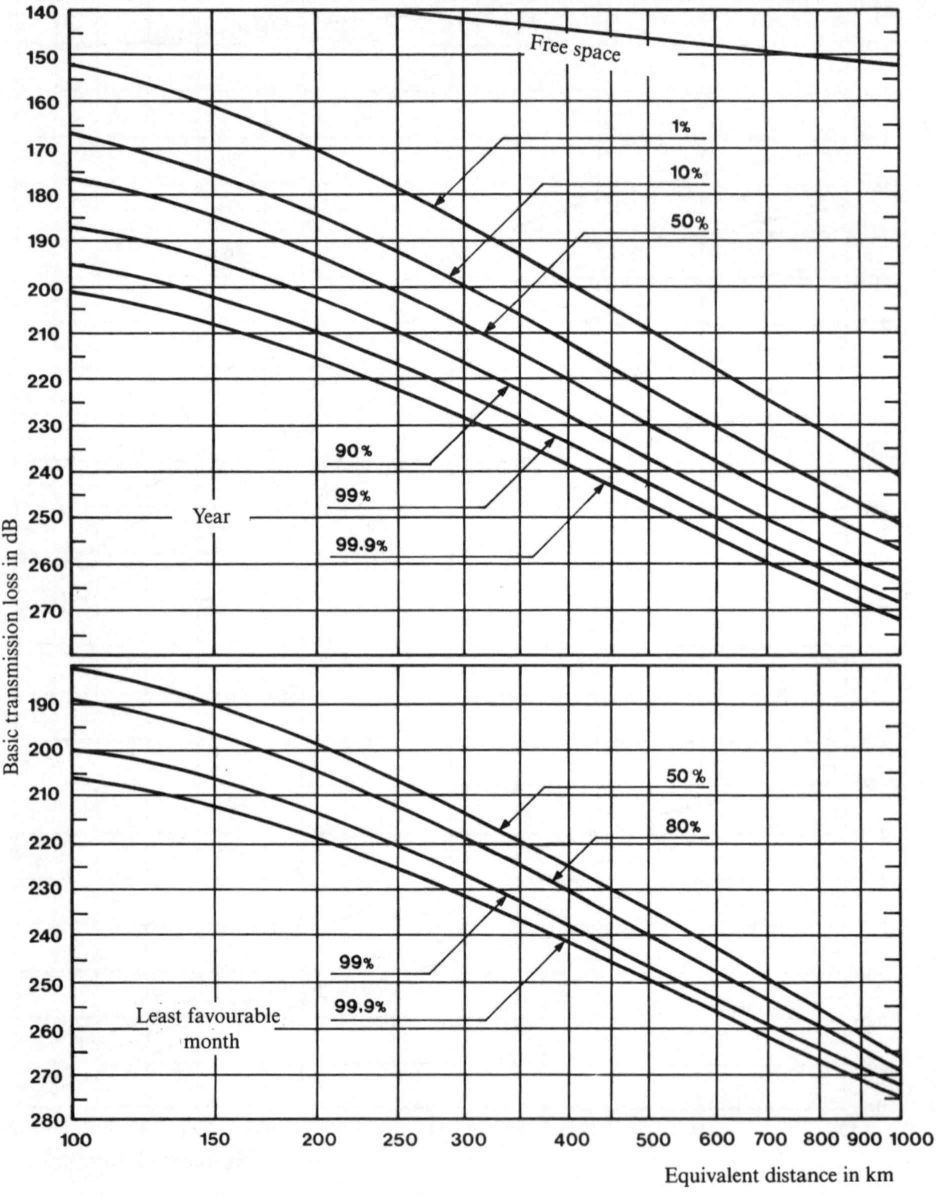

Fig. 8.2c. Mean attenuation in tropospheric scatter. 1GHz. Paths over land. Humid tropical climate.

the antenna beams in the common volume remains unchanged. If the angles of elevation θ_1 and θ_2 of the horizon seen from the antennas are given in milliradians and the distance in kilometres, we obtain the relation:

$$d_e = d + 8.5(\theta_1 + \theta_2)\,. \tag{8.2}$$

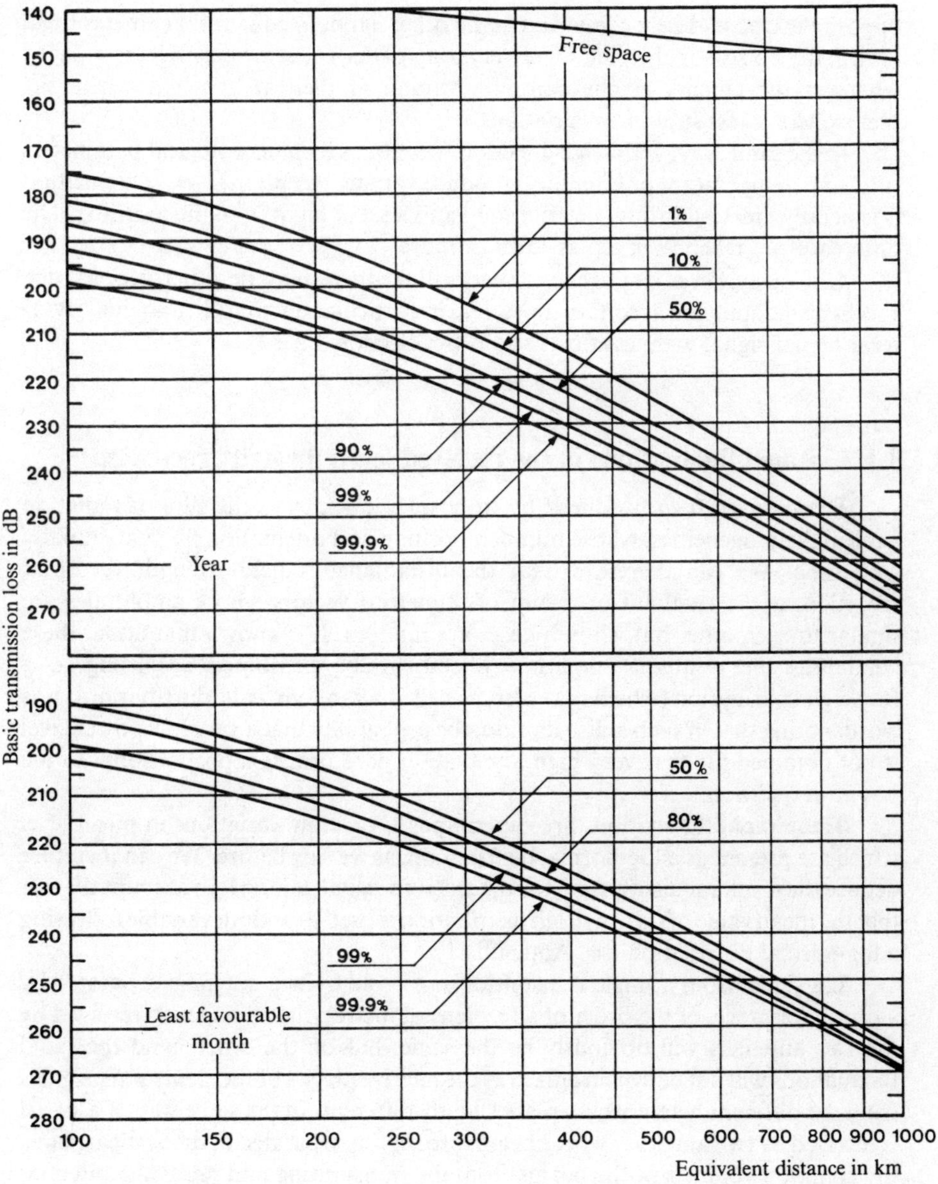

Fig. 8.2d. Mean attenuation in tropospheric scatter. 1GHz. Paths over land. Equatorial climate.

8.1.1.3. Seasonal and diurnal effects

Even within a single climatic zone, we generally find roughly significant seasonal variations, above all for short paths. In temperate climates, and for a path length of some 250km, attenuation in winter is 15dB greater than in summer. The opposite is true in desert climates. On the other hand, in equatorial climates there is virtually no seasonal effect. Over very long paths, the seasonal effect is reduced whatever the climate as the common volume is then at a height where the atmosphere is less subject to variations.

In the same way, within a 24h period, variations in meteorological parameters (pressure, temperature or humidity) produce varying reception levels. Attenuation is generally greatest during the afternoon and least at night or in the morning. The difference can reach as much as 15dB or more.

Obviously, on a tropospheric scatter link, there may be other propagation mechanisms applying at certain times, and in particular ducting phenomena, giving a far higher signal with less fluctuations (see Section 4.2.2.4).

8.1.2. Rapid fluctuations in the received level: diversity reception

The received field produced by tropospheric scatter is the sum of radiation from all heterogeneities, whose number, position and orientation fluctuate rapidly with time. We can therefore treat the instantaneous field strength vector at reception as the resultant of a sum of numerous vectors whose amplitudes are similar to each other but whose phases are random. It is known that under these conditions the resultant amplitude is statistically distributed according to a Rayleigh distribution (which may also be called a random walk distribution). It is worth noting that in such a distribution, the probability that a very low power level is not obtained (or that very high attenuation does occur) is proportional to the power in question.

These rapid fluctuations are superimposed on slow variations in mean level which are present as a log-normal distribution, as we saw before. We can therefore deduce the resultant distribution for the received signal, taking into account the fact that the mean value of the Rayleigh distribution is itself a random variable following a log-normal distribution (see Appendix 1).

If a signal from a single transmitter is received by two antennas separated by a certain distance, of the order of a few tens of metres, the mean levels received by the two antennas will obviously be the same, but on the other hand the rapid fluctuations will not be synchronized as the relative phases of the elementary signals from the different heterogeneities are slightly different. In the same way, if a signal is received by two antennas whose beams are slightly separated in the vertical plane, the common volumes of the beams from the transmitting and receiving antennas are slightly different on the two paths. Consequently, heterogeneities are not always the same and fluctuations are not always synchronized. Finally, if we use two different frequencies with the same antennas, phases over paths with the same heterogeneities are different, and if the difference in frequency is suffficient (of the order of 2% to 4%) fluctuations will not be synchronized.

In the three cases described before, we therefore have two signals carrying the same information but with fluctuations which are not always correlated. It is, furthermore, possible to have more than one receiver and therefore more than two signals.

As soon as we have different signals (generally two or four) from our receivers, we can obtain a resultant signal with far smaller fluctuations than any one signal. This method is called diversity reception. There are three types of diversity reception: space, frequency and angular. Angular diversity is less frequently used as the upper beam from the antenna generally receives a slightly lower mean level. Furthermore, this system makes it necesary to have a more complicated primary source for the antenna.

The simplest method of using the received signals is to select the strongest at any moment. However, there is an optimum solution which is to add signals from the receivers after having weighted each of them in direct proportion to the voltage of the signal and in inverse proportion to the power of the noise it contains. In this method, called optimum combination after detection, the resultant signal to noise ratio is the sum of the signal to noise ratio in each receptor (see Appendix 6.1).

There are also other methods which involve rephasing signals at intermediate frequency and adding them before detection.

8.1.3. Antenna gain degradation

Given these propagation mechanisms the wave in the aperture of the receiving antenna is not a plane wave since the amplitude and phase fluctuate. Consequently, summation of the various parts of this wave does not take place in phase at the antenna's primary source, and this effect grows as the aperture of the antenna itself grows, i.e. as its gain increases. This therefore leads to an apparent reduction in the gain of large antennas which is called 'antenna gain degradation' or sometimes 'aperture to medium coupling loss'.

These considerations have been applied to a number of theories concerning the microscopic structure of the atmosphere, but these theories have led to contradictory results, none of which has been confirmed experimentally (e.g. results concerning the influence of the length of the link or antenna gain).

Instead of studying deformations to the wave front on reception, it has been found to be far more efficient to consider the 'number' of refractive index heterogeneities in the common volume, since it is these heterogeneities which are the radiation source and therefore contribute to forming the resultant wave front. This allows us to apply the following simple reasoning.

If we replace the antennas on a link using tropospheric scatter propagation by antennas of greater gain, the heterogeneities in the common volume receive and scatter a power with increases in the same proportion as the antenna gain. But the dimensions of the common volume fall as a consequence of the increased gain of the antennas, so certain heterogeneities no longer contribute to transmission. Consequently, the increase in received level is less than would be produced by the increased gain of the antennas. What we find is apparent antenna gain degradation.

This leads to a further observation. In order to obtain the maximum received level, antennas must be oriented as low as possible on the horizon. This means that when we increase antenna gain, we only alter the upper limit of the common volume. As long as this limit is at a height where the number of heterogeneities is negligible, no gain degradation appears. We then see that gain degradation only appears above a certain value of gain.

Contrary to what was predicted theoretically, experiment has shown that antenna gain degradation does not depend in practice on the length of the path, between 150 and 500km, but almost exclusively on antenna gain, and begins to appear at gains of around 30dB.

Figure 8.3, obtained experimentally, allows us to calculate the effective total gain of antennas on a tropospheric scatter link (sometimes called 'path antenna gain'), as a function of the sum of the gains in free space between these two

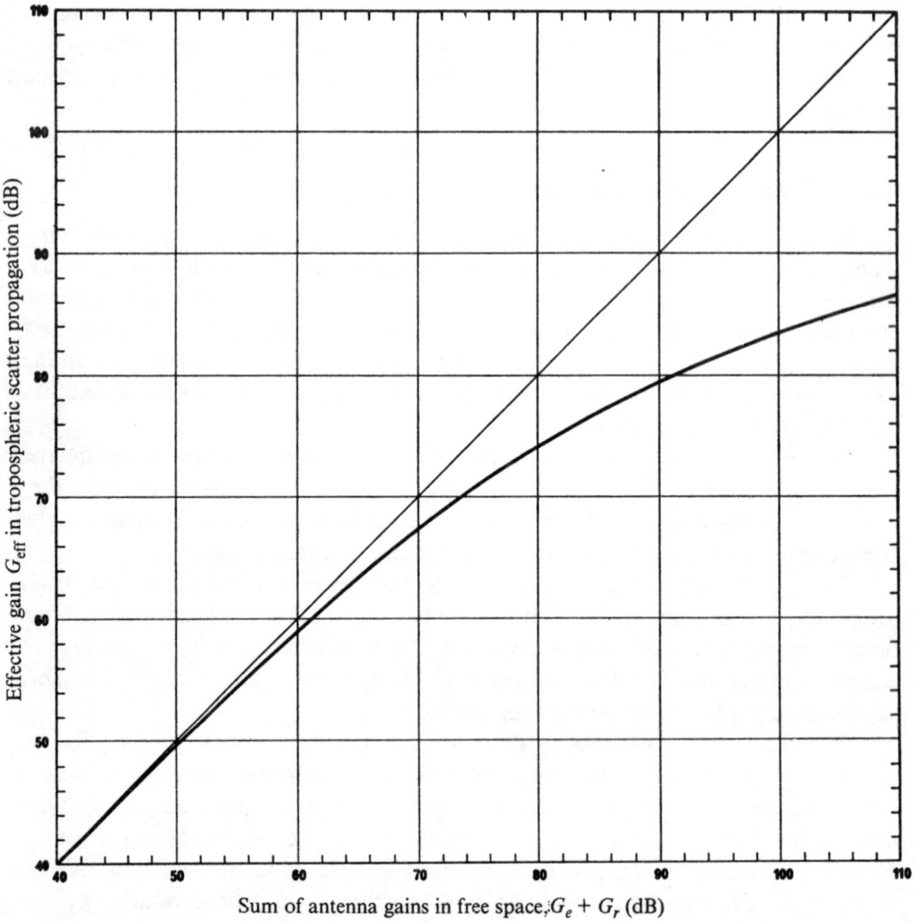

Fig. 8.3 Effective gain of an antenna in tropospheric scatter propagation.

antennas. The figure is valid on condition that the two antennas have similar gains, which is usually the case, as these links usually have identical antennas at each end. For quantitative calculations the curve in Fig. 8.3 can be represented by the following empirical relation:

$$G_{\text{eff}} = (G_e + G_r)\exp\left[-\frac{\left(\dfrac{G_e + G_r}{148}\right)^4}{1 + \left(\dfrac{G_e + G_r}{148}\right)^4}\right]. \tag{8.3}$$

This relation is valid for:

$$G_e + G_r < 120\ \text{dB}. \tag{8.4}$$

It can be seen that this phenomenon reduces the efficiency of large antennas considerably, in terms of received level. We however see (Section 8.1.5) that the use of large antennas is of value for other reasons.

We find that if we use very high gain antennas, the gradient of the curve in Fig. 8.3 tends towards 1/4. This value can easily be explained by the following very simple consideration.

Assume that at each end of the link there are identical antennas, of gain G and angular aperture α. We know that for a given type of antenna G varies with $1/\alpha^2$. In free space the ratio between the received power and transmitted power is proportional to G^2 and therefore to $1/\alpha^4$. In tropospheric scatter propagation we have to multiply this ratio by a coefficient representing the sum of the contributions of all heterogeneities, taking into account variations in the scatter angle between the top and the bottom of the common volume. If we are using very high gain antennas, the common volume is small and consequently all heterogeneities are of about the same size and have the same scatter angle. The coefficient is then proportional to the number of heterogeneities, and consequently to the size of the common volume and therefore to α^3. The received level is in this case proportional to $\alpha^3 \times 1/\alpha^4$, i.e. $1/\alpha$ instead of $1/\alpha^4$ in free space. As we can treat the effective gain of antennas as proportional to the received power, it is clear that on a logarithmic scale, this does indeed correspond to a gradient of 1/4.

It is theoretically possible to reduce the antenna gain degradation by using antennas with beams of different angular apertures in the vertical and horizontal plane to take account of the structure of heterogeneities. However, the complication introduced in this way generally seems too great for the results one can hope to obtain from it.

N.B: Here we are considering antenna gain degradation from the point of view of telecommunications, as it makes it possible to calculate the long term statistics for received levels, and therefore the quality of the link. Short term degradation in antenna gains have sometimes been studied by comparing the instantaneous received levels on two antennas and analysing the distribution of differences between these levels. This approach produces no valuable practical results.

8.1.4. Attenuation between antennas of given diameter: choice of frequency range

If antennas of a given diameter are selected for a given link, the long term variations of the received level depend on the carrier frequency because of the following three phenomena:

1. attenuation between isotropic antennas or 'basic transmission loss': it is generally accepted that this attenuation is proportional to the cube of the frequency;
2. the free space gain of the antennas used: for an antenna of given diameter, that gain is proportional to the square of the frequency;
3. the antenna gain degradation: for an antenna of given diameter, this degradation depends on the frequency, and can be calculated from Fig. 8.3.

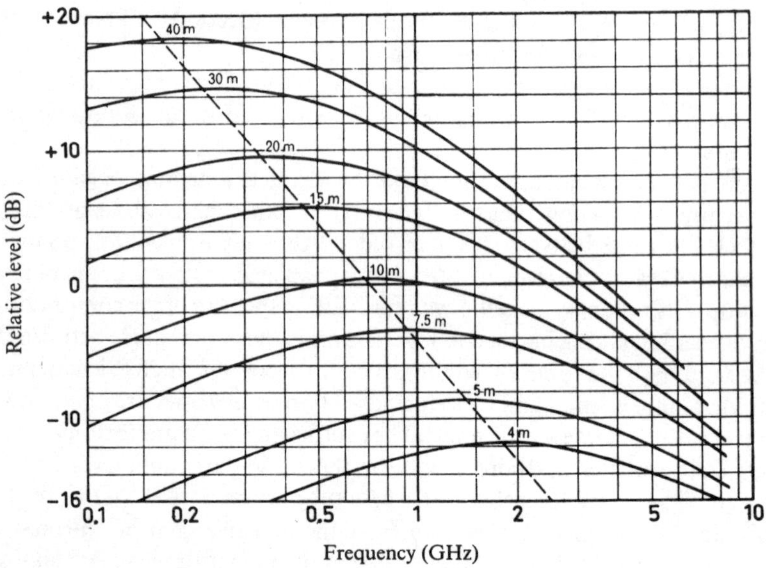

Fig. 8.4. Relative level received on a tropospheric scatter link using two antennas of given diameter.

The effects of variations of these parameters as a function of frequency are shown in Fig. 8.4 for antenna diameters between 3 and 40m. Figure 8.4 shows the relative level (or, by changing sign, the relative attenuation) obtained with two antennas of the same diameter on a transhorizon link; we have chosen as reference level (0dB) the level that would be obtained in the same conditions with two antennas of 10m diameter at 1GHz. As far as the length of the link is concerned, this figure has the same validity as that giving the antenna gain degradation, i.e. the link considered is assumed to be between about 150 and 500km in length.

It is clear that, for a given diameter, the relative level reaches a maximum for a certain frequency. It falls off on either side, for low frequency because the relative dimensions of the antenna measured in wavelengths falls and, consequently, its free space gain decreases, and for high frequencies because the antenna gain degradation increases more rapidly as the free space gain increases. The optimum operational frequency reaches from 300MHz for an antenna 30m in diameter to 3GHz for a 3m antenna; however, the maximum is very flat and deviation on either side up to a ratio of 2/1 leads to no significant loss in relative level.

8.1.5. Available bandwidth: transmission quality

All we have seen above concerns phenomena observed at a given frequency. In fact, however, transmission of information requires the use of a certain bandwidth. If amplitudes and relative phases of frequencies within this band are not strictly maintained during transmission, a certain distortion (in amplitude or phase) of the signal takes place and consequently quality is reduced. This degradation appears in different ways depending on the type of signal transmitted and the modulation used. For example, in transmission of an analog multiplexed telephone signal it appears in the form of intermodulation noise between channels which is added to the thermal noise associated with the received level. This intermodulation noise increases with the number of channels and the apparent effect is as if the propagation medium limited the available bandwidth.

To determine the available bandwidth, we have to consider the maximum level of the phase decorrelation that can occur between two propagation paths. Such phase decorrelation is linked with the dimensions of the common volume. However, by increasing the antennas gains, we only reduce phase decorrelation to the extent that the reduction in antenna beam width eliminates heterogeneities which were contributing to transmission. We saw before that it is precisely by eliminating 'effective' heterogeneities that we obtain antenna gain degradation. In other words, increased bandwidth is directly related to increased antenna gain degradation, and like that degradation, it only begins to appear for antenna gains greater than around 30dB.

This relation between antenna gain degradation and bandwidth is illustrated in Fig. 8.5 which gives two simultaneous recordings carried out in two series of experiments. At the receiving station the two antennas, which had very different gains, were positioned in such a way that they had a common axis, to avoid effects associated with difference in location. In the first series, with antennas of 9 and 29dB gain, antenna gain degradation was neglibible. It is clear from the figure that the received signals are almost totally correlated, showing that signals are produced by adding the elementary fields produced by the same heterogeneities. In the second series of experiments, carried out with two parabolic antennas of 34 and 52dB gain, it can be seen on the contrary that received signals are totally decorrelated and that the upper curve, corresponding to the larger parabola, contains less rapid fluctuations than the lower curve, showing that the number of heterogeneities contributing to the received signal on the 52dB parabola is less than for the 34dB gain antenna.

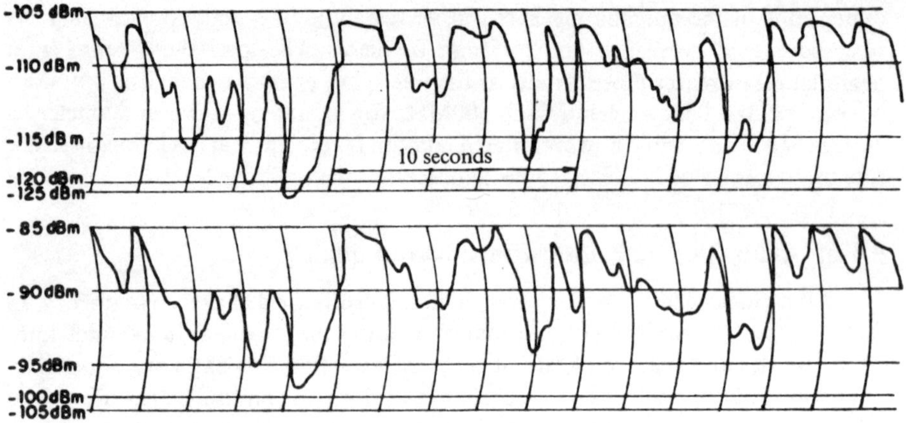

2170MHz

Antenna gain { upper curve 9dB
lower curve 29dB

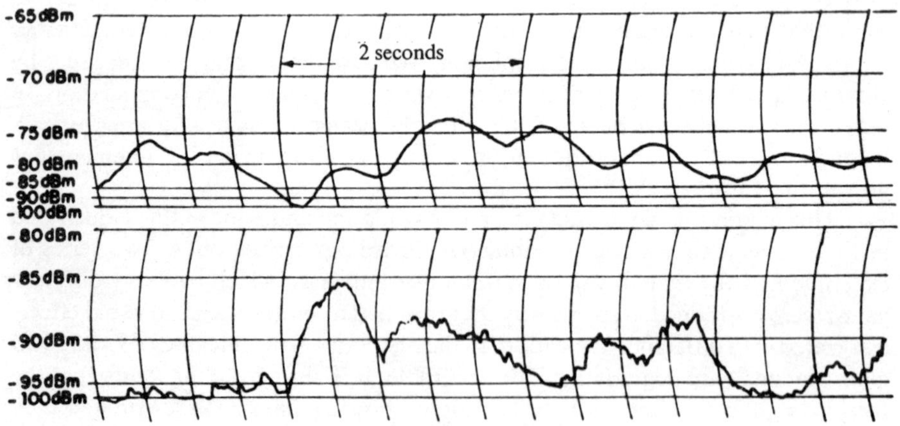

4000MHz

Antenna gain { upper curve 52dB
lower curve 34dB

Fig. 8.5. Comparison of rapid fluctuations on different gain antennas.

On this issue, we can use the following approximate reasoning. Assuming a zero starting angle, difference in length between the paths which pass through the top and bottom of the common volume respectively, can be calculated easily from elementary geometric considerations. We find that:

$$\Delta l = \frac{\alpha d}{2}\left(\alpha + \frac{d}{R}\right) \tag{8.5}$$

where:

d is the distance;
R is the Earth's radius;
α is the angular aperture (in radians) of the beam from each antenna.

For high gain antennas:

$$\alpha \ll \frac{d}{R}$$

and, consequently:

$$\Delta l \simeq \frac{\alpha d^2}{2R}. \tag{8.6}$$

If we want to keep the same bandwidth over any distance, the product αd^2 must remain constant, and consequently the antenna gain must be proportional to the fourth power of the length of the path.

It is difficult to calculate transmission quality. We only make a few general comments here. For any type of modulation, use of very high gain antennas is always useful, as this reduces phase decorrelation by reducing the dimensions of the common volume. In particular, with antennas of given diameter, the optimum frequency is always greater than one would expect from Fig. 8.4, if we take into account both thermal noise associated with the received level and intermodulation noise associated with the reduction in common volume. We can assume that the optimum frequency is between about 0.7 and 2GHz for antenna diameters between 6 and 20m and distances of the order of 200 to 400km.

For analog radio relay links using frequency modulation it is generally assumed that intermodulation noise increases with the fourth power of time delays in propagation within the common volume. Furthermore, if the mean level received varies by 1dB, thermal noise varies by 1dB and intermodulation noise by 0.7dB, whatever the order of diversity.

For digital radio relay links quality is limited in general not by thermal noise but by intersymbol interference due to multi-path. Consequently, the essential factor characterizing quality is the ratio between the mean square standard deviation of the time delay of the different paths and the duration of a binary element. The only way of obtaining a low error rate is by using a very high gain antenna, a high order of diversity (4 or 8) and adaptation systems for demodulation.

8.2. Ionospheric scatter

Just like heterogeneities in the troposphere, ionization heterogeneities in the ionosphere can be sufficient to create measurable scatter. However, the term scatter applied to the ionosphere encompasses several different phenomena.

8.2.1. Lateral scatter

The term lateral scatter is used for the fact that waves can be received well outside the circle containing the transmitter and the receiver. This phenomenon, which occurs on frequencies that are normally reflected, may be due to the fact that fronts of equal ionization or the Earth's surface are not perfectly spherical. This produces an effect covering 10° or more on either side of the plane of the circle. However, if the normal path contained in the plane of the circle undergoes a major change due to absorption, it is sometimes possible to discover paths differing by several tens of degrees.

8.2.2. Forward scatter

This type of scatter is found over frequencies which are not normally reflected, and particularly between 30 and 60MHz. It seems to be produced by two causes: meteoric ionization and turbulence in the lower part of the ionosphere at 85km in altitude. Turbulence increases with ionization, and therefore with the angle of elevation of the sun. Scatter reaches a minimum at around 9:00 pm (local time) as at that time both meteoric ionization and ionization within the ionosphere are low.

In general, the level received by scatter is at a maximum at the time of the solstice and at a minimum at the equinoxes. It is reinforced by sudden ionospheric turbulences because of the increased ionization of the lower layers.

As far as rapid fluctuations in the received level are concerned, we observe the same phenomena as for tropospheric scatter: the correlation between received values at two separate points decreases as these points move apart, making efficient diversity reception possible, using a separation of some 20 wavelengths in a direction perpendicular to the propagation path. There is also a phenomenon of antenna gain degradation with marked limitation of the effective antenna gain. Gain can only with difficulty exceed 15dB. In the same way, the available bandwidth cannot be greater than a few kilohertz.

To give an idea of the order of magnitude, attenuation between isotropic antennas can be taken to vary with $F^{7.5}$ between 30 and 60MHz. As in this frequency band the noise level varies approximately with $F^{2.5}$, the signal to noise ratio varies with F^{-5} if the power of the transmitter and the nominal gain of the antennas are held constant. For this reason, it is very difficult to work above 60MHz.

Transmissions by ionospheric scatter were used between 1950 and 1970 for telegraphic channels and high reliability data transmission channels over distances of 800 to 2000km, in particular for military applications. It was also attempted to make systematic use of meteoric traces using an auxiliary receiver which triggered the transmitter's modulation only at moments where it detected the presence of an ionized trace. These systems are now little used.

Chapter 9

Influence of the ionosphere on propagation

9.1. Structure and general characteristics of the ionosphere

In order to explain why the propagation of long waves used for transatlantic links was considerably more effective than had been expected, Kennelly (in the United States) and Heaviside (in Great Britain) suggested in 1902 that these waves might be travelling around the Earth by reflecting from ionized 'layers' a few hundred kilometres above the Earth's surface. As time went by this hypothesis became increasingly accepted and the layers were given the names of their inventors, or even abbreviated to 'KH layers'. Their use from 1922 onwards for decametric waves (called short wave), whose long distance propagation can only be explained by taking the ionosphere into account, completed the demonstration of its existence.

As reflection seemed to be taking place at clearly defined altitudes, it was assumed that several layers existed, for which Appleton used the letters *D*, *E* and *F*. The first layer discovered was given the first letter of the word Electric. When it became clear that there were other layers, letters *D* and *F* were chosen.

Later studies, and in particular rocket studies, showed that, contrary to appearances, the ionosphere was not made up of separate layers, but that the degree of ionization (the number of electrons per unit volume) was a continuous function of altitude, with maxima and points of inflection. These days we tend to talk of ionized regions (the *D* region, etc.). The word layer is, however, used for the altitude at which wave reflection seems to take place, i.e. generally the altitude of a maximum or of points of inflection of the ionization profile.

9.1.1. The source of ionization

The processes which produce ionization in the upper atmosphere are, in order of decreasing importance:

1. ultra-violet (UV) and X-ray radiation from the sun;
2. particle radiation from the sun;
3. meteor trails.

The sun therefore contributes the most, and consequently the characteristics of the ionosphere vary principally as a function of solar activity. In particular, we find a daily component to these variations, an annual component and a component with a period of around 11 years (mean solar activity period).

Electron density depends on the equilibrium between these electron creation processes and deionization processes, as follows:

1. recombination of positive particles with negative particles;
2. electron capture by a neutral atom. That atom becomes ionized but plays no role in propagation because of its mass.

When UV or X-ray radiation penetrate the atmosphere, it meets progressively denser layers, tending to increase the rate of electron production. But the intensity of the radiation decreases due to absorption. There is therefore an altitude where the electron production rate is at a maximum.

The first theory of electron production was developed by Chapman in 1931 based on very simple assumptions, including in particular monochrome radiation, a single ionizable gas and an atmosphere in thermal equilibrium. This theory already went some way towards explaining the morphology of the phenomenon, but could not predict the existence of all the various forms of reflection observed. The theory has since been considerably developed.

The particle radiation which makes up solar wind can lead to ionization by collision. Its influence appears primarily in certain relatively thin layers of the atmosphere in the polar regions. As the force lines of the magnetic field in these regions are virtually vertical, they do not prevent charged particles penetrating the atmosphere.

Ionized meteor trails are produced by small particles (0.1mg) at heights around 100km (shooting stars). Their individual life spans are very short (a few seconds).

9.1.2. Ionization profile

Rocket measurements give a fairly clear idea of the number of free electrons per unit volume as a function of height. Figure 9.1 shows maximum and minimum values, typical of the ionosphere by day and by night, respectively. We distinguish the following regions.

Region *D*, between 55 and 95km. Ionization is lower during the day (10^9 electrons/m^3) and fades almost entirely at night, as the recombination rate is high because of the high number of neutral particles at this altitude.

Region *E*, between 95 and 150km. Electron density is of the order of 10^{11} electrons/m^3 during the day and between 3 and 4.10^{10} electons/m^3 at night. Ionization is caused by X-ray and UV radiation with wavelengths around 100Å. At altitudes of around 125km, there is a point of inflection of the curve which may disguise a minor maximum.

Within region *E* there may be some cells, or clouds, with a very high level of intermittent ionization. These zones, which are generally fairly thin, are the sporadic *E* layers (E_s). There are many mechanisms by which these layers are formed and not all of them are fully understood.

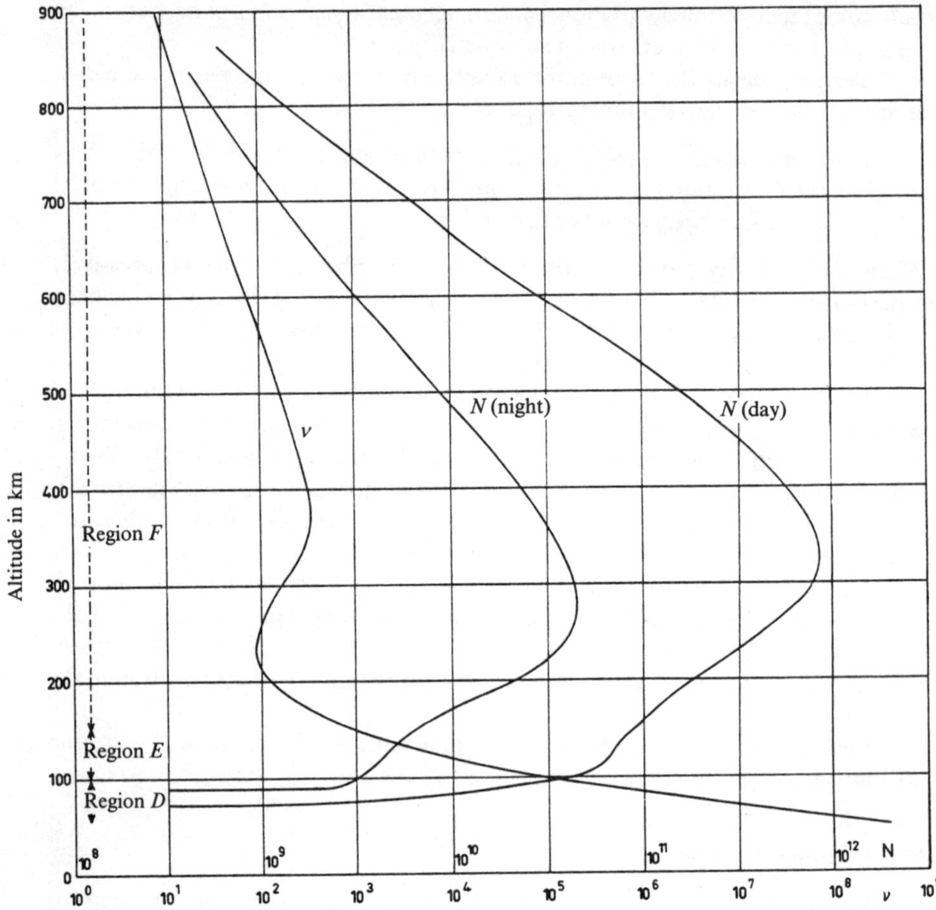

Fig. 9.1. Ionization N (electrons/m^3) and collision frequency ν (s^{-1}) profiles. Typical example for moderate latitudes.

Region *F*, above 150km. This is the region with the highest electron density, which may reach 10^{12} electrons per m^3 by day at an altitude of around 350km and 5.10^{10} electrons per m^3 at night at around 250km. Ionization here is caused by UV radiation with wavelengths between 200 and 600Å. During the day and in the summer region *F* can be divided into F_1 and F_2. Region F_1 is at 200km. The ionization profile rarely shows a maximum here, but rather a point of inflection.

The position and ionization during the day of regions *D*, normal *E* and F_1 seems to be rigidly dependent on the direction of the sun. Ionization depends mainly on the zenith angle χ of the sun, and is a function of $(\cos \chi)^n$ where n is of the order of 0.75. On the other hand, in region F_2 ionization is permanent and depends very little on the position of the sun. This is because of the slowness with which electrons recombine at this altitude, and because this region is fed with electrons from the lower regions by diffusion. Furthermore, whereas for the lower regions ionization is greater in summer than in winter, the opposite applies to region *F*. This phenomenon, called the winter anomaly of region *F*, is noticeable above all in periods of maximum solar activity.

9.1.3. Collision profile

Figure 9.1 also shows the number of collisions per second, ν, as a function of altitude. It can be seen that this number falls rapidly to around 200km, and then falls far more slowly. There is even a slight maximum at around 350km because of the high number of electrons at this altitude. The product $N\nu$, which affects wave absorption, reaches a maximum at around 70km.

9.2. Wave propagation in the ionosphere

In a vacuum the only possible current is the 'displacement current' given by:

$$\mathbf{i}_0 = \frac{\mathrm{d}\mathbf{D}}{\mathrm{d}t} = \varepsilon_0 \frac{\mathrm{d}\mathbf{E}}{\mathrm{d}t} \tag{9.1}$$

or assuming a sine wave and using complex notation:

$$\mathbf{i}_0 = \mathrm{j}\varepsilon_0\,\omega\mathbf{E}\,. \tag{9.2}$$

If the medium contains free electrons, they are set in motion by the electric field of the incident wave. This movement of charged particles produces a current **i** which has to be added to the displacement current. The effect is as though the permittivity of the medium had changed and taken a value η such that:

$$\mathbf{i} + \mathbf{i}_0 = \mathrm{j}\,\eta\omega\mathbf{E}\,. \tag{9.3}$$

To work out the movement of the particles we have to take account of the fact that in the ionosphere the Earth's magnetic field produces a magnetic induction $\mathbf{B}_0$. If we ignore thermal agitation (cold plasma), the movement of an electron is given by the equation of dynamics:

$$m\frac{\mathrm{d}\mathbf{v}}{\mathrm{d}t} = -\,e\mathbf{E} + e(\mathbf{v}\times\mathbf{B}_0) - m\nu\mathbf{v}\,. \tag{9.4}$$

9.2.1. Refractive index of an ionized medium in the absence of a magnetic field

Since the presence of the magnetic field introduces a major complication, we start assuming that it is zero, an assumption which already makes it possible to explain many of the phenomena involved. For a sine wave the previous equation can be written:

$$(j\omega + \nu)\,\mathbf{v} = -\frac{e}{m}\mathbf{E}\,. \tag{9.5}$$

As the velocity $\mathbf{v}$ is, in this case, in the direction of the electric field, we need only consider amplitudes. Let us assume that the mean free path of the electrons between two collisions is large and write N for the number of electrons per unit volume. The movement of these electrons produces a current:

$$i = Nev = \frac{Ne^2}{m}\,\frac{E}{j\omega + \nu}\,. \tag{9.6}$$

As this current is as i_0 in the direction of field E, the two currents can be added algebraically, and consequently Equation (9.3) becomes a scalar relation:

$$i + i_0 = j\eta\omega E \tag{9.7}$$

giving:

$$\eta = \varepsilon_0 - \frac{Ne^2}{m\omega}\,\frac{1}{\omega - j\nu} \tag{9.8}$$

which corresponds to a complex refractive index n such that:

$$n^2 = \frac{\eta}{\varepsilon_0} = 1 - \frac{Ne^2}{m\varepsilon_0}\,\frac{1}{\omega(\omega - j\nu)}\,. \tag{9.9}$$

Generally, the following parameters and normalized quantities whose physical meaning will become clear later, are used:

$$\omega_c^2 = \frac{Ne^2}{m\varepsilon_0}, \quad f_c = \frac{\omega_c}{2\pi}, \tag{9.10}$$

$$X = \left(\frac{\omega_c}{\omega}\right)^2 = \left(\frac{f_c}{f}\right)^2, \tag{9.11}$$

$$Z = \frac{\nu}{\omega}. \tag{9.12}$$

Under these conditions the expression for n^2 becomes:

$$n^2 = 1 - \frac{X}{1 - jZ}\,. \tag{9.13}$$

Let us write μ and χ for the real part and imaginary part of n, respectively, and in order to simplify the notation:

$$A = \frac{X}{1 + Z^2} \tag{9.14}$$

which gives:

$$2\,\mu\chi = ZA, \quad \mu^2 - \chi^2 = 1 - A \tag{9.15}$$

$$2\mu^2 = \sqrt{(1-A)^2 + Z^2A^2} + 1 - A\,, \tag{9.16}$$

$$2\chi^2 = \sqrt{(1-A)^2 + Z^2A^2} - (1-A)\,. \tag{9.17}$$

We now consider a few particular cases which provide a good approximation to real circumstances.

9.2.1.1. In the absence of collisions

This can be considered to be approximately the case in region F and even, as a first approximation, in region E for frequencies which are not too low:

$Z = 0$
$\chi = 0$

$$n = \mu = \sqrt{1-A} = \sqrt{1-X} = \sqrt{1-\left(\frac{\omega_c}{\omega}\right)^2} = \sqrt{1-\left(\frac{f_c}{f}\right)^2}\,. \tag{9.18}$$

The medium is then purely dielectric whose (phase) refractive index is less than 1 and vanishes for a certain frequency called the critical frequency or plasma frequency (strictly speaking, these two terms are not quite synonymous). The critical frequency of a layer is the plasma frequency corresponding to the medium electron density of the layer) given by:

$$f_c = \frac{\omega_c}{2\pi} = \frac{1}{2\pi}\sqrt{\frac{Ne^2}{m\varepsilon_0}}\,. \tag{9.19}$$

If we take the frequency in Hertz and N as a number of electrons per m^3, we obtain a very good approximation:

$$f_c = 9\sqrt{N}\,. \tag{9.20}$$

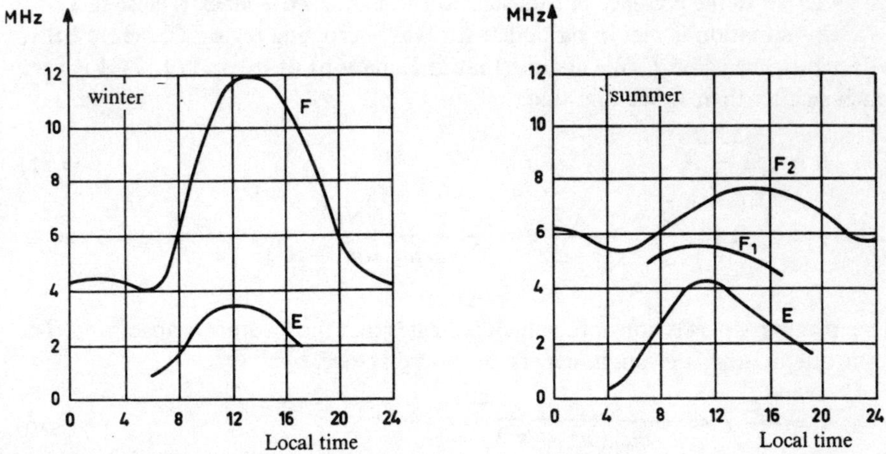

Fig. 9.2. Example of daily variation in critical frequencies.

The graphs in Fig. 9.2 give typical values of critical frequencies for a winter day and for a summer day in the temperate zone. Obviously, values measured any one day may vary considerably from mean values. Furthermore, they vary with the 11 year cycle of the sun, and increase considerably during periods of high solar activity. Finally, ionization decreases overall from the equator to the pole.

Propagation through the ionosphere is therefore only possible for frequencies greater than f_c. Lower frequencies are reflected. We see later that reflection can take place for frequencies greater than f_c if the wave meets the ionized layers at an oblique angle of incidence.

Note that the refractive index n which we have calculated is a phase refractive index, linked to the phase velocity of the waves by:

$$v_\varphi = \frac{c}{n} \,. \tag{9.21}$$

The phase velocity is therefore greater than c. On the other hand, the group velocity is lower than c. It is well known that we can define a group refractive index n_g associated with n using the general relation:

$$n_g = \frac{\mathrm{d}}{\mathrm{d}\omega}(n\omega) \tag{9.22}$$

which, for the ionosphere, gives:

$$n_g = \frac{1}{n} \tag{9.23}$$

or:

$$v_g . v_\varphi = c^2 \,. \tag{9.24}$$

9.2.1.2. In the presence of collisions when the refractive index is close to 1

This situation is met in particular for waves crossing region D, before being reflected from region F. We assume that in Equations (9.16) and (9.17) A is very much smaller than 1. We find that:

$$\mu \simeq \sqrt{1 - A} \tag{9.25}$$

$$\chi \simeq \frac{1}{2} ZA = \frac{1}{2} \times \frac{XZ}{1 + Z^2} = \frac{1}{2}\,\frac{\nu}{\omega}\,\frac{Ne^2}{m\varepsilon_0(\omega^2 + \nu^2)} \,.$$

The imaginary part is non-zero, which indicates that the medium is absorbing. The wave attenuation is given, in nepers per unit distance, by:

$$\alpha = \frac{2\,\pi}{\lambda}\,\chi = \frac{N\nu\,\mathrm{e}^2}{2\,cm\varepsilon_0(\omega^2 + \nu^2)} , \tag{9.26}$$

which can be rewritten using practical units as:

$$\alpha = 4.6\ 10^{-2} \frac{N\nu}{\omega^2 + \nu^2} \text{(dB/km)}. \tag{9.27}$$

If we furthermore assume that ν is very much smaller than ω, we obtain the general approximate relation:

$$\alpha \simeq 1.16\ 10^{-3} \frac{N\nu}{f^2} \text{(dB/km)} \tag{9.28}$$

where N is expressed as a number of electrons per m^3, ν as a number of collisions per second and f in Hertz.

Absorption reaches a maximum when crossing layer *D* at around 70km, as it is there that the product $N\nu$ is at a maximum. It occurs both for links crossing the ionosphere and for links involving reflection at layer *F*. As it is not associated with any major deviation, since n is close to 1, it is sometimes called non-deviative absorption.

We know that ν decreases as altitude increases. Equation (9.27) shows that α reaches a maximum at the altitude for which:

$\alpha = \omega$

9.2.1.3. In the presence of collisions when the refractive index is close to zero

This situation is met near the altitude at which the wave is reflected. Trajectories are very curved, phase velocity is very high and consequently group velocity is very low.

Starting from Equations (9.14) and (9.15) we can write:

$$\chi = \frac{ZA}{2\mu} = \frac{1}{2\mu}\frac{\nu}{\omega}(1 - \mu^2 + \chi^2). \tag{9.29}$$

From our assumption, μ^2 is small. From Equation (9.17), as χ^2 is always less the μ^2, it is also very small. We obtain:

$$\chi \simeq \frac{1}{2\mu}\frac{\nu}{\omega} \tag{9.30}$$

and then:

$$\alpha = \frac{2\pi}{\lambda}\chi = \frac{\nu}{2\mu c}. \tag{9.31}$$

Attenuation due to absorption, in this case called deviative absorption, is therefore inversely proportional to μ, i.e. it is proportional to the group refractive index. It does not affect links crossing the ionosphere, and it is often ignored for links by reflection, since the effect is relatively small.

9.2.2. Influence of the Earth's magnetic field

In the presence of the Earth's magnetic field H_0 the velocity of electrons is not generally parallel to the electric field because there is a vector product in Equation

(9.4). Permittivity and refractive index can still be calculated but because of the complexity of the calculations we simply give the results here. We need to define a certain pulsation ω_b called the cyclotron pulsation and the corresponding frequency called the cyclotron frequency, or more commonly called the gyrofrequency, by relation:

$$\omega_b = \frac{e}{m} B_0 . \tag{9.32}$$

In the ionosphere, the gyrofrequency is about 1.4MHz. We then introduce a further normalized variable:

$$Y = \frac{\omega_b}{\omega} = \frac{f_b}{f} . \tag{9.33}$$

If we ignore collisions ($Z = 0$) and if we write θ for the angle between the direction of propagation of the wave and the direction of the magnetic field, theory provides the expression:

$$n^2 = 1 - \frac{X}{1 - \dfrac{\dfrac{Y^2}{2}\sin^2\theta}{1 - X} \pm \left[\left(\dfrac{\dfrac{Y^2}{2}\sin^2\theta}{1 - X}\right)^2 + Y^2\cos^2\theta\right]^{1/2}} . \tag{9.34}$$

We can see that the expression for the refractive index depends on the frequency and the direction of propagation, and that it can take either sign. In other words, the medium is dispersive, anisotropic and birefractive. The term ordinary mode (written 0) is used for the mode of propagation corresponding to the positive sign and extraordinary mode (written x) for the mode corresponding to the negative sign.

As it is extremely complicated to study the general case, we consider two standard particular cases.

1. Longitudinal propagation: $\theta = 0$:

$$n^2 = 1 - \frac{X}{1 \pm Y} \tag{9.35}$$

which can be written:

$$n^2 = 1 - \frac{\omega_c^2}{\omega(\omega \pm \omega_b)} = 1 - \frac{f_c^2}{f(f \pm f_b)} . \tag{9.36}$$

It can be shown that in this case a circularly polarized wave will propagate in such a way as to preserve its circularity, but with a velocity which is different from right-hand and left-hand polarizations. We also know that a linearly polarized wave can be considered as the sum of two inverse circularly polarized waves. As these two waves have different velocities, the effect is as if the direction of polarization of the linearly polarized wave, which is the sum of the two others, turned while propagating through the medium. This rotation, which is the expression of the

Faraday effect, is proportional to the distance covered and to the difference in velocity of the two inverse circularly polarized components at the frequency considered.

The refractive index can vanish for two different frequencies f_1 and f_2 given by the positive roots of the following equations, respectively:

ordinary mode

$$f^2 + f_b f - f_c^2 = 0 \tag{9.37}$$

extraordinary mode

$$f^2 - f_b f - f_c^2 = 0 . \tag{9.38}$$

Let us first consider frequencies $f > f_b$. It is clear that:

1. if $f > f_2$, the two circularly polarized waves can propagate;
2. if $f_1 < f < f_2$, only one wave can propagate, while the other is reflected;
3. $f < f_1$, both waves are reflected.

In other words, the single critical frequency is replaced by two frequencies f_1 and f_2. The difference between f_1 and f_2 is equal to the gyrofrequency.

Now consider $f > f_b$. In this case propagation through the ionosphere is possible again, though frequency f is considerably less than the critical frequency. This wave does not play a large role in telecommunications, but it makes it possible to study the magnetosphere as it is ducted by the field lines of the Earth's magnetic field, enabling it to return to Earth at the magnetic conjugate of its starting point after moving away to distances several times greater than the Earth's radius. As this type of propagation involves considerable scatter, a complex radio wave, for example one generated by lightning, is received in the form of a whistling of variable frequency.

Note: We derived Equation (9.35) from Equation (9.34) by setting θ immediately to zero. If instead we allow θ to tend gradually towards zero (quasi-longitudinal propagation), we find that, beyond $X = 1$, it is the extraordinary mode wave which corresponds to the positive sign (Z wave) while the ordinary mode wave corresponds to the negative sign.

2. Transverse propagation: $\theta = \pi/2$. The magnetic field is perpendicular to the direction of propagation:

$$n^2 = 1 - \frac{1}{1 - \dfrac{Y^2}{2(1 - X)} \pm \dfrac{Y^2}{2(1 - X)}} . \tag{9.39}$$

Depending on which sign we choose we can obtain the following equations:

$$n^2 = 1 - X , \tag{9.40}$$

$$n^2 = 1 - \frac{X}{1 - \frac{Y^2}{1 - X}}. \tag{9.41}$$

Equation (9.40) corresponds to the ordinary mode wave and we can see that it is identical to the one obtained in the absence of a magnetic field. It can be shown that polarization under these conditions is linear and lies in the direction of the magnetic field. The vector product in Equation (9.4) is therefore zero.

Equation (9.41) corresponds to the extraordinary mode wave, which is polarized elliptically in the plane perpendicular to the field B_0. For $Y < 1$, i.e. for frequencies greater than the gyrofrequency, the graph of function $n^2(X)$ shows two different branches for the extraordinary mode wave. For $Y > 1$ there is only one branch (Z wave).

3. General case: Without going into detail, we give a few guidelines for cases where angle θ can take any value and the number of collisions is non-zero. From Equations (9.40) and (9.41), n^2 vanishes for 3 values of X, one corresponding to f_0 and the others to frequencies f_1 and f_2 defined by Equations (9.37) and (9.38). In practice, however, we find that only f_c and f_2 can be identified in most cases. Frequency f_1 only appears exceptionally. It gives the Z wave or Z component.

If collisions are taken into account, n^2 is complex and its modulus never vanishes. It can be shown that the extraordinary mode wave is always attenuated more than the ordinary mode wave.

9.2.3. Non-linear effects of the ionosphere

In 1934, after putting a powerful long wave transmitter (Radio Luxembourg) into service, it was discovered that waves reflected from the ionosphere in the region of this transmitter were slightly modulated by it. This is why the effect was at first called the 'Luxembourg effect'. The following is a simplified explanation.

We can assume that at a given point in the ionosphere the mean free-path of electrons is independent of their velocity and, consequently, the frequency of collision grows in proportion to the velocity of the electrons. This means that the attenuation of the wave due to energy losses by collisions also increases proportionally with this velocity [Equation (9.28)].

Let ν_0 be the mean frequency of collision in the absence of any wave. If a wave is being propagated in the ionosphere, its electric field $E(t)$ gives electrons a suplementary velocity and consequently the collision frequency becomes a function $\nu(t)$. If the wave is modulated, function $\nu(t)$ cannot follow the sinusoidal variations of $E(t)$ because of the specific inertia of the medium. It therefore simply takes a mean value ν slightly greater than ν_0. There is therefore a small increase in absorption due to the passage of a wave, but the effect is slight.

If the radio wave is amplitude modulated at a sufficiently low frequency (acoustic frequencies), the amplitude E(t) includes a component at this frequency, and $\nu(t)$ can follow these variations. The effect is therefore as if attenuation varied in rhythm with the modulation. This phenomenon has two consequences.

1. If only one modulated wave is being propagated, then, as the medium's attenuation increases at the modulation peak, the modulation factor is reduced.
2. If two waves of very different amplitude are being propagated, the wave with the greater amplitude modulates the attenuation due to the medium in rhythm with its own modulation. The amplitude of the second wave therefore undergoes this amplitude modulation which adds to its own modulation. This cross-talk phenomenon is called ionospheric crossmodulation. It is clear that crossmodulation increases not only with the power of the interfering transmitter but also with the mean attenuation of the medium. It is therefore particularly significant with kilometric and hectometric waves (LF and MF) reflecting from region *D* of the ionosphere. The Earth's magnetic field also plays a role. This is because, as the extraordinary mode wave is attenuated most, it is particularly effective when it comes to modulating the ionosphere's attenuation if the perturbation frequency is close to the gyrofrequency. If frequency increases, the effect of crossmodulation diminishes according to a function which tends towards f^{-2}.

The explanation above works well for observations carried out on amplitude modulated waves, which are virtually the only ones used for ionospheric links. However, it is also true that a powerful transmitter also leads to a variation in the ionosphere's permittivity, which could influence waves using other types of modulation.

9.3. Links penetrating the ionosphere

As a first approximation we ignore collisions between electrons and neutral molecules, as well as the magnetic field, and assume that the ionosphere is locally planar, i.e. that equal index surfaces are horizontal. A trajectory from the Earth's surface and reaching the ionosphere with an angle φ_0 is defined by Snell's law:

$$n_h \cos \varphi = \cos \varphi_0 \tag{9.42}$$

where n_h is the refractive index at height h, which may be expressed as a function of the critical frequency at that height by Equation (9.18):

$$n_h = \sqrt{1 - \left(\frac{f_c}{f}\right)^2} \tag{9.43}$$

from which we deduce:

$$\sin \varphi = \sqrt{\frac{\sin^2 \varphi_0 - \left(\frac{f_c}{f}\right)^2}{1 - \left(\frac{f_c}{f}\right)^2}} \tag{9.44}$$

which gives angle φ at any point on the trajectory. That trajectory only exists if:

$$f > \frac{f_c}{\sin \varphi_0}. \tag{9.45}$$

For geometrical reasons which we consider below, angle φ_0 can only take very low values, and consequently only high frequency waves will be able to penetrate the ionosphere.

If the magnetic field is taken into account, we have seen that a wave with a frequency below the gyrofrequency may also penetrate the ionosphere, but this mode of propagation is of more interest to geophysical research than telecommunications. Such a mode is characterized by major group delays, extremely variable attenuation and complex interaction between waves and charged particles. If will not be studied in detail here. We only consider the case of higher frequencies because of their importance in satellite links.

9.3.1. Angle of elevation errors

Since the refractive index is not the same everywhere, the trajectory tends to be curved and consequently there is a difference between the real direction of a source

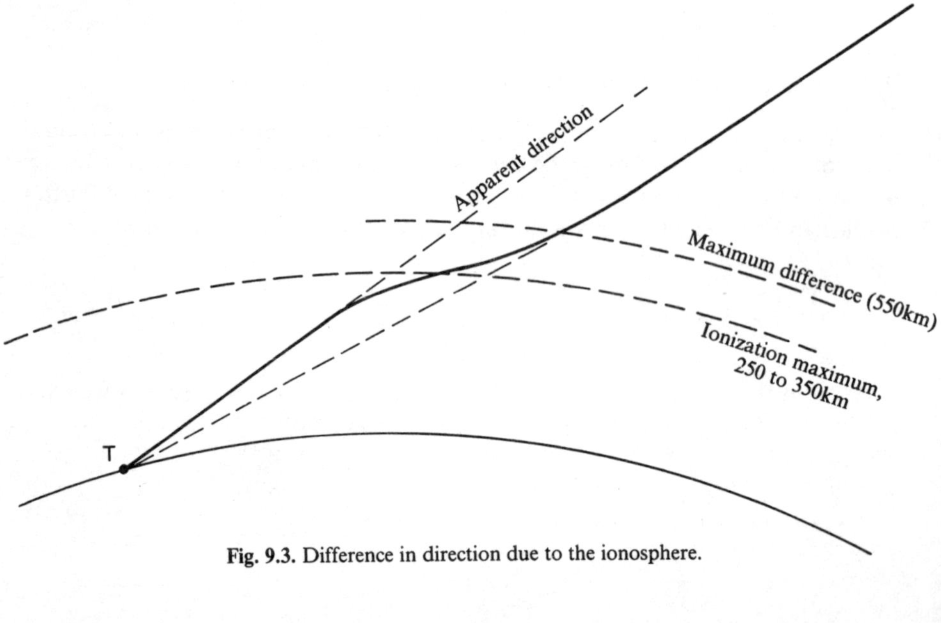

Fig. 9.3. Difference in direction due to the ionosphere.

and the apparent direction from which the waves arrive. We have already seen from the case of the troposphere that in a spherically symmetrical medium the radius of curvature of trajectories is proportional to the vertical gradient of the refractive index. In the ionosphere the vertical gradient of the refractive index changes sign at the altitude when ionization is maximum; consequently, trajectories tend to be S-shaped, but the resulting total bending is always turned downwards (Fig. 9.3).

For a given ionization profile the difference in the angle of arrival depends on the frequency, on the angle of elevation and on the altitude of the source (satellite). The difference reaches a maximum for an altitude of around 550km, a little greater than that of the ionization maximum, and for an angle of elevation of 4 to 5 °. It then reaches around 0.35° during the day for a frequency of 100MHz. The difference falls approximately inversely with the square of the frequency. It falls as the angle of elevation increases and, naturally, vanishes for a vertical path. Beyond about 200MHz the angular difference due to the ionosphere is always less than that due to the troposphere.

9.3.2. Propagation time and distance errors

Just as for the troposphere, the error in the distance is given, for all practical purposes, by the expression:

$$\Delta l = \int (n_g - 1)\,\mathrm{d}s \tag{9.46}$$

but here n_g is not equal to the phase refractive index. If we consider frequencies considerably greater than the critical frequency, we obtain approximately:

$$n_g = \frac{1}{n} = \left[1 - \left(\frac{f_c}{f}\right)^2\right]^{-1/2} \simeq 1 + \frac{1}{2}\left(\frac{f_c}{f}\right)^2 \tag{9.47}$$

and consequently:

$$n_g - 1 \simeq \frac{1}{2}\,\frac{81\,N}{f^2} \simeq 40\,\frac{N}{f^2} \tag{9.48}$$

and hence:

$$\Delta l \simeq \frac{40}{f^2}\int N\,\mathrm{d}s\,. \tag{9.49}$$

The integral represents the total electronic content, i.e. the total number of electrons contained in a prism whose cross-section has unit area and contains the trajectory (the units used are Hertz and metres). For a vertical path during the day the integral takes a value of around 30.10^{18} electrons/m^2. For a frequency of 100MHz we find that $\Delta l = 1200$m.

For a zero angle of elevation the figure would be about three times greater. This effect is clearly of major importance.

As Δl depends on the frequency, the propagation time is distorted leading to deformation of transmitted signals, and consequently to a limitation in the available bandwidth. This limitation is not generally a problem for frequencies greater than a few hundred Megahertz.

9.3.3. Polarization rotation

We have seen that the presence of the Earth's magnetic field leads, because of the Faraday effect, to the appearance of a double refractive index, and therefore to two different propagation velocities, depending on the polarization of the waves. If n_1 and n_2 are the two refractive indexes, the difference between phase shifts is equal to:

$$\Delta \Phi = \frac{2\pi}{\lambda} \int (n_2 - n_1)\, \mathrm{d}s \tag{9.50}$$

and the angle of rotation is equal to half the phase shift angle.

The quantity $n_2 - n_1$ depends essentially on angle θ between the direction of propagation and the direction of the magnetic field. For non-transverse propagation the angle of rotation in radians is given by an approximate expression:

$$\Omega = \frac{A}{f^2} \int N H_0 \cos\theta\, \mathrm{d}s = \frac{A}{f^2} \int \frac{N H_0 \cos\theta}{\sin\varphi}\, \mathrm{d}h \tag{9.51}$$

where h is the height, φ is the angle formed with the horizontal by the tangent to the trajectory and A is a constant. Figure 9.4 gives the angle of rotation as a function of the frequency for a zero angle of elevation by day and by night. For a vertical path the angle should be divided by about 3.5. It clear that this angle of rotation is extremely large for metric and decimetric waves, making it absolutely essential to use circularly polarized links with satellites. On the other hand, beyond a few Gigahertz linear polarization can be used, even for longitudinal propagation. However, if we want to use the same frequency on both orthoganal polarizations, we need to use frequencies greater than 10GHz to obtain sufficient decoupling between the two polarizations.

9.3.4. Attenuation

On penetrating the ionosphere waves undergo attenuation associated with the imaginary part of the refractive index, which is itself due to collisions between electrons and neutral molecules [Equation (9.9)]. As the frequencies considered are high, their values in Hertz are greater than the number of collisions per second. We, in addition, ignore the magnetic field. This means that the collisions are the same as for non-deviative absorption examined in Section 9.2.1.

Specific attenuation given by Equation (9.28) has to be integrated over the

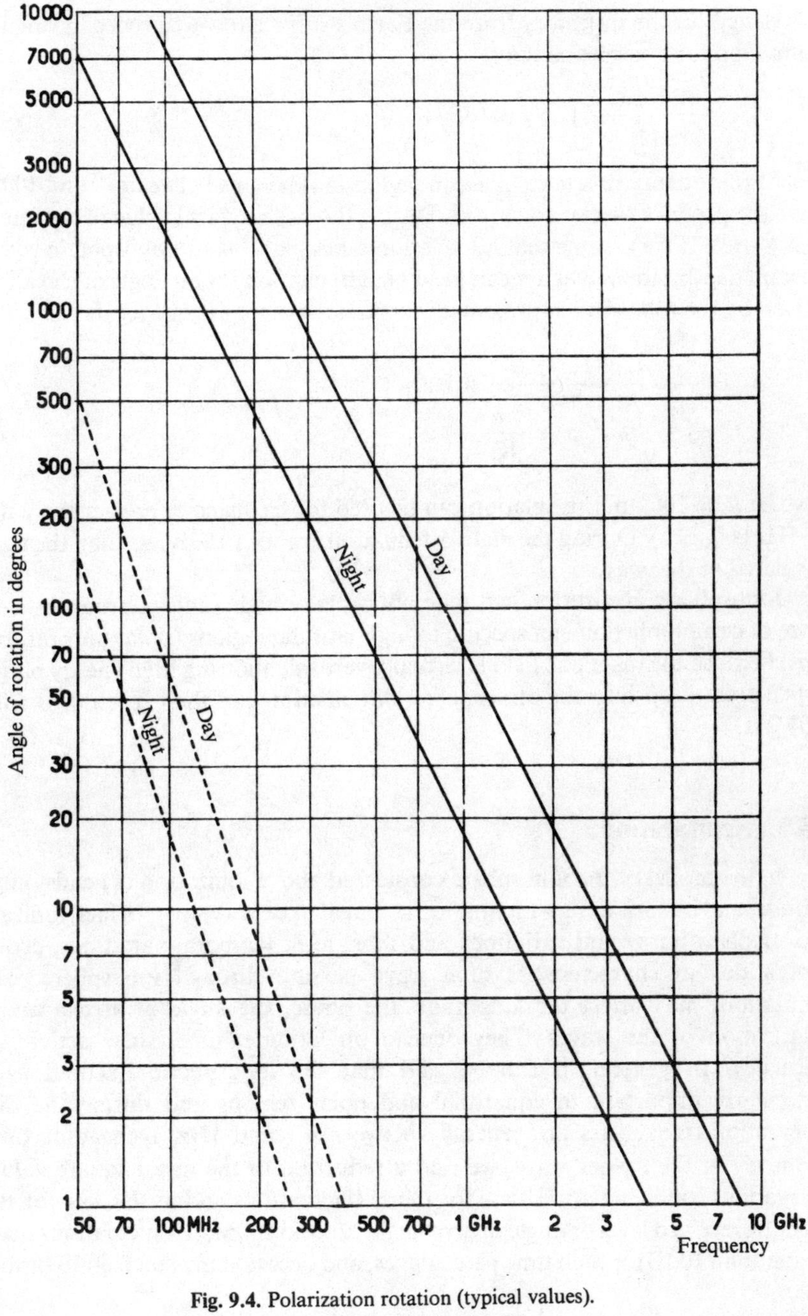

Fig. 9.4. Polarization rotation (typical values).
——— longitudinal propagation.
– – – transverse propagation.

entire length of the trajectory from the Earth's surface towards space, giving total attenuation for a vertical link of:

$$A_v = \frac{1.16\ 10^{-3}}{f^2} \int Nv\ ds\,(\text{dB})\,. \tag{9.52}$$

Most of this attenuation takes place in region D at heights between 80 and 100 km where the product Nv can reach 10^{17}. During the day a typical value of the integral is 1.8×10^{18}. If we assume that attenuation takes place in an atmospheric layer of constant depth, situated at a mean height h, attenuation for an angle of elevation φ is given by the following approximate expression where f is in Megahertz:

$$A_\varphi = \frac{2{,}2\ 10^3}{f^2 \sqrt{\sin^2 \varphi + \dfrac{2\,h}{R}}}\ (\text{dB})\,. \tag{9.53}$$

If we let h be 100km, this relation can be used for frequencies greater than about 50MHz (Fig. 9.5). During the night attenuation is about 15dB less than the figures calculated in this way.

Ionospheric absorption can take abnormally high values during the occurrence of certain phenomena specific to high latitude regions (polar auroras, polar caps) because the magnetic field is virtually vertical, allowing high energy particles to penetrate deep into the atmosphere. Attenuation can then reach 4 to 5dB at 100MHz.

9.3.5. Scintillation

In the model of the ionosphere considered above ionization depends only on altitude, and its variations with time are assumed to be very slow. In fact, ionization also involves horizontal variations and short term temporal variations, creating fluctuations in characteristics of a wave passing through ionosphere. These fluctuations may affect the amplitude, the phase, the angle of arrival and the polarization of the waves. They depend on latitude, time, solar activity and frequency. In general, it can be said that the ionospheric fluctuations are particularly important in equatorial and polar regions and during the night. Fluctuation frequencies are generally between 0.1 and 1Hz. Depending on the frequency of the carrier wave, we find a reduction in the mean square value of fluctuations following an F^{-1} law for phase fluctuations and an F^{-2} law for those affecting received level or angle of arrival. At around 100MHz, fluctuations may be greater than 10dB for high time percentages, and occasionally reach 30dB or more.

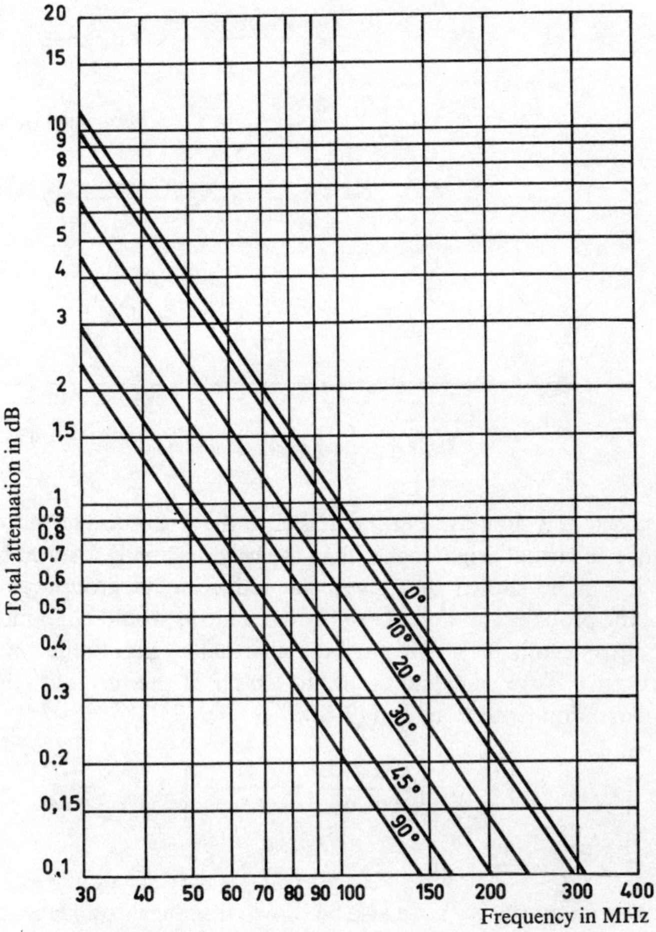

Fig. 9.5. Attenuation on crossing the ionosphere by day. Parameter: angle of elevation with respect to the Earth's surface.

9.4. Links by means of ionospheric reflection

As in the previous case, let us ignore collisions between electrons and neutral molecules and also the magnetic field, and assume that the ionosphere is locally planar. Equation (9.44) ceases to be valid at a certain altitude for which we have:

$$\sin \varphi_0 = \frac{f_c}{f}. \tag{9.54}$$

At this altitude the angle φ fades, and the trajectory becomes horizontal and then turns down again towards the ground. This is called ionospheric reflection, though the physical phenomenon is in fact refraction.

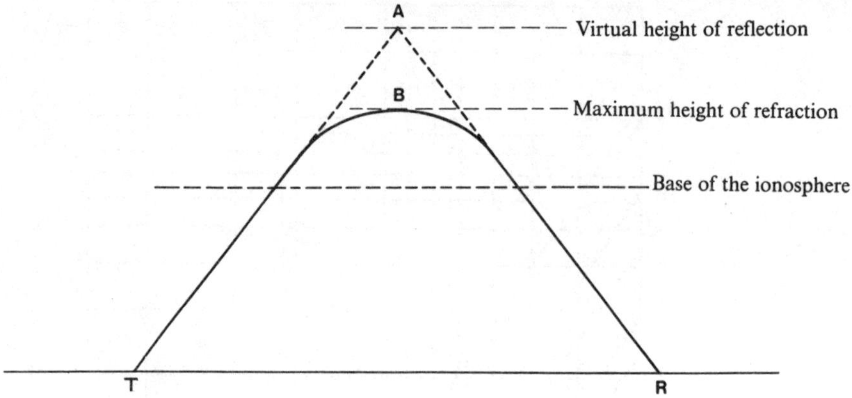

Fig. 9.6. Virtual height of reflection.

The trajectory is shown in Fig. 9.6. It is clear that reflection seems to take place from a plane at a virtual height higher than the height actually reached by the wave. However, it can be shown that given the reduction in group velocity in the ionosphere the propagation time is exactly the same as would be found in the case of reflection from a mirror in a vacuum at the virtual height of reflection (Breit and Tuve's theorem). If we calculate the radio length of the link, we can make the following derivation from Equation (9.42):

$$L = \int n_g \, \mathrm{d}s = \int \frac{\mathrm{d}s}{n} = \int \frac{\cos \varphi \, \mathrm{d}s}{\cos \varphi_0} = \frac{1}{\cos \varphi_0} \int \mathrm{d}x = \frac{TR}{\cos \varphi_0} = \tag{9.55}$$

$$= TA + TR .$$

Using the same assumptions it can also be shown that the virtual height for a wave of frequency f reaching the ionosphere at an angle φ is the same for a wave of frequency $f \sin \varphi$ which is propagating vertically (Martyn's theorem).

9.4.1. Ionospheric trajectories and distances reached by ionospheric reflection

In the frequency range concerned (see Section 9.4.2) the Earth's surface is generally a good reflector, a path between two points at the Earth's surface may include several successive reflections, alternately from the ionosphere and the ground. There may even be successive reflections between two ionospheric layers. Figure 9.7 gives the shapes of possible links between two points and the conventional symbols used for these links. A negative sign shows that reflection from the layer takes place upwards.

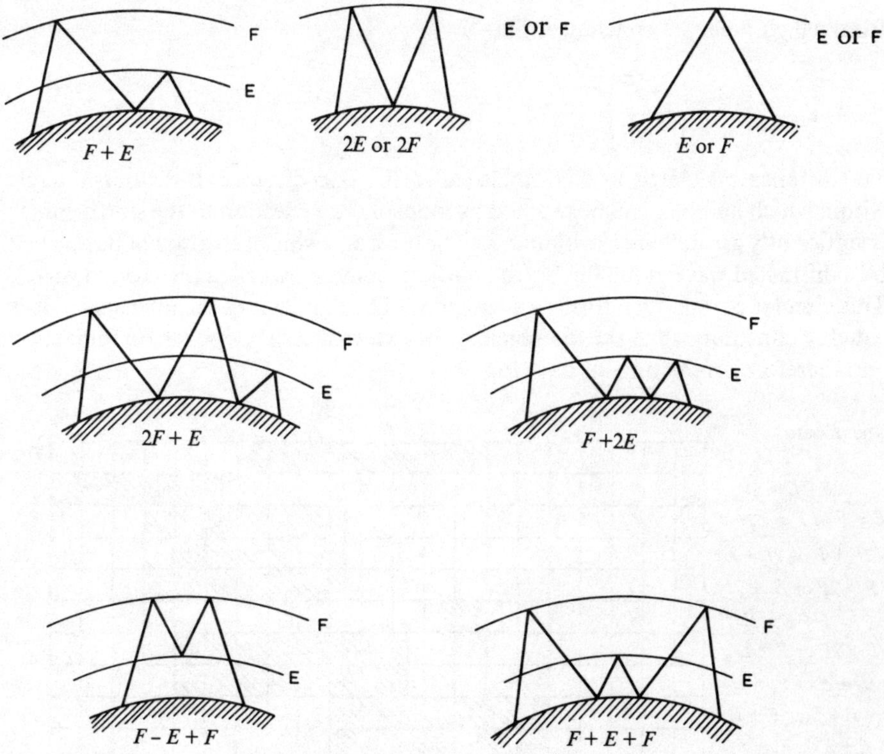

Fig. 9.7. Various types of ionospheric path.

The maximum distance reached in a single hop is limited for purely geometrical reasons by the height of the reflecting layers and by the Earth's curvature. Even if we take account of refraction of trajectories during their passage through the troposphere, this distance seldom exceeds 3500km.

Figure 9.8 shows the most likely paths possible for links less than 10,500km long. Most of these paths contain no more than 3 reflections from ionospheric layers.

As a link may include several successive hops there is no real maximum distance. On the other hand, for a given frequency f_1 greater than f_c there is a minimum distance. For this distance, angloe φ_0 reaches its maximum value given by the relation:

$$(\sin \varphi_0)_{\max} = \frac{f_c}{f_1}. \tag{9.56}$$

As this is a short distance, the Earth's curvature can be ignored and:

$$d_{\min} = 2\, h \operatorname{cotg} \varphi_{0\,\max} \tag{9.57}$$

from which it can be concluded that:

$$d_{\min} = 2\,h\sqrt{\left(\frac{f_1}{f_c}\right)^2 - 1}\,. \tag{9.58}$$

The distance calculated in this way is called the skip distance. It delimits a circle within which no point can be reached by ionospheric reflection. If the skip distance is sufficiently great, there is within this circle a corona which can also not be reached by a diffracted wave along the Earth's surface (sometimes called the ground wave). This circular corona then forms a silent zone. The diameter of the internal circle is roughly constant, while the diameter of the external circle depends on ionization and therefore on the time of day (Fig. 9.9).

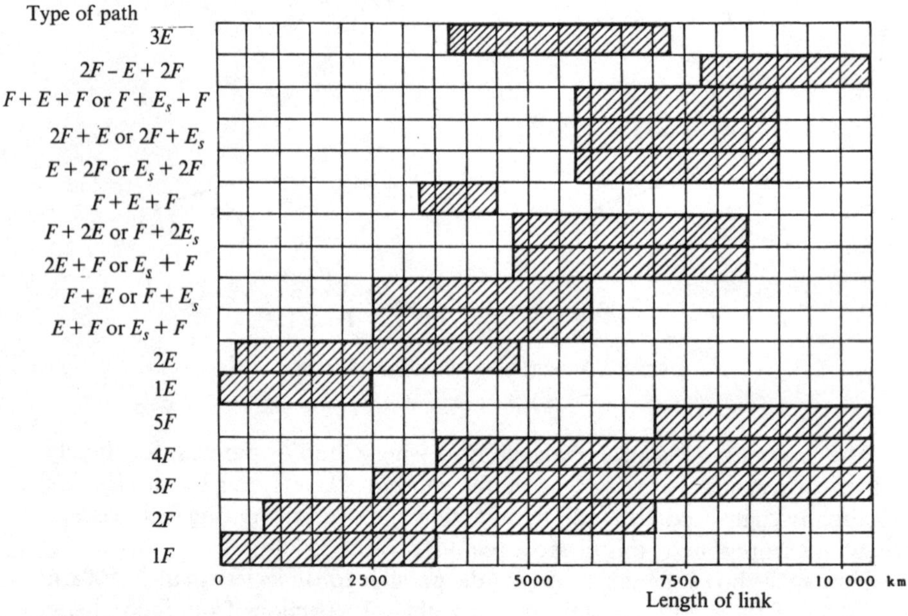

Fig. 9.8. Paths to consider depending on the length of a link (a minus sign indicates upward reflection).

9.4.2. Range of frequencies used for an ionospheric reflection link

We have seen that above a certain frequency depending on the degree of ionization and the angle of incidence a wave will penetrate the ionosphere, which gives a fairly sudden upper cut-off point to the range of frequencies usable for reflection links. We also saw that for progressively lower frequencies attenuation increases constantly, giving a lower limit to the usable frequencies, but here the cut-off point is far less abrupt. Since these two limits can vary fairly widely with time, the establishment of an ionospheric link chiefly requires that we find the most suitable frequency range, given geographical location, distance, date and time.

The upper limit of frequencies beyond which there is no further reflection is called the maximum usable frequency (MUF) and the lower frequency below which the received level is no longer sufficient is called lowest usable frequency (LUF). In almost every case these frequencies lie between 2 and 30MHz, which means that the range of waves used for ionospheric reflection links is practically the same as the range of decametric waves (HF band).

9.4.2.1. Maximum usable frequency (MUF)

If we write f_M for the highest frequency that can be reflected by an ionized layer of critical frequency f_c if the direction of arrival of the wave forms an angle φ_0 with this layer, these quantities are linked by Equation (9.54).

If we consider a hop of length d obtained by reflection from a layer at virtual height h, and if we write R for the Earth's radius (Fig. 9.10), simple geometrical calculations will give us the following approximate relation:

$$\tan \varphi_0 = \frac{d}{4R} + \frac{2h}{d}. \tag{9.59}$$

The minimum value of φ_0 is obtained when the link is tangential to the Earth's surface at the start:

$$(\varphi_0)_{\min} \simeq 2 \text{ Arc sin} \sqrt{\frac{h/R}{2(1 + h/R)}} \simeq \sqrt{\frac{2h}{R}}. \tag{9.60}$$

In this way we find that φ_0 is greater than about 20° for reflection from layer F and 11° for reflection from layer E. Eliminating φ_0 from Equations (9.54) and (9.59) gives an expression for the quantity f_M/f_c as a function of the length of hop and the height of the ionized layer. This quantity is called the transmission factor or the distance factor for the ionized layer considered:

$$\frac{f_M}{f_c} = \frac{1}{\sin \varphi_0} = \sqrt{\frac{4Rd}{d^2 + 8Rh} + 1}. \tag{9.61}$$

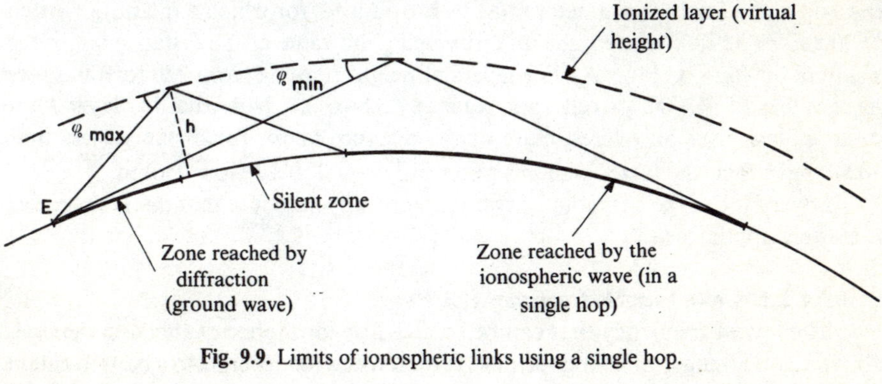

Fig. 9.9. Limits of ionospheric links using a single hop.

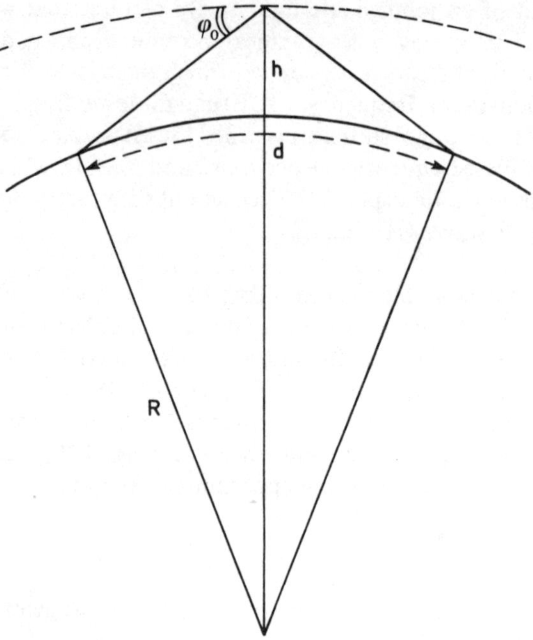

Fig. 9.10. Geometric quantities affecting ionospheric reflection.

Taking into account the real (or modelled) curve of ionization as a function of height, the distance factor differs slightly from the $1/\sin \varphi_0$ law, but the difference is always small on condition that a suitable hypothetical value for the representational height h is chosen.

Layer E: It is generally accepted that the distance factor can be represented by a single curve for which a good approximation can be deduced by using the preceding equations with the following values: $R = 6800$km and $h = 145$km. For example, we obtain a factor of 3.2 for 1000km and 4.7 for 2000km.

Layer F: As the shape of this layer is extremely variable, a single curve is not enough, and a further parameter must be introduced, for which a quantity written $M\,3000F_2$ or $M\,3000$ is used, the quantity being the value of the distance factor for a skip of 3000km on layer F_2. For layers of height running from 220 to 500km we have values of $M3000F_2$ from approximately 3.8 to 2.2. Note that for layer F the $1/\sin \varphi_0$ law gives an overestimate of the distance factor for distances less than 1000km. In fact, the factor remains practically equal to 1 below 250km.

Layer E_s: For the sporadic E layer it is generally accepted that the $1/\sin \varphi_0$ law is entirely applicable.

9.4.2.2. Lowest usable frequency (LUF)

The lowest frequency that can be received by ionospheric reflection depends above all on attenuation along the link. It therefore also depends to a certain extent on the performance of the equipment used (transmitter power and antenna gain). Attenuation along the link is due to several causes which are described separately.

9.4.2.2.1. Spreading loss

This attenuation is due to the fact that transmitted energy diverges from the antenna (propagation in free space), but a certain convergence is produced by the spherical shape of the ionized layers, which in this way acts as a concave mirror. Spreading loss therefore reaches a maximum at a certain distance.

If the field at 1km from the transmitter is taken as the reference point, and if the fact that the virtual height h is small compared with the Earth's radius is taken into account, the spreading loss A_s for a skip is given by the following approximate relation in which d, R and h are expressed in kilometres:

$$A_s = 10 \log \left[(d^2 + 4h^2) \frac{1 - \dfrac{d^2}{8Rh}}{1 + \dfrac{d^2}{8Rh}} \right] \text{(decibels)}. \tag{9.62}$$

In this expression the second term on the right-hand side, which represents the convergence effect, is deduced from the calculations given in Chapter 4 for the divergence factor with the assumption made here that heights h_1 and h_2 are negative and equal to opposite of the height of the ionized layer. It is known that the figures obtained in this way will be underestimated as we approach the limiting distance d_l obtained for a zero starting angle:

$$d_l = \sqrt{8Rh}\,. \tag{9.63}$$

Spreading loss is shown in Fig. 9.11. Correct values can be obtained from Equation (9.62) if $R = 6800$km and $h = 110$km for layer E and 240km for layer F, on condition that we do not come too close to the limiting distance.

9.4.2.2.2. Attenuation due to reflection

If a link is established with several hops, we have to take account of attenuation due to ground reflection. Attenuation reaches a maximum of the order of 1dB for grazing incidence even on low conductivity ground. On the other hand, for higher angles of reflection, attenuation increases up to about 2dB over sea, 3 to 4dB over moderate conductivity ground and more than 10dB over very low conductivity ground (frozen continents).

9.4.2.2.3. Attenuation due to ionospheric absorption

This type of attenuation limits ionospheric transmission most severely since it varies with time and frequency, and can take very high values. We have seen that attenuation is due to collisions between electrons and neutral molecules, and that attenuation in dB/km is generally proportional to the number of collisions per

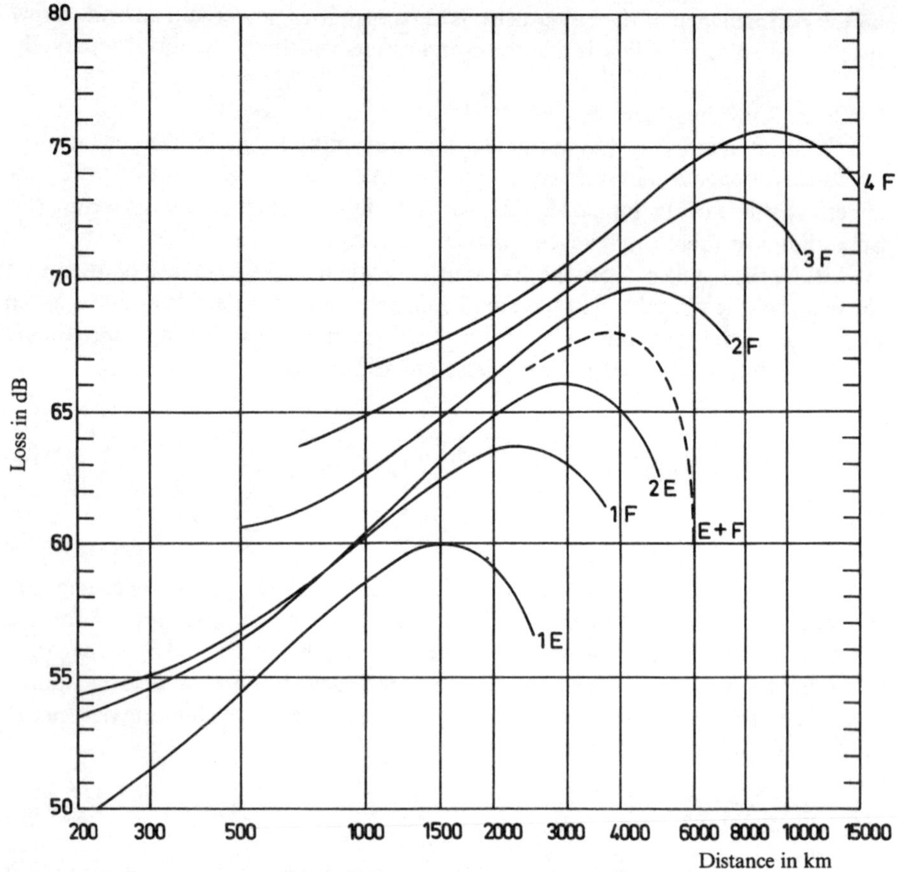

Fig. 9.11. Ionospheric spreading loss.

second, and inversely proportional to the square of the frequency, except for very low frequencies.

To evaluate absorption on a given link we have to carry out trajectory calculations based on models of the ionosphere, i.e. expressions giving the number of electrons per cubic metre and the number of collisions per second as a function of height. By way of illustration, Fig. 9.12 shows the median values obtained for a 4200km link (Paris-Dakar) for two typical periods in a year of low solar activity. The link involves intermediate reflection from the Earth's surface. It is clear that absorption which is very low during the night increases considerably during the day, particularly in summer.

9.4.2.2.4. Auroral absorption

If trajectories penetrate zones of high latitude where there are polar auroras,

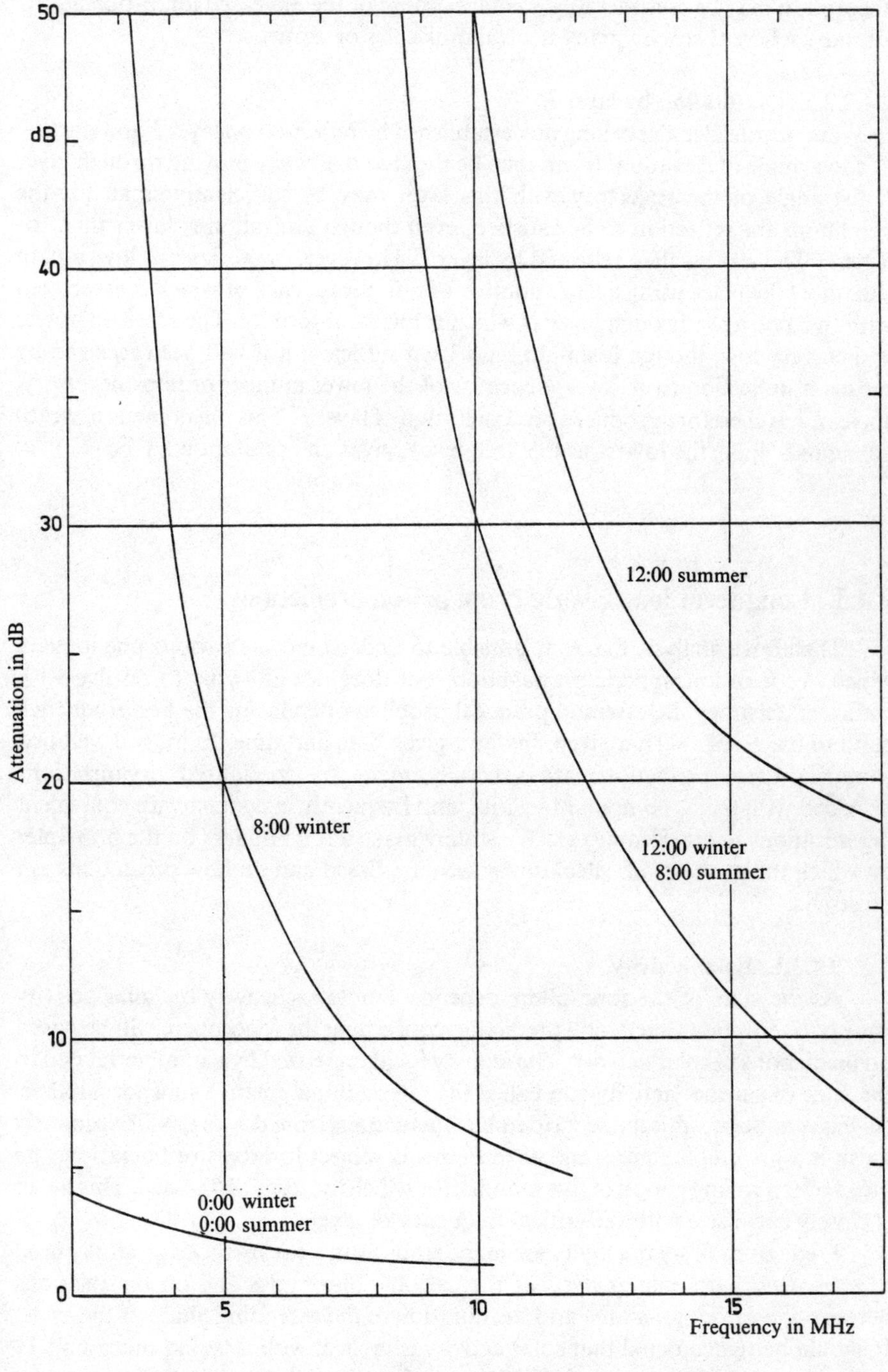

Fig. 9.12. Ionospheric absorption on a 4200km link (Paris-Dakar).

absorption may be considerably greater because of the increased ionization at low altitudes where there are many neutral molecules or atoms.

9.4.2.2.5. Occultation by layer E

Let us consider a very long link established by reflection on layer F and starting at a low angle of elevation. It can then be the case that while moving through layer E the angle of the trajectory with this layer may be sufficiently small for the conditions for reflection to be satisfied, even though ionization is lower than for layer F. The wave is then reflected by layer E. However, as this layer is low, a path can only take place using a large number of reflections, each of which is associated with two passages through layer D which is highly absorbing. The received power is then very low, though it should have been sufficient if it had been received by means of reflection from layer F because of the lower number of hops necessary. Layer E has therefore produced an occultation of layer F. This phenomenon, which sometimes limits the lowest usable frequency, gives an occultation LUF.

9.4.3. Long term ionospheric propagation predictions

The above analysis makes it possible to understand the various phenomena which occur in ionospheric propagation, but does not allow us to resolve with sufficient accuracy the essential practical problem of choosing the best frequency band to use to establish a given link at a given date and time. In most developed countries these predictions are generally made by specialized organizations provided with major computer facilities, and frequently in contact with equivalent organizations in other countries. We simply make a few remarks on the principles on which the methods of calculations used are based and on how predictions are presented.

9.4.3.1. Solar activity

As the state of the ionosphere depends almost exclusively on solar activity (sunspots, eruptions, etc.), any prediction concerning the ionosphere will be based on predictions of solar activity. The activity is characterized by a number related to the scale of sunspot activity and called the international relative sunspot number. As this number is subject to considerable fluctuations from day to day, the monthly mean is a preferable figure, and as even this is subject to excessive fluctations, its mean over a rolling period of five months, for which the term IR5 is used. This mean is closely correlates with the critical frequency of layer F_2.

Predictions of solar activity are made empirically. The methods generally used are based on harmonic analysis of the various solar cycles and on the relations between the value of maxima and the duration of the ascending phase of the cycle. It should be remembered that solar activity is cyclical with a period of around 11 years, but with significant major differences from one cycle to another.

9.4.3.2. Determination of basic ionospheric characteristics

There are four such characteristics:

1. the critical frequency of the normal layer *E*;
2. the critical frequency of the sporadic layer *E*;
3. the critical frequency of layer F_2;
4. the $M\,3000F_2$ factor.

There are major daily variations in these characteristics, and efforts to determine them are limited to the prediction of median monthly values. This involves two operations: on the one hand prediction of characteristics at one location where there is a station, and on the other hand the extrapolation of these results to the whole of the Earth's surface.

The median value of the critical frequency of the normal *E* layer undergoes little fluctuation, and consequently it can be calculated using empirical equations valid for at least one solar cycle.

The other characteristics, whose fluctuations are far greater, are obtained from median values, measured at corresponding periods of previous years, taking into account the predicted development of solar activity. Difficulties arise because ionospheric observation stations are very unevenly distributed over the Earth's surface and because allowance in measurements has to be made for irregular fluctuations. Special analysis procedures have been established to make this possible.

9.4.3.3. Presentation of long term predictions

To satisfy the needs of various users, several ways of presenting monthly statistical predictions concerning the ionosphere are used by the organizations responsible in various countries. We briefly consider the methods used in France.

9.4.3.3.1. Predictions for standardized geographical regions

A number of geographical regions have been defined to take account of the space variations in the characteristics of the ionosphere. The prediction for a geographical region is valid as long as the two ends of the link are within the region. It is also reasonably valid as the halfway point of the path is within the region. If the ends of the link are within two contiguous regions, such as Europe and the Mediterranean, it is possible to take a mean between predictions corresponding to the two regions in question. But, over longer links, it is no longer enough to examine the ionospheric situation in a single region or in two contiguous regions.

The predictions are established for distances up to 3000km, using data from one, two or even three ionospheric measurement stations. They are presented in the form of graphs showing the approximate statistical limits to the operating frequency to be used in the 2 to 35MHz band as a function of time, of the place within the region and of the available operational resources (Fig. 9.13). The two upper curves give the MUF limits at 30% and 90% probability, i.e. exceeded by values of the real MUF, respectively some 9 days out of 30 and 27 days out of 30.

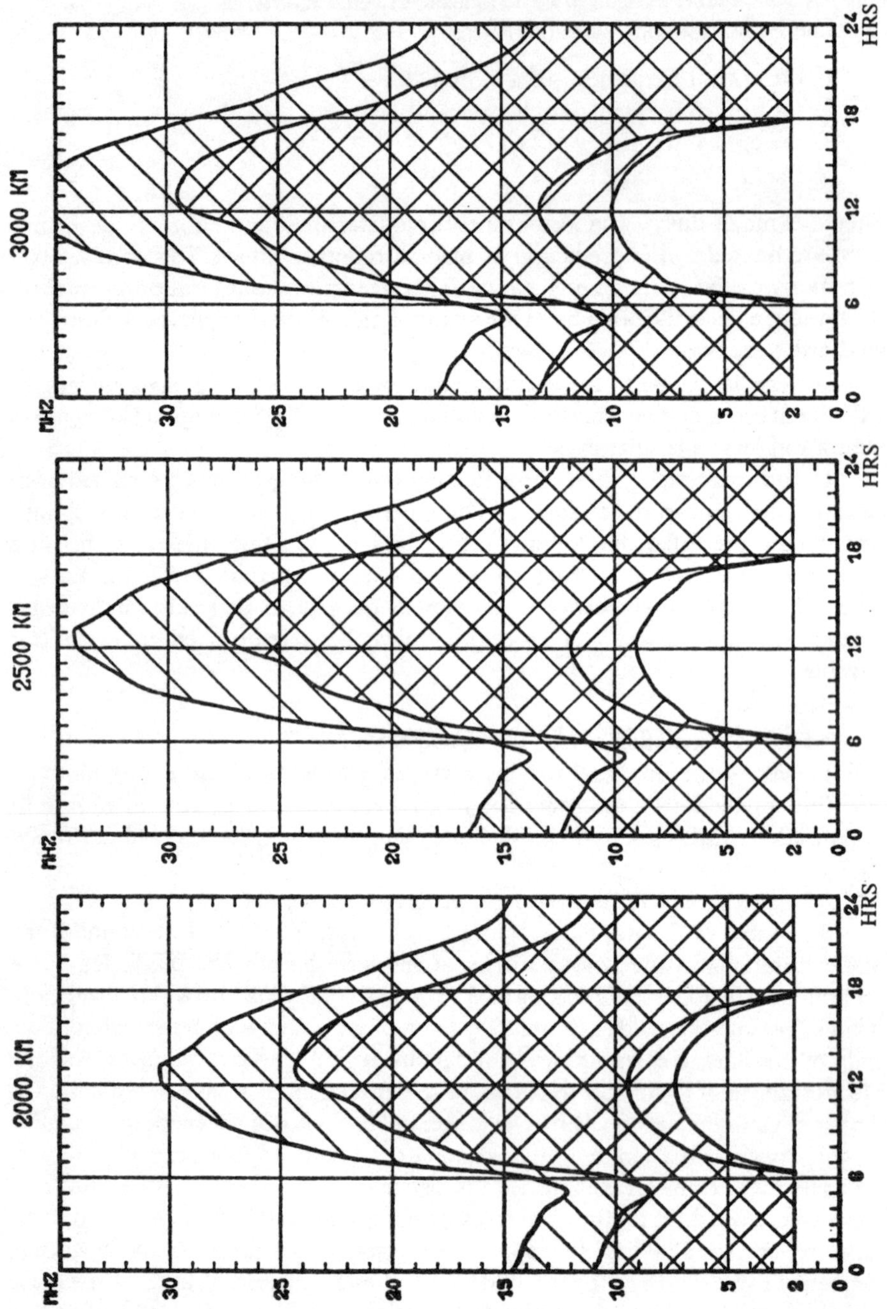

Fig. 9.13. Example of prediction by zones, March 1979.

The two lower curves give the LUF 90% and 30% probability limits, i.e. those exceeded by values of the true LUF, respectively some 3 days out of 30 and 21 days out of 30. If the calculated LUF is less than or equal to 2MHz, it is not shown on these graphs.

The zone in which the probability of establishing a link is greater than 90% (more than 27 days out of 30) is cross-hatched. The area where the probability of establishing a link is greater than 30% but less than 90% (more than 9 and less than 27 days out of 30) is single-hatched.

In general, the four curves do not cross. But if the solar activity index is low, the LUF curves may cross the MUF curves. If the curve giving the 90% probability LUF crosses the curve giving the 90% probability MUF, no frequency will provide the link with a probability greater than or equal to 90%. The graph will then only contain zones with single hatching and the maximum probability will lie between 90% and 30%. At times where the calculated value for the 30% LUF is greater than that for the 30% MUF, no curve appears.

The numbers shown on the upper curves are mean values, taken over 24 hours, of the starting angle for the radio link determining the MUF.

9.4.3.3.2. Point to point predictions

Certain particularly important paths may be the subject of specific predictions. These paths are generally very long, and predictions require a knowledge of the ionosphere over almost the entire globe. These point to point predictions are usually presented as graphs.

For links less than 10,500km long numeric tables give an estimate of the starting angles of the radio link giving 90% MUF and LUF. If the starting angle setting the MUF is less than 4°, another upper frequency limit is represented for the replacement link, giving a frequency limit less than the MUF, but with a starting angle greater than 4°. Users who do not have antennas capable of radiating at an angle less than 4° ought to consider this frequency limit as an 'operating MUF'.

For links longer than 10,500km no table of estimated starting angles is provided. These angles are always low and most often lower than 4°. If the length of the link is greater than 16,000km, a prediction is also drawn up for the link by the major arc.

Point to point predictions can also be presented in the form of a numeric table giving, for every hour UTC (universal time coordinated) and for all integer frequencies from 2 to 35MHz, the transmission loss for a given probability.

9.4.3.3.3. Point to zone predictions, also called fixed-mobile or mobile-fixed predictions

These predictions are intended for links in which one of the ends has a known and well-defined position, and the other end may be in an unknown variable place. This is the case, for example, of a link between a plane or a ship and its base.

These predictions are presented as follows: the surface of the Earth is divided into a grid, separated by 15° intervals in longitude and 7° 30′ in latitude, and for each of the 'rectangles' obtained in this way, the MUF and LUF for the month in question are given.

9.4.4. Short term ionospheric propagation predictions

This is the term used for predictions concerning only the next 24 hours or the next week. The essential aim of such predictions is to take account of temporary ionization anomalies due to perturbations in the ionosphere which cannot be considered for long term predictions more than a month in advance. The anomalies are either abnormally low or abnormally high ionization levels, where the latter may be accompanied by excessive absorption in the lower layers of the ionosphere. By way of example, Fig. 9.14 shows measured and predicted values of MUF 3000 for three consecutive days (short term predictions). For two of these three days the prediction was entirely correct. For the third, on the other hand, a transitory ionospheric anomaly produced a spectacular drop in MUF, making links using normal frequencies impossible. It is obviously useful to have at least one day's notice of such anomalies.

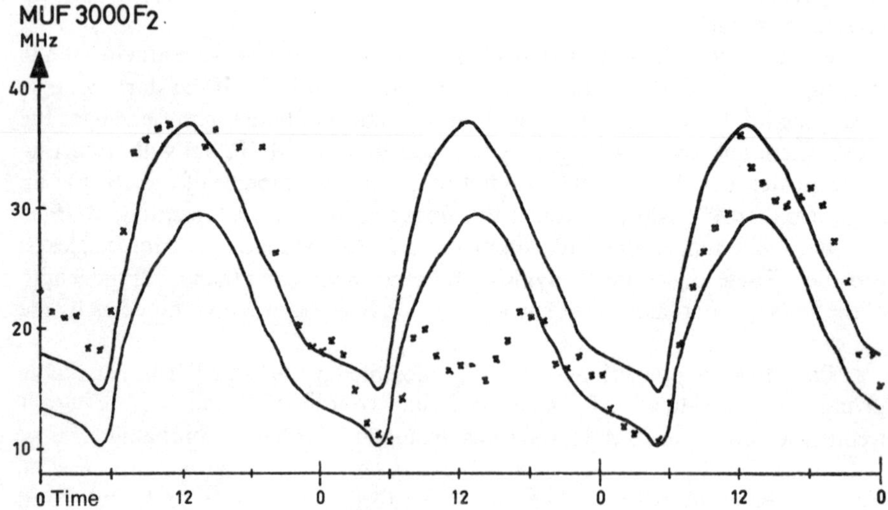

Fig. 9.14. Example of an ionization anomaly.
——— Predicted values
× × × Measured values

All ionospheric perturbations have the same cause, which is the appearance of a temporary perturbation in solar activity. Short term prediction techniques are therefore based on detecting and indentifying as quickly as possible all solar and geomagnetic phenomena likely to precede ionospheric perturbation, and evaluating the scale of such perturbations.

9.4.4.1. Solar and geomagnetic phenomena

We know that from time to time localized zones on the surface of the sun become the sites of temporary phenomena, in particular of eruptions accompanied by the ejection of particles and electromagnetic radiation, including X-rays and UV radiation. These eruptions are generally associated with a wholly visible phenomenon, the appearance of sunspots, which, consequently provide a visual indicator of solar activity.

X- and UV-rays reach the Earth 8 minutes after the start of this phenomenon. When they penetrate deeply into the atmosphere they produce heavy ionization in the relatively dense layers of the *D* region. The result is a very considerable increase in the number of collisions and therefore in absorption. This phenomenon is known as sudden ionospheric disturbance (SID), and affects all or part of the illuminated hemisphere and can last for about half an hour. Absorption may sometimes be sufficient to interupt all decametric wave links. On the other hand, the transmission of kilometric waves is slightly improved but with phase anomalies.

High energy particles, which arrive a few hours after the start of the phenomenon, may penetrate deeply into the ionosphere in regions close to the magnetic poles where the lines of the magnetic field are nearly vertical. They then produce intense ionization, leading to absorption in the polar caps which may last several days.

Lower energy particles produce the solar wind. Their increase at the time of a solar eruption may produce a modification and heating of the ionosphere, leading to the occurrence of magnetic storms which produce sudden variations in the electron density of the *F* region, lasting as long as several days.

9.4.4.2. Short term prediction techniques and presentation

Short term prediction techniques involve identifying the solar and geomagnetic events, amongst those that are constantly observed, that are precursors or ionospheric disturbance. This first of all requires rapid collection of a large quantity of data from various observatories scattered throughout the world. It is then necessary to analyse the data in order to identify and understand the geophysical phenomena as they emerge. The possible consequences for radiocommunication can then be worked out with reference to similar situations that appeared earlier.

Rapid exchange of data across the world is made by the international ursigram service which collects data, as it is analysed, from solar and ground observatories around the world. These data are transmitted by telex from the observatories to world centres which in turn distribute them, again by telex, to subscribers to the service. A large quantity of information can be exchanged in this way in the shortest possible time.

Information received by the short term prediction organizations can be divided into two categories:

1. results of observations of the principal solar parameters (description of sunspots observed by white and monochromatic light, chromospherical eruptions, measurements of radio flux from the sun at different wavelengths and radio bursts), as well as geomagnetic parameters (magnetic activity, deduced from magnetograms and micropulsations) and ionospheric parameters (deduced from vertical soundings, radio propagation indexes deduced from test links);
2. predictions for solar and geomagnetic parameters (radio flux, X-ray radiation probability, and magnetic activity indexes) made by specialists in these different fields, attached to the observatories.

Weekly prediction messages take the form of an uncoded text giving both a description of the state of the ionosphere and the geomagnetic field observed during the previous week, and a description of the likely changes in that state for the following week. These predictions are purely qualitative and concern the mean state of the ionosphere.

Daily prediction messages concern the following 24 hours. They take the form of a prediction for parameters MUF and LUF for six hour periods. Messages are coded in a simple way with reference to the geographical region prediction curves made available to each user. These messages are therefore supplied daily as corrections to the permanent region prediction.

9.4.5. Ionospheric propagation for frequencies less than 1500kHz

We have been almost exclusively concerned with frequencies greater than about 1500kHz, for which propagation takes place principally through the F region where variations are fairly irregular and require careful prediction. At lower frequencies propagation takes place chiefly in the lower ionospheric regions whose characteristics depend far more on the position of the sun than on its activity. Moreover, absorption plays a more important role.

9.4.5.1. Frequencies between 150 and 1500kHz

Propagation of these frequencies takes place chiefly in region E. Absorption while crossing region D is very considerable during the day but often much lower at night, and consequently ionospheric propagation of these frequencies is practically only possible at night. This is why, in particular, the range of sound radio broadcasting transmitters operating in this frequency band is very great at night, though by day it is reduced to the area that can be reached by spherical diffraction (ground wave).

Semi-empirical formulas have been used to calculate the field created by a transmitter. The CCIR, in particular, has proposed the following formula:

$$E = 80.2 - 10 \log d - 0.001\,8\, f^{0.26}\, d + \Delta H + \Delta I, \tag{9.64}$$

where:

E is the median field strength at midnight, expressed in decibels with respect to $1\mu V/m$, created by a transmitter whose field at 1km over an infinitely conducting ground would be $3.10^5 \mu V/m$;
d is the distance in kilometres;
f is the frequency in kilohertz between 150 and 1500kHz;
ΔH is a correction factor to take account of the local time half way along the path. The field strength is reduced by around 10dB at 19.00 hours and 5.00 hours, with respect to its value at midnight;
ΔI is a correction factor to take account of magnetic inclination half way along the path. This correction is large (and negative) only for distances greater than 1500km and inclinations greater than 62°.

Furthermore, a certain distribution of values of the field strength about the median value calculated in this way must be taken into account. Variations in solar activity only have a small influence, not greater than a few decibels.

9.4.5.2. Frequencies below 150KHZ

At lower frequencies we find the following changes.

1. The diffraction wave (ground wave) extends further, for example up to 1000 or 2000km (see Chapter 7).
2. The ionospheric wave penetrates less deeply into the ionosphere, and we can assume for all practical purposes that it is reflected from the 'lower surface of the ionosphere' at 70km during the day and 90km at night.
3. The influence of variations in solar activity reduces.
4. For all these reasons the propagation of such waves shows great stability in amplitude and in phase, which explains their use for radio navigation systems over very great distances (Omega system) and for time signal transmissions. Vertical polarization is used almost exclusively.

The propagation of these waves can be studied in two different ways.

9.4.5.2.1. Theory of propagation by hops
This is in fact ray theory from optics. We study possible trajectories with reflections from the lower surface of the ionosphere and from the ground, taking into account the ionospheric reflection coefficient, the convergence effect due to the curvature of the reflecting surface and the nature of the ground. For short distances, we add the diffracted wave effect. This method is used above all for kilometric waves and may be used for myriametric waves at distances less than about 1000km.

9.4.5.2.2. Mode theory
For myriametric and longer waves, and for distances greater than 1000km, it is best to consider the Earth-ionosphere space as a large wave duct capable of transmitting several modes, whose number falls with the frequency. This method is fairly accurate but leads to complicated calculations, particularly if the lack of homogeneity of the wave duct and the existence of the Earth's magnetic field are considered.

9.4.6. Artificial modifications to the ionosphere

Since 1970, high power transmitters have been used, installed on the Earth's surface and operating in the decametric wave band, with the intention of modifying the electronic density and the speed of electrons in region *F* of the ionosphere at around 300km. To do this, a frequency is chosen close to the critical frequency at the altitude where the ionosphere is to be modified, in order to obtain maximum absorption of the transmitted power, and consequently the greatest possible degree of modification to the ionosphere at that point. In the experiments carried out, the power flux intensity of the wave in the region to be modified was of the order of 1 to 50μW/m^2. Given the collision frequency, the time constant for the loss of energy is of the order of 10 seconds in region *F*, which leads to a significant heating of the electrons (i.e. electron speeds are increased).

For radio signals crossing the disturbance region, whether on reflection links or Earth/space links, fades increases in both depth and number. Moreover, given that the ionization fluctuations created tend to be oriented along the field lines of the Earth's magnetic field, there is a reflecting effect comparable with that of a radar obstacle of large cross-section (10^5 to 10^9m^2).

Reflecting and scatter properties increased in this way have been used for various transmissions between terminals on the ground separated by thousands of kilometres and at frequencies running from the decametric to decimetric wave band. Given the influence of the Earth's magnetic field, the effect is highly directed, and the zone containing the points between which a link is possible may reach 4000km in the east-west direction, while it is only a few hundred kilometres across in the north-south direction. A few experiments and modification in the *E* region at around 110km show a much smaller directional effect.

Obviously, such modifications to the ionosphere, though they allow links which would be impossible under normal conditions, are also a source of interference and are to be used with care.

9.4.7. Ionospheric sounders

As well as obvious applications in telecommunications, wave reflection provides a convenient method for studying the ionosphere using ionospheric sounders. An ionospheric sounder or 'ionosounder' works like a radar. Pulses, lasting around a tenth of a millisecond at a power of a few kilowatts, are transmitted upwards. The carrier frequency for these pulses varies cyclically between 0.5 and 20MHz, with a period of the order of a minute.

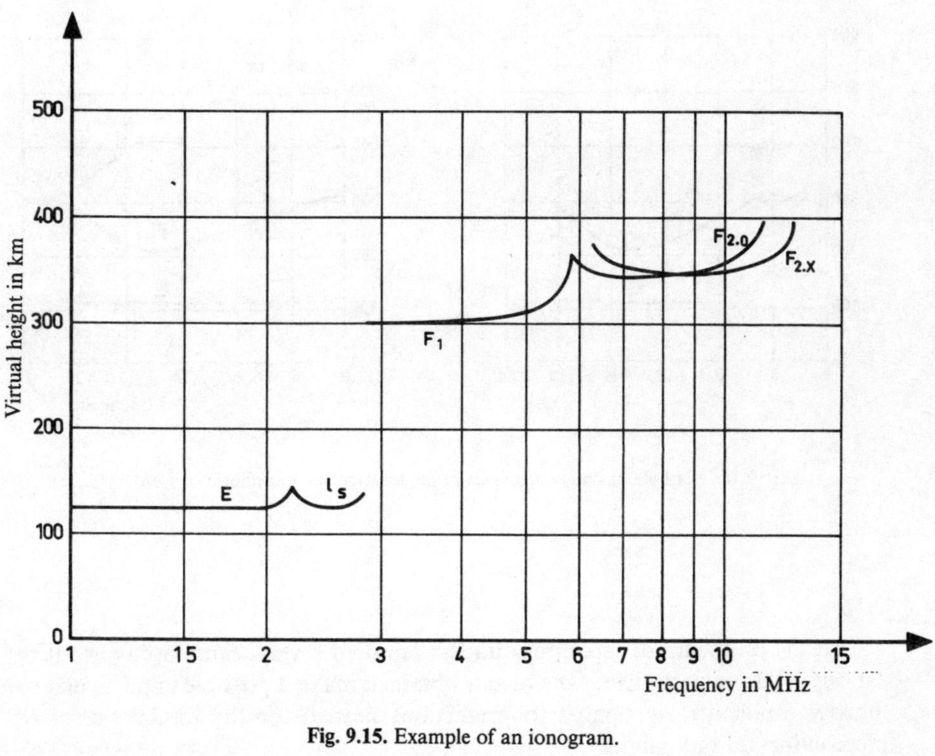

Fig. 9.15. Example of an ionogram.

After reflection from the ionosphere these pulses are received using a receiver tuned to a frequency that follows the transmitter frequency. The result is usually presented on a cathode ray tube with horizontal scanning proportional to the frequency (or, more generally, to the logarithm of the frequency). Vertical scanning is proportional to the time the pulse takes to complete the return journey. It therefore gives the virtual height which can be taken as equal to the real height if not too close to a critical frequency. Numerical processing and recording are associated with the visual presentation.

Ionograms may vary enormously with the time of day, with solar activity and with latitude. A simple example is given in Fig 9.15 where 'layers' corresponding to E and E_s, F_1, F_2 ordinary mode and F_2 extraordinary mode are shown.

Beyond 3MHz layer F_1 is no longer visible as it is occulted by layer E or E_s.

Obviously, the ionospheric image given by ionograms is very different from that provided by a direct measurement of ionization carried out by rocket, and this explains why the original concept of a layer was replaced by that of an ionized region. Nevertheless, the concept of a layer is still useful for telecommunications. Figure 9.16 gives an idea of the mean variations in the height of these layers as deduced from ionograms.

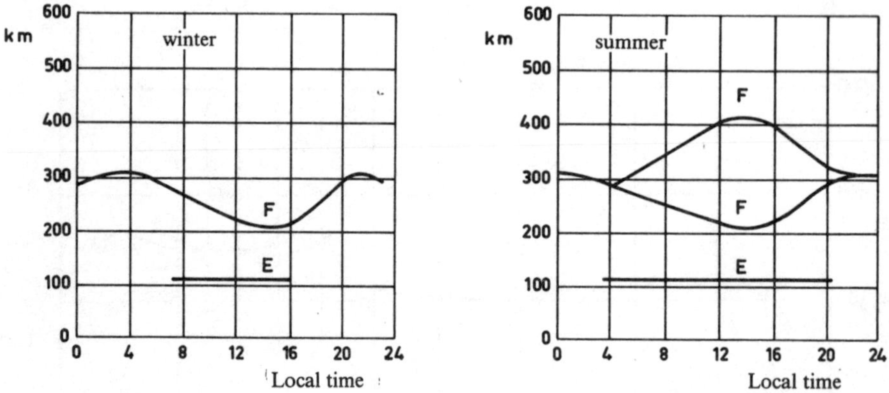

Fig. 9.16. Example of daily variations in the altitude of ionospheric regions.

At the moment, ionospheric sounders are used several times a day at many stations in various countries. The results obtained make it possible in particular to improve ionospheric propagation predictions needed for the establishment of intercontinental radio links.

Since the advent of space techniques, it has been possible to carry ionospheric sounders on satellites, allowing the upper part of region F to be studied. This has revealed the ordinary and extraordinary components as well as the characteristic frequencies of the ionosphere (plasma frequency and gyrofrequency) at the altitude at which the satellite travels.

Chapter 10

Frequency sharing and frequency re-use, coverage area, interference

10.1. Introduction

The large number of radio users means that very detailed regulations have to be applied to their use, and, in particular, to the allocation of frequency bands to different radio services. That allocation must take into account propagation phenomena and the technology available.

However, as shown in Appendix 3, most frequency bands are shared by different services which have to be able to operate without interference, even under propagation conditions that favour interference. Moreover, within a given service, we may be obliged to reuse the same frequencies several times while avoiding such interference. We study this problem from the point of view of wave propagation.

To establish a continuously operating radio link, we have to make use of propagation mechanisms whose characteristics are stable except, perhaps, during a small proportion of the time. This is possible either using a fixed frequency, or, if absolutely necessary, by changing frequency on a cyclical basis, as is the case for ionospheric links.

We have also seen that there are propagation mechanisms which operate for very short times, but which, when they occur, can give very high signal levels. These intermittent phenomena cannot generally be used to establish links, but they can, on the other hand, lead to interference over very great distances. See Chapter 11 for definitions.

When planning a link or an entire telecommunication service, whether at the national or international level, all the propagation mechanisms described in the preceding chapters must be taken into account. This chapter is particularly concerned with practical considerations and general caluclation methods, some of which have been adopted internationally by the CCIR, in order to resolve problems concerning coverage areas and interference for various types of service and radio links.

10.2. Factors to take into account when sharing or reusing frequencies

Radio Regulations simply allocate frequency bands to a given service or services. It is up to the authorities concerned to determine, in conjunction with the CCIR, the practical rules to allow the greatest possible number of users and the lowest possible level of interference. A certain number of factors have to be taken into account to solve this problem.

10.2.1. Radio channels

The concept of a radio channel has been introduced for frequency bands above 30MHz, to assist in planning television, mobile services and radio relay systems. The frequency band allocated by Radio Regulations is subdivided into a number of consecutive radio channels for which other characteristics such as polarization may also be fixed. This makes planning simpler and reduces the number of possible causes of interference.

10.2.2. Polarization

We know that to any radio wave, there corresponds another wave with orthogonal polarization, allowing the two waves to propagate over the same path and be received independently of each other. Common examples are a vertically polarized wave associated with a horizontally polarized wave, or two waves with circular polarization in opposite senses of rotation.

The use of orthogonal polarization therefore makes it possible, in theory, to double the transmission capacity of a channel or to discriminate between two links, at least insofar as propagation phenomena do not cause too great a level of transpolarization (see Sections 4.3 and 5.4).

At this point it is worth giving the definitions of certain terms used in polarization problems.

Transpolarization is the appearance (may be due to equipment) of a component with polarization orthogonal to the expected polarization irrespective of the cause of the phenomenon. The word cross polarization is commonly used to describe this concept. However, this word has other meanings, which can lead to ambiguity. The use of transpolarization (similar to transmodulation) instead avoids such ambiguities.

Polarization discrimination is the ratio at a receiving point between received power in the expected polarization and received power in a polarization orthogonal to it from a wave transmitted with a given polarization.

Transpolarization isolation is the ratio at a receiving point between received power from a wave in the polarization expected for that wave and the received power from another wave transmitted from the same point as the first and with the same power but with orthogonal polarization.

We can therefore see that polarization discrimination is defined on the basis of a singe transmitted wave, while polarization isolation involves transmission of two waves, and therefore corresponds directly to interference with one signal by another with orthogonal polarization. In practice, the values of these quantities may not be very different.

Depolarization is a phenomenon in which all or part of the power of a radio wave transmitted with a given polarization no longer has a defined polarization after propagation. Note that the word is sometimes incorrectly used to mean a change in polarization, leaving the wave polarized nonetheless.

If polarization is taken into account in neighbouring frequency channels, additional concepts have to be defined. In particular, radio channel arrangement with alternate polarization, in which transmission in two adjacent channels have orthogonal polarizations, is often used.

10.2.3. Antenna directivity

Antenna directivity is an efficient protection against interfering transmission from a direction that is different to the signal direction. In particular, as radio relay links became increasingly dense, the required characteristics for antennas concerning the side lobe level became increasingly demanding.

10.2.4. Distance between stations using the same frequency

Increased distance between an interfering station and a station likely to suffer from interference is the most effective way to reduce the chance of interference. However, the minimum distance at which a given frequency may be reused has to be estimated very carefully, since adopting too great a distance leads to the use of a higher number of frequencies for a given service, with the consequence that less efficient use is made of the frequency spectrum.

To determine the chances and level of interference, the critical factor is therefore the received levels which are only exceeded for small time percentages. To determine these levels, we must take account of propagation phenomena leading to small losses at the frequency and for the path considered.

10.3. Classification and characteristics of radio communications

Radio communications can be divided into three categories called point to point, point to zone and point to multipoint communication, respectively.

In point to point communication, corresponding in particular to fixed service links, the characteristics and locations of the ends of the path are entirely known. Fixed service links operating above 30MHz notably include radio relay links and satellite links. As this type of link involves relays, frequency reuse is standard, which therefore leads to the possibility of interference.

In point to zone communication, where the two major examples are radio broadcasting (sound and vision) and mobile services (terrestrial, aeronautical or maritime), only one of the ends (transmitter or base station) has a fixed location and known characteristics. The other end could be anywhere within a certain zone, covered by the fixed station. In general, one fixed station serves several stations, either unilaterally (radio broadcasting) or bilaterally (mobile service). Unlike the previous case, it is impossible to protect against interference by means of a shift in angle between the direction of the desired signal and the interfering signal. Consequently, study of the coverage area of a station and that of the zone in which it can cause or undergo interference must be taken together, for example by using the same set of propagation curves. A distinction should be made between coverage area whose definition is purely technical and service area whose definition also implies administrative procedures and regulations. The two areas may be different. For example, for a coastal station of the maritime mobile service, the service area is purely maritime though the coverage area may include an overland part.

There is another type of communication which appears both in radio broadcasting and fixed services, which is called point to multipoint communication. This is the case, for example, of a television transmitter serving a certain number of relay transmitters whose locations are well known. It is also the case of satellite links using time division multiple access (TDMA) and with the use of digital modulation the system can even be extended to terrestrial links. From the point of view of propagation, point to multipoint communication can be treated in the same way as point to point communication, since the characteristics of the path are known.

The distinction between point to point and point to zone communication is of value principally for terrestrial links because for these links propagation conditions are very different, depending on the distance between the base station and the point considered in the zone. On the other hand, for Earth-satellite links, the distance and propagation conditions are practically the same for all points in the zone. And the distinction between these two types of communication becomes less clear, at least as far as propagation is concerned.

From the point of view of propagation, terrestrial links associated with point to point communication and those associated with point to zone communication must be treated differently for the following reasons.

1. Services using point to point communications must transmit a signal from a given point to another given point, possibly using intermediate relays, and transmission towards any other point is regarded as undesirable. On the other hand, the aim of services using point to zone communications is to obtain complete coverage of a region, dividing it into contiguous zones, sometimes called 'cells', such that the radio station allocated to each zone provides adequate coverage of that zone and does not cause interference with others. In general, neighbouring cells use different radio channels and possibly different polarizations.
2. In point to zone communications, one of the ends of each link is at an

arbitrary location not specified in advance, and some of these locations are more favourable than others at the same distance. We must therefore introduce a further concept which is the percentage of locations for which given transmission characteristics apply. Statistics obviously must only consider locations for which the service in question is possible. For example, for terrestrial mobile services and radio broadcasting in vehicles, only locations in roads or streets should be taken into account. Most quality objectives are based on 50% of locations.

3. For point to point communications working at frequencies above a few hundred Megahertz, the time percentage used to define quality is always very high, of the order of 99.9% or greater, and consequently interference must not appear more than about 0.1% of the time. On the other hand, for point to zone communications the time percentages considered are much closer to the middle of the distribution (10% to 90%) and are often limited to the median value (50%).

   This difference has a major practical consequence. For point to point communications, we must use curves or formulas which differ depending on whether we are calculating low levels allowing adequate reception or high levels which can cause interference, since the propagation phenomena are different in the two cases. On the other hand, for point to zone communications, as the percentages considered are around 50%, the same systems of curves can be used to determine the field in the service area and interference at greater distances in neighbouring zones.

4. The most important propagation phenomena are not the same in point to point communications and in point to zone communications.

   Point to point communications operating at frequencies lower than about 30MHz make use of, in particular, propagation by ionospheric reflection. At higher frequencies, they either use line-of-sight propagation for relatively short paths, or propagation by tropospheric scatter for longer paths.

   Point to zone communications are seldom line-of-sight. They mainly use propagation by diffraction, either around the sphere of the Earth for relatively low frequencies (hectometric and longer wavelengths), or around obstacles of different sizes and shapes for higher frequencies. Note, incidentally, that major frequency bands allocated to point to zone communications (radio broadcasting and mobile services) are in that part of the radio spectrum where diffraction phenomena play a major role (metric and decimetric waves).

10.4. Point to point communication: interference

When planning point to point links, a study of the received level and the quality of transmission has to be carried out for each case. Chapters 6, 8 and 9 describe the procedures to use. However, it remains to study problems of interference between these links themselves, as the same frequency is shared between several services, or is reused several times by the same service.

The problem of interference on line of sight links has increased in importance since the appearance of telecommunication satellites, which often use the same frequency bands as terrestrial links. At the international level (that of the CCIR) relatively complex procedures have been established to check that a planned new station will not cause or undergo more than a predetermined maximum level of interference. As these procedures are complex a new concept has been introduced for the sake of simplicity: the coordination area is defined as the area around the station outside which the probability of interference is regarded as negligible. We therefore have only to check the possibility of interference within the coordination area.

The coordination area is determined using formulas similar to those allowing interference levels to be calculated, but choosing pessimistic conditions. For example, it is assumed that the ground is relatively flat and that antennas are directed towards each other.

The main propagation mechanisms to be taken into account to determine the possibility of interference are:

1. Super-refraction and propagation by tropospheric ducts (ducted propagation).
2. Scatter by hydrometeors.
3. Reflection or scatter due to particular ionospheric phenomena.

The first two mechanisms only affect frequencies higher than 100MHz for the first, then about 1GHz for the second, while the third phenomenon only appears between about 30 and 100MHz.

Finally, there is the phenomenon of tropospheric scatter (see Chapter 8) which always occurs, and occasionally gives a sufficient field to create interference at great distances, particularly in the 0.5 to 2GHz range. This phenomenon needs to be taken into consideration particularly for estimating interference during relatively high time percentages, e.g. greater than 5%.

On a link a few hundred kilometres long, operating at a frequency greater than a few hundred Megahertz, the distribution of received levels during periods where they are high typically takes the shape shown in Fig. 10.1, where the ordinate is a linear scale in decibels and the abscissa is a gaussian scale. The distribution curve goes through two points of change of gradient. Point A corresponds to the passage from the tropospheric scatter field (to the right) to the super-refraction field (to the left). Its abscissa is in general at around 0.1 to 0.5% over land and 5 to 10% over sea. There may even be higher time percentage values above hot seas. Point B corresponds to passage from the super-refraction field to a free-space field. There may be fields higher than the free-space field, but in this zone the distribution curve has a very low gradient, while in zone AB corresponding to super-refraction the gradient is far higher.

10.4.1. Interference due to super-refraction and ducting

We saw in Chapter 4 that if the vertical gradient of the refractive index is below –40N/km, super-refraction takes place, and when it is below –157N/km, ducting can occur.

However, for ducting actually to occur, two further conditions must be fulfilled:

a. The height of the duct must be sufficient for the frequency considered;
b. The trajectory must include a certain number of successive return trips between the upper and lower boundaries of the duct.

In the case of a surface duct, the first condition is almost always satisfied for decimetric or shorter waves. The second condition is always satisfied over sea, but hardly ever satisfied over land because of the low value of the reflection coefficient (except perhaps for paths over very large flat humid regions). This can be appreciated by the fact that there is generally no big problem due to reflection on radio relay paths entirely over land, while when they pass over sea, reflections always cause major difficulties.

To summarize, for decimetric or shorter waves, the phenomenon of ducting practically only occurs over sea or over very flat wet areas of land. For most terrestrial paths susceptible to super-refraction conditions, the radio trajectory is limited to a single 'arc'. Chapter 4 gives certain typical values for these arc lengths.

10.4.1.1. Maritime links

As reflection from the sea is virtually perfect and the atmosphere is homogeneous over a very large area, each trajectory may include many successive reflections. The received level diminishes slowly as distance increases and there may be very high levels over very long distances, if the duct is long enough. Over sea the influence of propagation therefore depends above all on the extent of the zone over which ducting takes place at any given time. In general, the zone covers a few hundred kilometres but it may under exceptional conditions reach several thousand kilometres, for example in the trade wind region for the east-west direction. It has also been found that such a zone can extend over much of the Mediterranean or the Persian Gulf. The maximum extent of the ducting zone at any given moment depends little on the time percentage considered and it is generally accepted that it increases slightly as the time percentage considered is reduced.

Although most available measurements have been obtained for television transmission at frequencies below 800MHz, it has been possible to draw up a method of prediction taking account of the following simplifying factors:

1. As we are only interested in fairly long distances, generally greater than 100km, we can assume that in Equation (4.50) the exponential term is preponderant with respect to the $1/d$ term and that, consequently, attenuation in decibels is virtually a linear function of distance.

2. To take account of the angle of elevation of antenna beams, we replace geographical distance by an 'angular' distance, deduced from the angle between the beams of the antennas, as we did for propagation by tropospheric scatter.
3. The effect of frequency is deduced from spherical diffraction theory, as the theory applies under super-refraction conditions, and it may be generally accepted that it also applies to cases of ducted propagation.
4. All seas and oceans can be divided into two categories depending on their latitude, 'cold seas' at latitudes greater than 30° and 'warm seas' at latitudes less than 30°. However, seas like the Mediterranean and the Black Sea must be classified as warm seas despite their high latitudes. Moreover, it is usually accepted that the probability of the field exceeding a given value is three times greater over warm seas than over cold seas.

The following equations have been drawn up on these assumptions, and the coefficients have been determined experimentally. They correspond to the AB section of the curve in Fig. 10.1.

$$A = U + 20 \log F + d_a V F^{1/3} + d(\gamma_0 + \gamma_v) \tag{10.1}$$

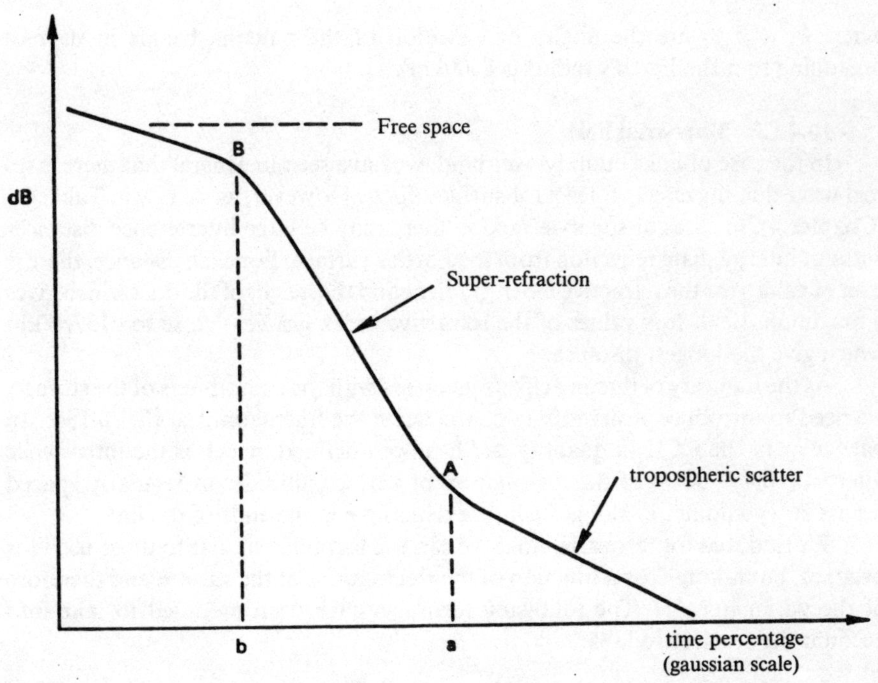

Fig. 10.1. Typical form of distribution of received levels for small time percentages.

where:

A is the attenuation in decibels between isotropic antennas;
F is the frequency in GHz;
d is the geographical distance in kilometres;
d_a is the angular distance.

γ_0 and γ_v are specific attenuation in dB/km due to oxygen and water vapour, respectively (see Chapter 5); U and V are functions of probability p, given by the following relations:

cold seas:

$$U = 130 + 10 \log p + 1.5 (\log p)^2 \tag{10.2}$$

$$V = 0.075 + 0.01 \log p \tag{10.3}$$

warm seas:

$$U = 130 + 10 \log p/3 + 1.5 (\log p/3)^2 \tag{10.4}$$

$$V = 0.075 + 0.01 \log p/3 \tag{10.5}$$

The relation between d and d_a is as follows:

$$d_a = d + 148 (\theta_1 + \theta_2) \tag{10.6}$$

where θ_1 and θ_2 are the angles of elevation of the antenna beams in degrees (assuming that the Earth's radius is 8500km).

10.4.1.2. Terrestrial links

In the case of links entirely over land, we have seen in general that there is no real wave ducting effect, at least for surface ducts. However, as we saw in Table 4.1 (Chapter 4), in cases of super-refraction there may be large interference distances without intermediate reflection from the Earth's surface. For each distance, there is a set of values for the refractive index gradient and the height of the duct which gives a maximum field. It is values of the refractive index gradient close to –157N/km which give the longest distance.

As the intensity of ducting effects increases with the smoothness of the surface, we need to introduce a quantity to characterize the fluctuations at the surface. In particular in the CCIR a quantity ΔH has been defined, which is the inter-decile interval (10% – 90%) of the distribution of surface altitudes at regularly spaced points every kilometre, along the link, excluding e.g. the ends of the link.

We find that for terrestrial links we can use formulas similar to those used for over sea, but taking U as a function of the fluctuations at the surface and therefore of the parameter ΔH. The following formulas have been modified to take into account the results available:

$$U = 170 - 40/(1 + \Delta H/50) + 20 \log p + (1.5 \log p)^2 \tag{10.7}$$

$$V = 0.11 + 0.01 \log p \tag{10.8}$$

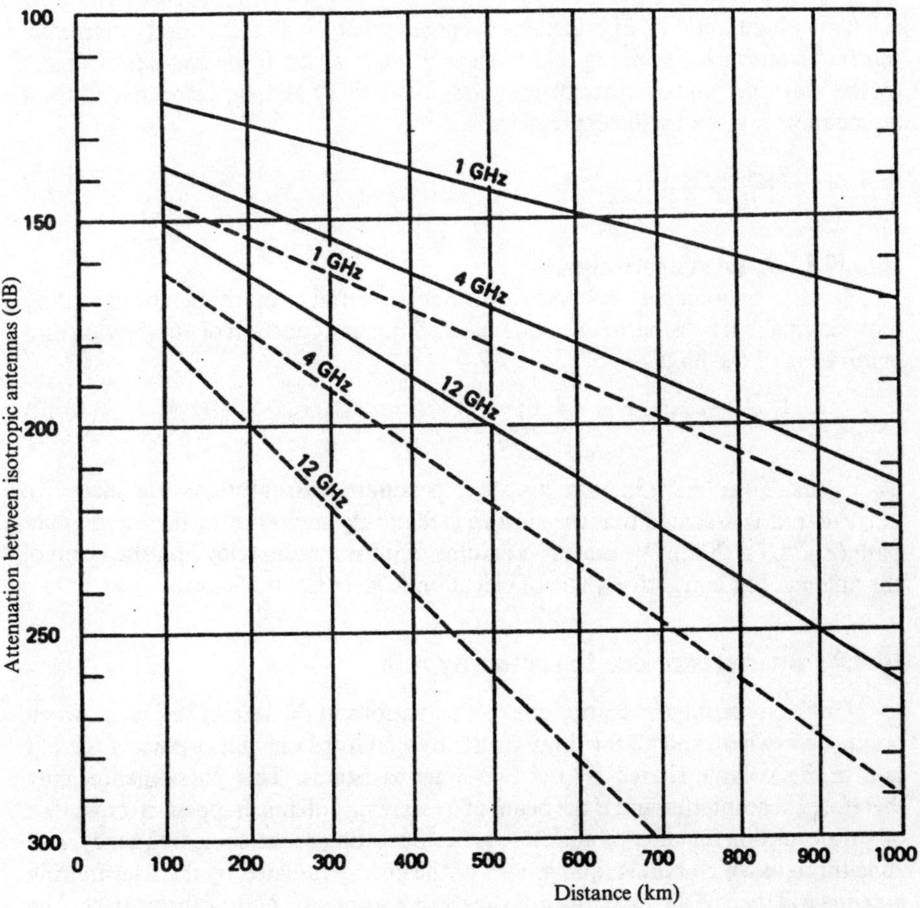

Fig. 10.2. Attenuation not exceeded for 0.01% of the time (interference estimates).
——— maritime links (cold seas).
– – – terrestrial links (ΔH =50m).

Some examples of these calculations are given in Fig. 10.2 (for 0.01% of the time).

10.4.1.3. Links over land and sea

These links often raise problems. In general, this type of link is made of a maritime section to which a land section, including the transmission and reception stations, is added at each end. If the land section is very smooth and at low altitude, ducting conditions can extend into the hinterland. It is also possible that the refractive index gradient above land is sufficient at certain times for the sea to be in line of sight of the ends of the link. Then the link behaves as if it was entirely maritime. In many other cases, a mixed link behaves more like an entirely terrestrial link, particularly if the land part has an irregular surface.

The following approach allows us to estimate attenuation on a mixed link. We calculate attenuation A_1 obtained by supposing that the link is entirely maritime, then attenuation A_2 assuming that it is entirely terrestrial. If the angular distances of the maritime and terrestrial links are equal to d_1 and d_2, respectively, total attenuation is given by the expression:

$$A = \frac{A_1 d_1 + A_2 d_2}{d_1 + d_2} \tag{10.9}$$

10.4.1.4. Interference distance

In cases of homogeneous paths, whether terrestrial or maritime, the preceding formulas make it possible to calculate the distance as a function of attenuation and parameters of the link:

$$d = \frac{A - U - 20 \log F - 148\,(\theta_1 + \theta_2)\, V F^{1/3}}{V F^{1/3} + \gamma_0 + \gamma_v} \tag{10.10}$$

To calculate the coordination distance, pessimistic assumptions are made. In particular, it is assumed that the surface is relatively smooth over the interference path (e.g. $\Delta H = 20$m). We may also assume that for a radio relay link the beam of the antenna has a negative angle of elevation (e.g. $148\,\theta_1 = -50$km).

10.4.2. Interference due to scatter by rain

Unlike tropospheric scatter due to fluctuations in the refractive index, which occurs everywhere and all the time, scatter by raindrops only takes place if there is rain in the volume shared by the two antenna beams. This phenomenon may therefore cause interference if the beam of a receiving antenna happens to cross that of a transmitting antenna of another service operating at the same frequency, in a zone through which rain is falling. Part of the energy radiated by the transmitting antenna will then be directed towards the receiving antenna of the other service. The result depends very much on the ratio between the common volume of the antenna beams and the volume occupied by the scatter particles.

For tropospheric scatter (Chapter 8) the volume shared by the antenna beams is always entirely within the scatter volume. For scatter by raindrops between a radio relay antenna and a satellite telecommunication antenna, the beam of the latter passes entirely within the scatter zone, which it crosses over a length d, while, in general, the other antenna has a far larger beam which covers the entire scatter zone (Fig. 10.3). This is the only case we consider.

If we write V for the scatter volume, G_1 and G_2 for the antenna gains, r_1 and r_2 for the distances to the centre of the scatter volume (Fig. 10.4), σ_D for the total scatter area per unit volume, A_i and A_e for coefficients less than 1 representing losses due to the rainfall within and outside the scatter volume, respectively, the ratio of received power to transmitted power is given by:

$$\frac{P_r}{P_t} = G_1 \cdot \frac{V.\sigma_D}{4\,\pi r_1^2} \cdot \left(\frac{\lambda}{4\,\pi r_2}\right)^2 \cdot G_2\; A_i\, A_e. \tag{10.11}$$

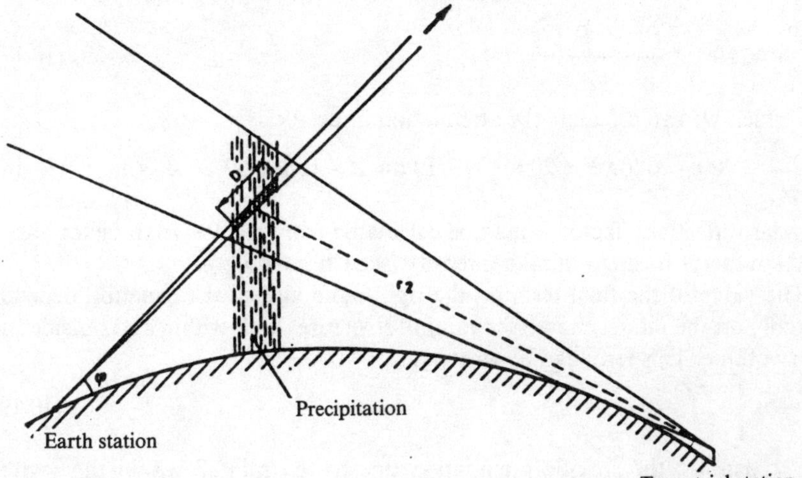

Fig. 10.3. Interference due to precipitation between a terrestrial station and an Earth station. ▦ Scatter zone.

From Equations (5.41) and (5.43):

$$\sigma_t = 0.93 \frac{\pi^5}{\lambda^4} \cdot Z . \tag{10.12}$$

Where Z is related to rainfall intensity R by the empirical Equation (5.45).

Moreover, if we write Ω for the angle of the beam from the very high gain antenna, we obtain the following approximate relation:

$$V = \Omega r_1^2 D . \tag{10.13}$$

But Ω is related to the gain G_1 of the antenna by the approximate relation:

$$G_1 = \frac{8}{\Omega} . \tag{10.14}$$

Consequently:

$$G_1 V = 8 r_1^2 D \tag{10.15}$$

and:

$$\frac{P_r}{P_t} = \frac{0.93 \pi^2}{8} \cdot \frac{D.Z.G_2}{\lambda^2 r_2^2} . \tag{10.16}$$

If we express D and r_2 in km, Z in mm^6/m^3 and frequency in GHz:

$$\frac{P_r}{P_t} \simeq 10^{-20}\frac{F^2.Z.D.G_2}{r_2^2}. \tag{10.17}$$

From which we can calculate the attenuation in decibels:

$$10\log\frac{P_r}{P_t} = 200 - 20\log F + 20\log r_2 - 10\log Z - 10\log(D\ G_2\ A_i\ A_e) \tag{10.18}$$

The radar reflectivity factor Z may be calculated from the Rayleigh factor and a correction factor for high rainfall intensity (see Chapter 5).

The value of the final term on the right-hand side of the equation depends essentially on the model chosen for rainfall structure, both within and outside the scatter volume. This term can be written as:

$$-\log\left[A_e \int_0^D G_2\, e^{-\gamma_i l}\ dl\right] \tag{10.19}$$

where γ_i denotes the specific attenuation due to the rainfall within the scatter volume (i.e. see Chapter 5):

$$\gamma_e = K\,R^{\alpha} \tag{10.20}$$

If the 0^0 isotherm passes through the scatter zone, we have to calculate attenuation in each part of the volume separately.

Calculation of attenuation A_e within the scatter volume depends on the extent chosen for the rain zone and on the horizontal profile of rainfall intensity.

Note that the choice of a model of spatial structure of rainfall is more critical when it comes to estimating propagation by scatter than to estimate attenuation on a given link. This is because the effect of all the raindrops on attenuation is cumulative over the entire link and errors within the model can compensate for each other. On the other hand, when it comes to scatter, each particle in the scatter volume has two opposite effects, on the one hand scatter which produces the field strength and on the other hand losses which reduce it. Moreover, the field strength is reduced by losses outside the common volume. The result is a differential effect, and is therefore extremely sensitive to the spatial profile of the rainfall adopted as a model for calculation.

We must also take account of the random time characteristics of the rainfall. The coefficient of coupling by scatter and total loss over the link are two random variables that are not entirely correlated, since much of the loss occurs outside the coupling zone. We must therefore consider the combined probability of coupling and loss.

The CCIR has proposed (Report 569) a prediction method based on a rainfall structure such that the scatter volume is always within the maximum precipitation intensity zone, with intensity decreasing exponentially moving away from the zone. Moreover, loss outside the scatter volume is taken into account by an approach based on equi-probability rather than combined probability. The result obtained is probably an overestimate of scatter.

To calculate the coordination distance the method is the same in general, but the formulas have been simplified in a way that gives an increased estimate of interference. Moreover, the various rainfall rate distribution that appears around the world has been brought together into five standardized models. Sets of curves are provided by the CCIR.

10.5. Point to zone communication: space service area, interference

Prediction of propagation effects is generally based on curves drawn up by mainly empirical methods. The CCIR has adopted sets of curves established first of all for radio broadcasting and then for the mobile land service and the mobile aeronautical service. The curves give either the field strength (radio broadcasting) or the basic transmission loss (mobile service) as a function of distance, for certain antenna heights and for various time percentages or locations. To reduce the number of curves the heights selected are only those considered representative of real situations:

1. 10m for a television antenna;
2. 1.5m and 3m for a land-based vehicle antenna;
3. 15m for a base station antenna in a mobile aeronautical service;
4. 1524m, 7620m, 18,300m (5000 feet, 25,000 feet and 60,000 feet) for a flying station in a mobile aeronautical service.

10.5.1. Radio broadcasting

For radio broadcasting of hectometric or longer waves the planning of networks is based on the diffraction curves (ground wave) given in Chapter 7 (Figs 7.4 to 7.7). For decametric waves we must also take account of ionospheric propagation (Chapter 9).

For metric and decimetric waves planning is based on diffraction and tropospheric scatter. By way of example, Fig. 10.4 shows a set of curves (drawn from a CCIR report) used to plan radio broadcasting services (television) between 450 and 1000MHz. To allow comparison we have used dotted lines on the same figure to show values calculated using curves from Chapter 7 for diffraction and those in Chapter 8 for tropospheric scatter. It is clear that for short distances the predicted field strength is considerably smaller than the field strength due to diffraction above a smooth Earth. On the other hand, at great distances, where the field is due to tropospheric scatter the difference is much smaller.

In general, sets of curves give values corresponding to 'average' conditions, and calculations are completed using empirical formulas or correction tables to take account of different circumstances. This applies in particular to the percentage of locations considered, the irregularity of the surface and the density of buildings in the area studied.

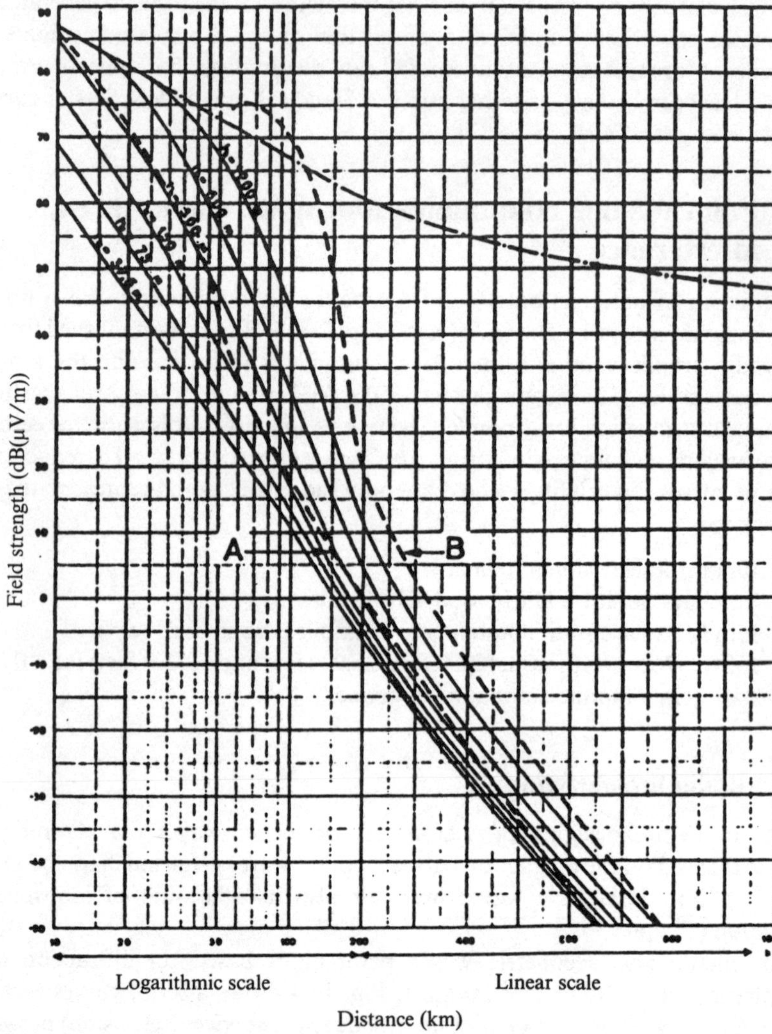

Fig. 10.4. Field strength (dB(μV/m)) for 1kW effective radiated power (CCIR) 450 to 1000MHz (band IV and V). Land 50% of time.
50% of locations. h_2 = 10m. Δh = 50m.

— · — · —— free space field strength
— — — — median field strength

A $\begin{cases} h_1 = 37.5\text{m} \\ h_2 = 10\text{m} \end{cases}$ B $\begin{cases} h_1 = 1200\text{m} \\ h_2 = 10\text{m} \end{cases}$

To characterize the irregularity of the surface we again use the parameter ΔH defined previously for point to point communication. However, radio broadcasting services (CCIR Rec. 370) only calculate this parameter for that part of the link between 10 and 50km from the transmitter. It should also be noted that for point to zone communications the irregularity of the surface influences not only statistics for received levels at a given distance but also statistics for locations with a given reception quality. If the surface is very irregular, many points of reception are likely to be hidden by obstacles.

10.5.2. Land services

10.5.2.1. General considerations

A land-mobile service intended to reach across a country has a ‘cellular’ structure. Each cell is served by a base station, and all the base stations are interconnected. Cells generally have a radius of 5 to 10km in urban areas and 20 to 50km in the country. Adjacent cells operate on different frequencies. Vertical polarization is used, as it requires a very simple vehicle antenna (vertical quarter wave antenna) which is omnidirectional in the azimuthal plane.

Another important characteristic of links in the mobile service is the asymmetry of the link from the point of view of antenna height. The vehicle antenna is very close to the ground (1.5m to 3m), and is therefore near obstacles, at least in urban areas. The base station antenna, on the other hand, is generally at a high point, some tens or hundreds of metres above the vehicles, and is not near obstacles. This asymmetry leads to different antenna height effects at the two ends of the link (see below).

10.5.2.2. Propagation modelling and prediction

Given the distances and frequencies generally used, the atmosphere plays no role and multi-paths due to reflection and refraction effects are stable with time. The random element is caused by the movement of the vehicle in question and other vehicles, which produce temporal variations or spatial variations in the field strength due to phase differences between different paths. The situation is therefore very similar to that created by tropospheric scatter and we can expect similar propagation laws to apply, in particular:

1. If we consider the mean value of field strength measured over a few tens of metres in order to eliminate short period fluctuations, these mean values for a given distance to the base station will follow a log-normal distribution.
2. Fluctuations in the field stength about the mean value follow a Rayleigh distribution.
3. Loss between isotropic antennas increases with the cube of the frequency.

Experimental measurements carried out, particularly in Japan and New Zealand, have confirmed these conclusions. On the other hand, the influence of distance and that of antenna height must be determined from other considerations.

To determine loss during propagation measurements must be made with respect to the free space level. However, that level does not play an important role in the case of mobile stations, and it seems more interesting to take measurements with respect to the level which would be received if the link passed over a smooth spherical Earth, for the same distance and the same antenna heights.

To determine the usual orders of magnitude let us consider a path between a base station whose antenna is at a height of 100m and a mobile station whose antenna is at a height of 1.5m for the three frequencies 150MHz, 450MHz and 900MHz. Figure 10.5 shows attenuation between isotropic antennas under these conditions for distances between 0.5 and 100km. As the horizon is at 46km the link is in line of sight for short distances and depends on diffraction for greater distances, which leads to a rapid increase in loss beyond that distance.

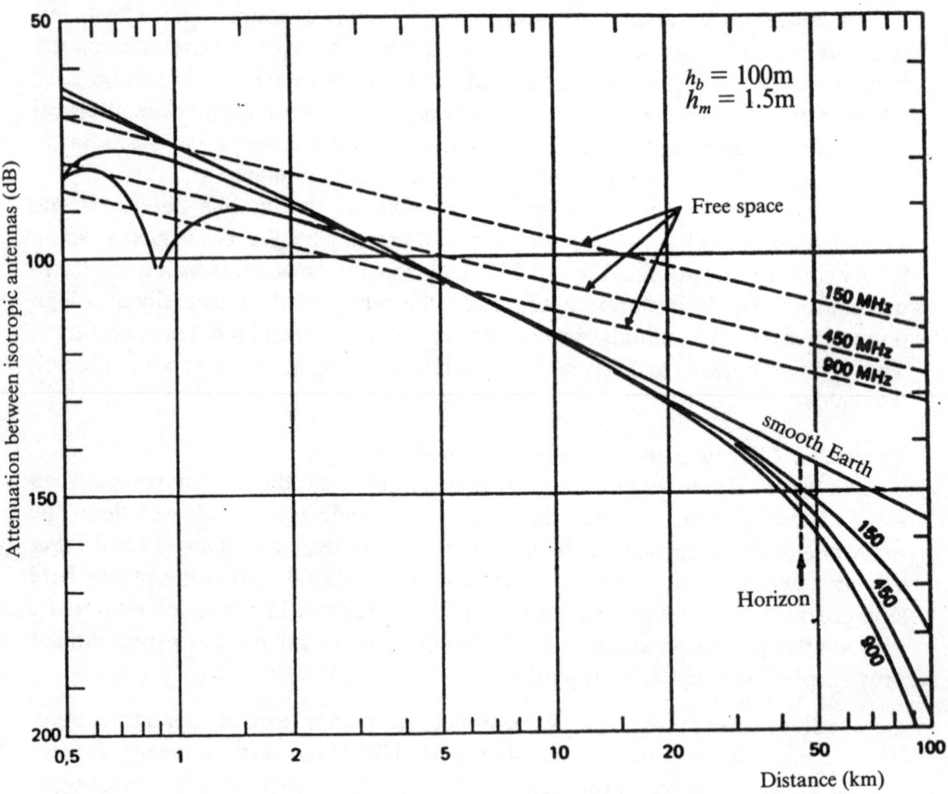

Fig.10.5. Attenuation as a function of distance

We find that over great distances loss is practically the same for all three frequencies considered, and that the gradient of the curves is close to twice the free space level. This can be easily demonstrated, as it is a case for the application of the approximate Equation (3.39). The ratio between the power received and the power received in free space is therefore:

$$\frac{P_r}{P_0} = \left(\frac{4\pi h_1 h_2}{\lambda d}\right)^2 \tag{10.21}$$

and if we take free space attenuation between isotropic antennas into account:

$$\frac{P_0}{P_t} = \left(\frac{\lambda}{4\pi d}\right)^2 \tag{10.22}$$

from which we can deduce that the attenuation between isotropic antennas on the link:

$$\frac{P_r}{P_t} = \left(\frac{h_1 h_2}{d^2}\right)^2 \tag{10.23}$$

is independent of the frequency and varies with the fourth power of the distance.

This approximation, which is sometimes called the 'smooth Earth' approximation, is usable between a distance:

$$d_1 = \frac{2\pi h_1 h_2}{\lambda} \tag{10.24}$$

and a distance d_2 a little short of the horizon. Beyond d_2 the Earth's curvature introduces an exponential factor and attenuation therefore increases more rapidly than the fourth power of distance. Below d_1 attenuation shows maxima and minima, but its mean value varies with d^2.

The smooth Earth approximation is suitable for uses by mobile services as its field of validity corresponds for all practical purposes to the range of distances currently used by this service. Moreover, it is found experimentally that the median extra attenuation due to obstructions on the link by obstacles is almost independent of the distance, for distances shorter than d_2. The general median attenuation therefore varies with d^4 to a good approximation between distances d_2 and d_2. The power of distance is higher if we include distances greater than d_2, and it is lower if we include distances less than d_1.

As far as the effect on antenna height is concerned, the symmetry of Equation (10.23) suggests that the effect is the same at both ends of the link. In fact, this symmetry does not exist, as we saw earlier. Experience shows that if the antenna of the base station is not near obstacles, the median field strength received increases proportionally with the height of that antenna, as suggested by Equation (10.23). On the other hand, if the vehicle antenna is at a lower height than that of the surrounding obstacles, which is the case generally in towns, the median field strength received increases approximately as a function only of the square root of the height of the antenna. For example, the level increases by 3dB approximately when the antenna height is increased from 1.5 to 3m.

If we bring together all the results mentioned above, a prediction formula for median attenuation between isotropic antennas is obtained:

$$\frac{P_r}{P_t} = \frac{K\, h_b^2\, h_m}{d^4 F^3} \tag{10.25}$$

where h_b is the antenna height at the base station and h_m is the vehicle antenna height, F is the frequency, d is the distance and K is a coefficient depending on the environment, in particular on the height of obstacles (buildings, trees), the width of streets, the extent of buildings, etc. The coefficient can only be determined experimentally.

If we express heights in metres, distances in kilometres, frequencies in MHz, attenuation is given in decibels by the following expression:

$$A = a + 40 \log d + 30 \log F - 20 \log h_b - 10 \log h_m \tag{10.26}$$

where the value of a can be taken as equal to 65dB for links over large towns. The standard deviation of the log-normal distribution is then between 5 and 10dB.

For rural areas conditions are very different. On the one hand, antennas at base stations are higher in order to have a greater coverage area, and at the edges of the area we generally reach distances for which spherical diffraction begins to be significant. On the other hand, additional attenuation due to obstacles on the link to the vehicle antenna is far smaller.

10.6. Interference due to ionospheric phenomena

Between about 3 and 30MHz we have seen that propagation by ionospheric reflection makes it possible to reach almost any point on the Earth's surface, and consequently in this frequency range it is impossible to separate interference problems from problems related to the use of these frequencies.

Beyond about 30MHz, on the other hand, reflection from the ionosphere is much less common, which makes it possible in theory to reuse the same frequency at a certain distance. However, there are ionospheric phenomena which can create abnormal ionization conditions, for periods lasting from a few minutes to hours, allowing propagation up to a distance of several thousand kilometres on frequencies significantly above 30MHz, which can therefore lead to interference.

Table 10.1, drawn from a CCIR Report, summarizes the main characteristics of the ionospheric phenomena likely to create interference at large distances and frequencies greater than 30MHz. It is clear that the maximum frequency at which they can be observed is less than 100MHz. There are however two phenomena which can be observed at far higher frequencies. These are reflection from ionization columns in regions of polar auroras and refraction from meteor ionization trails. The latter phenomenon occurs at all latitudes and has even been used for transmissions using special devices, for example for telegraphic transmission in a few second bursts, at the moment that ionization due to meteors appears.

Table 10.1. The principal causes of interference from the ionosphere between 30 and 300MHz.

Origin of interference	Latitude	Intense Interference time	Maximum frequency of intense interference (MHz)	Maximum frequency of interference (MHz)	Approximate distance (km)
Reflection by normal *F*	Medium	Day, winter Equinox, maximum solar activity	50	60	E-W links 3000-6000 or N-S links 3000-10,000
	Low	Afternoon and end of evening, maximum solar activity	60	70	
Reflection by Sporadic *E*	Aurora	Night	70	90	500-4000
	Medium	Day and evening, summer	60	80-135	
	Equatorial	Day	60	90	
Scatter by sporadic *E*	Low	Evening to midnight	60	90	to 2000
Reflection by meteor ionization	All latitudes	During meteor showers	Any frequency between 30 and 300MHz		to 2000
Reflection by aurora ionization	Aurora	End of evening and night			
Scatter in *F* region	Low	Evening to midnight, Equinox	60	80	1000-4000
Special effects of transequatorial propagation	Low	Evening to midnight	60	80	4000-9000

Interference can also be caused by abnormally high ionization (solar activity maximum) and by reflection from the sporadic *E* layer under certain conditions. For example, it has frequently been reported that certain European television transmissions have been received in central Africa in the frequency band allocated to that services between about 50 and 60MHz.

Chapter 11

Noise and interference

11.1. Wanted and unwanted signals

Any physical phenomenon used to convey information constitutes a signal. The same phenomenon may therefore be or not be a signal, depending on circumstances. For example, light produced by an electric lamp is not a signal if it is only being used as a source of lighting, but it is a signal if it is used to indicate danger or the functional state of a piece of equipment. Transmission of information is therefore merely transmission of a signal in a form appropriate to the system being used.

The end of the link receives not only the wanted signal but also, in general, other unwanted signals which are not necessarily related to the wanted signal. If the link is implemented by radio, the level of these unwanted signals depends on their propagation up to the receiving antenna.

The problem of unwanted signals became clear in the early days of radio technology, and there emerged a vocabulary in this field which is not always very consistent and in which, in particular, the words interference, noise, and disturbance are often used interchangeably depending on habits adopted in the technical field in question. It is therefore worth attempting to define the initial idea behind these terms more clearly, in order to deduce logical rules for their use, even if those rules cannot always be strictly respected in view of established practices.

11.2. Significance of the terms disturbance, interference and noise

11.2.1. Disturbance

In general usage disturbance is defined as being the interruption of tranquillity (*OED*), peace, rest or settled conditions. In telecommunications a disturbance corresponds to the appearance of some difficulty in the operation of a system, caused by an unwanted phenomenon. If the phenomenon is an electromagnetic field, the disturbance is said to be electromagnetic (or radioelectric) disturbance. In this case the trouble may take the form of interference, jamming of operations, the untimely triggering of an operation, modification to calibrations, etc.

Two important remarks need to be made:

1. Although the term disturbance is defined as being the effect of a

phenomenon it is sometimes used for the phenomenon itself, for instance an interfering signal or a noise.

2. As interference is the most usual kind of electromagnetic disturbance these two terms are often used interchangeably.

11.2.2. Interference

Etymologically, 'to interfere' means 'to strike against each other'. In telecommunications an interference corresponds to a 'collision' between two signals. For this to happen two signals have to be in the same frequency band. If the two signals are purely sinusoidal at the same frequency the interference is called phase interference.

Given these comments, an interference may be defined as an electromagnetic (or radioelectric) disturbance due to the superimposition of an unwanted signal on a wanted signal.

Various levels of interference have been defined by the Radio Regulations:

1. Permissible interference corresponds to a level of interference which, under given conditions, implies degradation in reception quality to an extent which is regarded as insignificant. The value of this level generally appears in the recommendations of the CCIR or other international agreements and must be taken into account in system projects.
2. Accepted interference corresponds to a higher level of interference than in the previous case, implying moderate degradation in reception quality which, under given conditions, has been agreed between the authorities concerned.
3. Harmful interference corresponds to a level of interference which 'seriously degrades, obstructs or repeatedly interrupts a radiocommunication service'. This definition is obviously very vague and corresponds to no precise level, but it is very old and appears in numerous regulatory texts, where its vague nature is undoubtedly an advantage.

11.2.3. Noise and random noise

The word noise has been borrowed from acoustics where it is used to mean 'non-musical'. But what does 'non-musical' mean? This concept has developed through time and space: however, in its current form it is applied to the acoustic phenomenon due to the superimposition of harmonic vibrations. It is worth mentioning in passing that to designate frequencies that are multiples of a fundamental frequency, we do in fact use a word which suggests 'harmony', i.e. in general, the opposite of noise. From the physical point of view, the distinction between a noise and a musical sound therefore depends on a comparison of the spectra of these two types of signals.

Even though the word noise does not in itself give an idea of quality or of value, it is often implicitly assumed that a musical sound is more pleasant to hear than a noise, and noise therefore appears as a source of disturbance when listening to sounds. This is, however, far from being universally the case, and, for example, it is more pleasant to hear the noise of a waterfall than the sound of a fire siren. On the other hand, if you are bothered by your next door neighbour playing a symphony on his stereo (but over powerful) record player, your inclination is to tell him, 'stop that noise'. This example points out the ambiguity of the word noise in everyday language.

The idea of a noise moved quite naturally from the acoustic field to the electrical and radio fields, with the appearance of microphones and loudspeakers. A new extension of the word noise again took place when we began to transmit non-acoustic signals electrically, and this led us to speak, for example, of noise on a television picture. These various extensions led to the establishment of a theory of noise (often called random noise) without adopting any frequency limitation.

The final step was taken when it became possible to consider useful signals taking the form of noise. This is the case in radio astronomy and also in telephony, where we simulate a multiplex signal with many channels using a noise generator.

There is, however, one constant factor running through all these successive extensions to the word, which is that in telecommunications the idea of noise is principally associated with the characteristics of the frequency spectrum of the signals.

In consequence of these considerations a 'random noise' (usually abbreviated to 'noise') may be defined as a signal of random character generally transporting no deliberate information, and, by extension, the effect of this signal. A noise may be characterized by the random process which generates it, e.g. gaussian noises, by the physical phenomenon with which it is associated, e.g. thermal noise, shot noise, or by its spectrum, e.g. white noise, $1/f$ noise. Noise often produces interference in the reception of a wanted signal, but in certain cases, e.g.. in radio astronomy, the wanted signal itself has the characteristics of random noise.

11.3. Transmission quality

The performance of a transmission circuit is generally characterized by the ratio of two quantities measured at a specified point of the circuit. One of these quantities is the power of the wanted signal and the other is the power of unwanted signals such as harmonic, noise, interfering signals, etc. In the case of radio circuits there are two possible specified points on the circuit.

In point to point radiocommunications, such as a fixed service, the specified point of the circuit is the output of the receiver, and the quality is characterized for example by the 'signal to noise ratio' S/N.

In point to area radiocommunications, such as mobile services and radio broadcasting services, the characteristics of the receiving stations are not known and consequently the specified point of the circuit is the input to the receiver. The quantity considered in this case is the 'protection ratio', which is defined by the Radio Regulations as 'the minimum value of the wanted to unwanted signal ratio at receiver input, determined under specified conditions such as a specified reception quality of the wanted signal which is achieved at the receiver output'.

The conditions specified include, among others:

1. the nature and characteristics of the wanted signal;
2. the nature and characteristics of the unwanted signals;
3. the characteristics of the receiver;
4. the propagation conditions.

11.4. Theoretical remarks concerning thermal noise in one-port and two-port networks

To study noise, thermal noise is a convenient standard to choose, i.e. noise due to thermal agitation of electrons, as its statistical properties and the physical laws which govern it are relatively simple. We start, therefore, with a few theoretical comments on this subject.

11.4.1. Exchangeable power, available power

Noise takes the form of an electric signal whose amplitude is random. Only the mean power is well determined. Furthermore, we know that the exchangeable power of a generating one-port network of given internal impedance is the stationary value (maximum or minimum) of the active mean power. This one-port network can provide to an external charge if we make the impedance of that charge vary. If the one-port network has an internal impedance whose real part is positive, exchangeable power is called available power, and the stationary value is a maximum.

11.4.2. Noise temperature

The real signal can only appear at the end of a resistance. But any resistance is a noise generator, whose available power is given by the Nyquist approximation as:

$$\mathrm{P} = k.\ T.\ \Delta f \qquad (11.1)$$

Where k is the Boltzmann constant (1.38 10^{-23} J/^{0}), T is the absolute temperature of the resistance and Δf is the bandwidth considered. In the case of an antenna the resistance at the ends of which the signal and noise appear is the radiation resistance. It is not localized at one point but is associated with the radiation of the antenna in all directions.

Equation (11.1) also makes it possible to express a noise power within a given frequency range by an equivalent noise temperature or internal noise temperature, which is the physical temperature of a resistance at whose ends this noise power would appear.

In the same way we define the internal noise temperature of a two-port network as the noise temperature of a hypothetical one-port network which, if it were connected to the entry point to the two-port network, itself assumed to be free of all noise sources, would give the same noise power at the output as the internal noise sources of the two-port network.

11.4.3. Attenuator noise temperature

An attenuator at physical temperature T and giving an attenuation a (as a power ratio) is a two-port network with an internal noise temperature T_i:

$$T_i = T(a - 1)\,. \tag{11.2}$$

If a is small and expressed in decibels by the relation:

$$\alpha = 10 \log a\,, \tag{11.3}$$

and if T is equal to 290K, approximately:

$$T_i = 70\,\alpha \tag{11.4}$$

From which we can derive the following practical rule: every tenth of a decibel of loss increases the noise temperature by 7K.

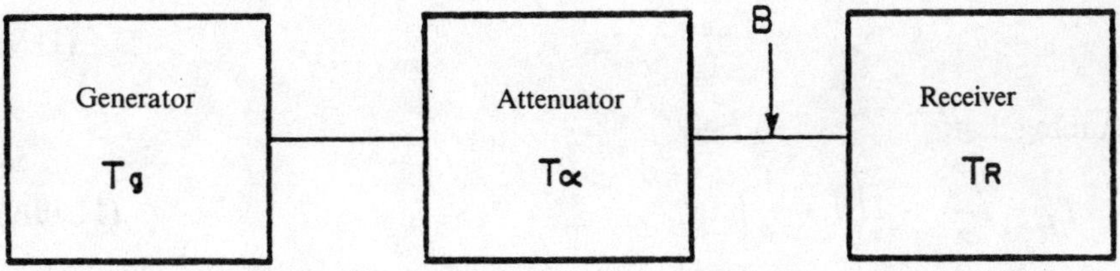

Fig. 11.1. The influence of an attenuator on noise.

Let us therefore consider the network formed by a generator, an attenuator and a receiver, with respective noise temperatures T_g, T and T_R (Fig. 11.1).

The noise temperature of the whole, measured at point B between the attenuator and the receiver, equals:

$$T_B = \frac{T_g}{a} + T\left(1 - \frac{1}{a}\right) + T_R\,. \tag{11.5}$$

The attenuator therefore leads to an additional noise temperature T_S:

$$T_S = T\left(1 - \frac{1}{a}\right). \tag{11.6}$$

11.4.4. Antenna noise temperature

In the case of a radio receiving station, T_g is the antenna noise temperature T_A, which is defined by assuming that the noise power received by the antenna equals:

$$P = kT_A \, \Delta f. \tag{11.7}$$

The noise temperature of the antenna T_A can be calculated by assuming that a noise temperature can be associated with each direction in space, equal to that of a black body giving a noise radiation equal to that obtained in that direction. It is then possible to prove that:

$$T_A = \frac{1}{4\pi} \int\int G.T.\mathrm{d}\Omega, \tag{11.8}$$

where G is the gain of the antenna in a direction where the noise temperature is T, and $\mathrm{d}\Omega$ is the elementary solid angle around this direction. These calculations assume that we know the radiation pattern of the antenna throughout space, or, at least, that we know its 'envelope pattern', which gives a slightly pessimistic result. It should however be noted that T_A is not an intrinsic characteristic of the antenna, and its value depends on the direction in which it is oriented and on the external noise sources present.

The terms a and T represent all the losses occurring in the antenna and its associated parts (radome, feeder lines, filters, etc.) as well as losses due to mismatching. Therefore:

$$T_B = T_E + T_R \tag{11.9}$$

assuming that:

$$T_E = \frac{T_A}{a} + T\left(1 - \frac{1}{a}\right). \tag{11.10A}$$

where T_E is the equivalent noise temperature of the antenna and associated circuits.

11.4.5. Figure of merit

The ratio of signal to noise on reception for a satellite link depends on the radio characteristics of the Earth station, because of the factor G/T which is called the figure of merit of the station. G is the isotropic gain of the antenna (assumed to be lossless) in the direction of the received signal and T is the total noise temperature measured at the receiver input, i.e. T_B.

11.4.6. Noise factor

If we take as a standard for noise that produced by a black body at ambient temperature T_0 (generally standardized to 290K), we can define the noise factor of a two-port network as being the ratio, generally expressed in decibels, of the total noise power measured at its output to that part of the power which depends only on the source connected to its input. The noise factor is 0dB if the two-port network adds no noise. The noise factor F of a two-port network is associated with its noise temperature T by the relation:

$$F = 1 + \frac{T}{T_0}. \tag{11.10B}$$

If the source of the signal has a noise temperature equal to the ambient temperature (e.g. in cases of a terrestrial link), the noise factor is an exact measure

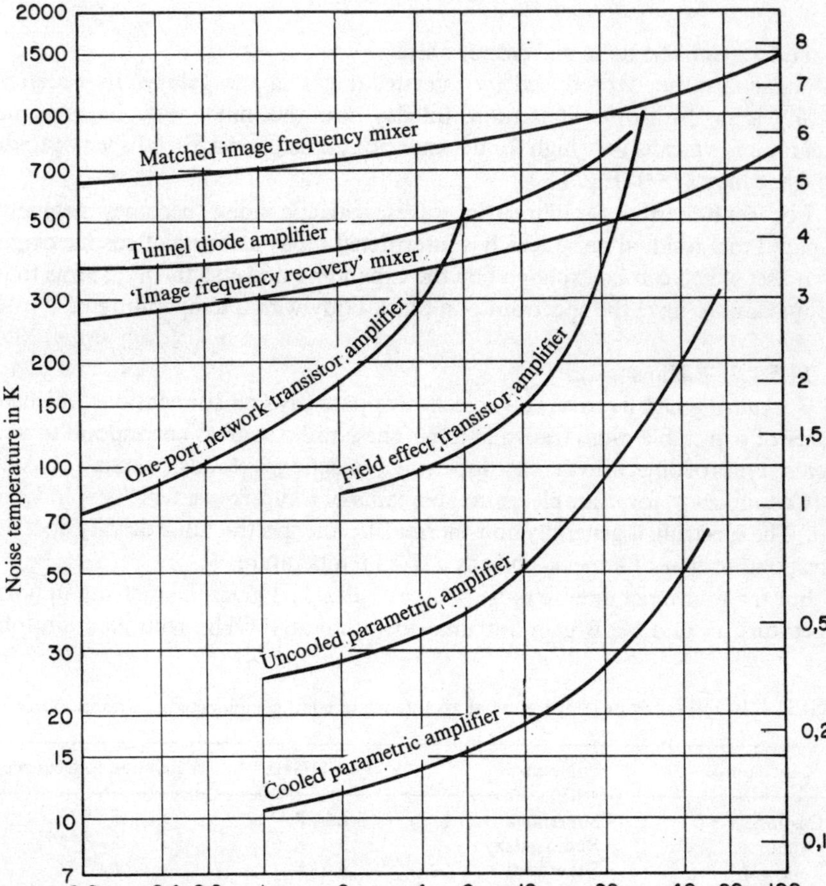

Fig. 11.2. Performances of the best low noise devices.

of the degradation of the signal to noise ratio due to the two-part network. If the noise temperature of the source of the signal is smaller than the ambient temperature (e.g. a satellite link), the degradation of the signal to noise ratio is greater than the noise factor. Figure 11.2 gives the noise temperatures and the noise factors of a number of low noise amplifiers.

11.5. External noise sources

External noise sources can be classified into two groups depending on whether they are terrestrial or extra-terrestrial.

11.5.1.. Extra-terrestrial noise

This is above all galactic noise, noise radiated by celestial radio sources and noise radiated by the bodies of the solar system.

11.5.1. Galactic noise and cosmic noise

Galactic noise is produced by interstellar gas in the galaxy. Its radiation intensity is not isotropic, it is more intense near the milky way. Its brilliance temperature, which is very high for decametric and metric waves, falls very rapidly with frequency, ($f^{-2.6}$) (Fig. 11.3).

For centimetric or millimetric waves, galactic noise becomes negligible compared to a residual noise which is interpreted today as being of cosmic origin. This noise, which can be explained by the expanding universe theory, seems to be isotropic and to have the spectrum of a black body with a temperature of 2.76K.

11.5.1.2. Radio sources

This term generally covers all objects of apparently small dimensions which are sources of detectable radio transmissions. These radio sources correspond to very different types of objects in the sky: supernova galactic gas clouds, galaxies, quasars, etc. (Cassiopeia *A*, for example, is the remnants of a supernova which exploded in 1750.) The spectrum is generally non-thermal, i.e. the spectral noise density does not correspond to that of a black body at a fixed temperature.

For the four most intense radio sources Table 11.1 gives the increase in noise temperature T_A of a 60dB gain antenna operating at 11GHz, with its main lobe

Table 11.1. Increase in noise temperature of an antenna of 60dB gain directed at a radio source.

Radio source	Nature	T_A at 11GHz	Relation to frequency
Cassiopeia *A*	Supernova	7.4 K	$F^{-0.87}$
Cygnus *A*	Radio galaxy	3 K	$F^{-1.06}$
Taurus *A*	Supernova	9.6 K	$F^{-0.26}$
Virgo *A*	Radio galaxy	0.76 K	$F^{-0.87}$

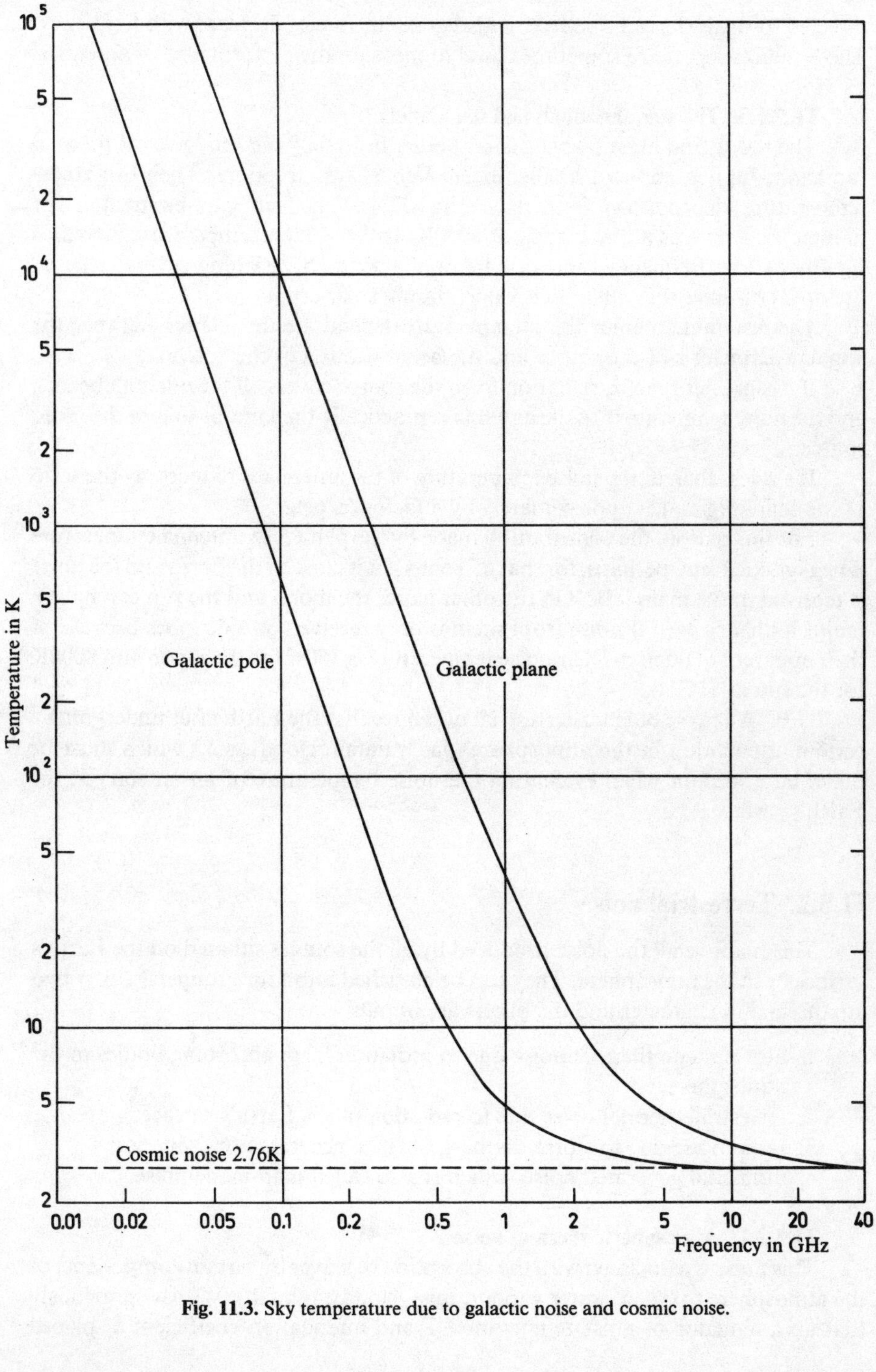

Fig. 11.3. Sky temperature due to galactic noise and cosmic noise.

directed towards the radio source; it also gives the rate of decrease with frequency. These radio sources are sometimes used to measure the performance of antennas.

11.5.1.3. The sun, the moon and the planets

The moon and most planets, solar bodies heated by the sun, give off thermal radiation. Jupiter, and to a smaller extent Venus, are exceptions. Their brightness temperature depends on frequency (Fig. 11.4). The sun can be treated for millimetric waves as a black body at 6000K. Its brightness temperature increases rapidly at low frequency, and can become very high, reaching several tens or hundreds of times the 'quiet Sun' value, during solar eruptions.

The resultant antenna noise temperature depends on the relation between the angular diameter ε of the source and the beam width α of the antenna.

If ε is greater than α, radiation from the source covers all the antenna beams, and the noise temperature of the antenna is practically the same as that of the radio source.

If ε is less than α, the noise temperature of the antenna is reduced by the ratio of the solid angles, i.e. approximately by a factor $(\varepsilon/\alpha)^2$.

For this reason, the contribution made by the planets to antenna temperature is negligible except, perhaps, for that of Venus if it is close to the Earth and the noise is received in the main lobe. On the other hand, the moon and the sun can have a major influence even if noise from them is only received on side lobes because of their apparent diameter. Their noise temperature is 150K for the moon and 8000K for the sun at 11GHz.

N.B.: All noise of extra-terrestrial origin reaches the Earth after undergoing a certain attenuation in the atmosphere (gas, rainfall, clouds, etc.) which must be taken into account when evaluating the noise temperature of an antenna at the Earth's surface.

11.5.2. Terrestrial noise

This is above all the noise produced by all the sources situated on the Earth's surface or in the atmosphere. They can be classified into four groups: the first two are thermal in character and the others are impulsive.

1. atmospheric thermal noise due to radiation from absorbing bodies in the atmosphere;
2. terrestrial thermal noise due to radiation of the Earth's surface;
3. radio noise due to storm discharges which generate atmospherics;
4. industrial generated noise, which is also called man-made noise.

11.5.2.1. Atmospheric thermal noise

This noise is associated with the absorption of waves by certain components of the atmosphere (oxygen, water vapour, mist, clouds, rainfall). We saw previously that an attenuator of noise temperature T and attenuation coefficient a, placed

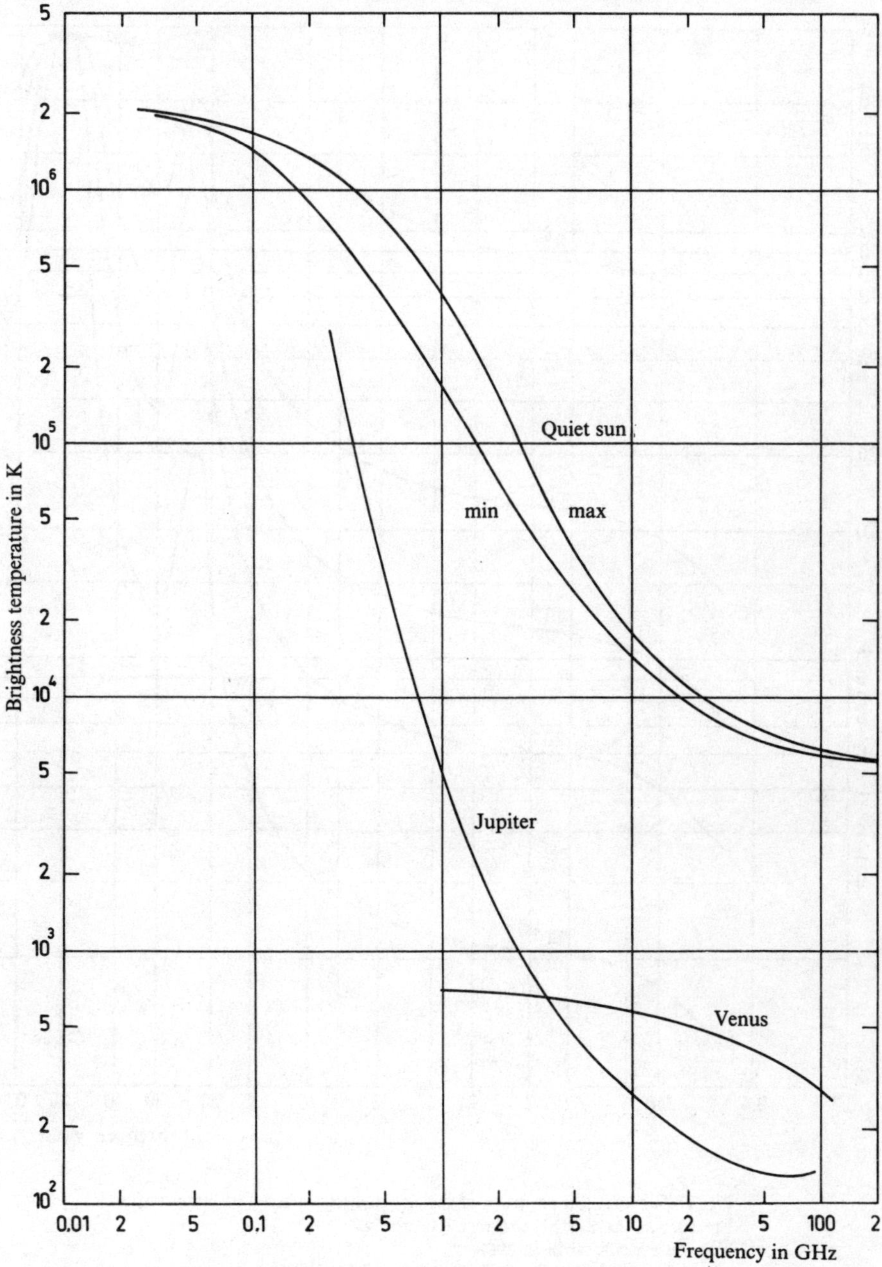

Fig. 11.4. Brightness temperature of the sun, Jupiter and Venus.

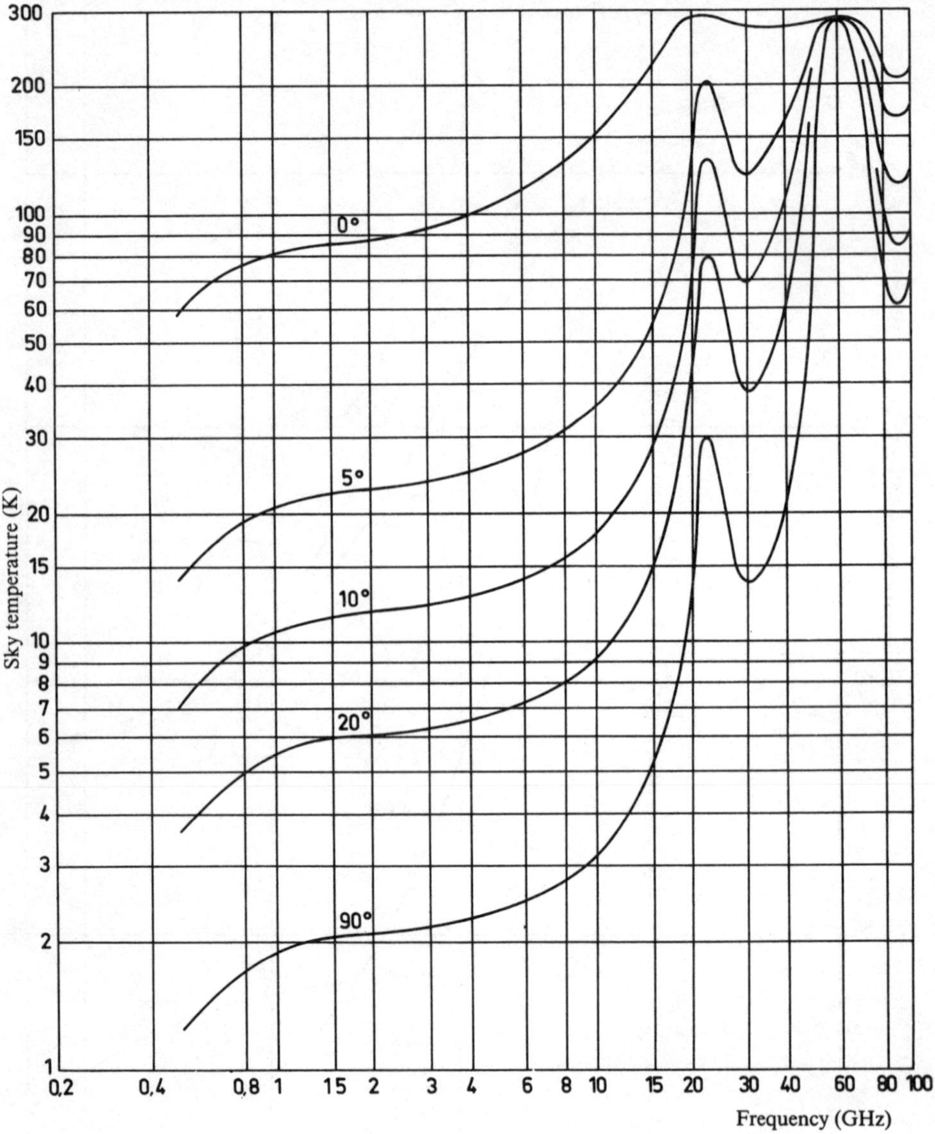

Fig. 11.5. Noise temperature of the sky due to oxygen and water vapour.
Pressure at the Earth's surface: 760mmHg.
T at the Earth's surface: 290K.
Humidity at the Earth's surface: 7.5g/m^3 ($R_H = 50\%$).

between a source of noise temperature T_g and a receiver of noise temperature T_R gives a noise temperature at the input of the receiver which can be calculated by Equation (11.5).

In the case of the atmosphere, the noise temperature T and attenuation a vary along the propagation path between the satellite and the receiver. If we write $\alpha(r)$ for the attenuation per kilometre at a distance r from the receiver, the previous expression can be written as an integral:

$$T_s = \int_0^{\infty} T(r).\alpha(r).\exp\left[\int_0^{r} \alpha(r')\,\mathrm{d}r'\right] \mathrm{d}r\,. \tag{11.11}$$

Assuming that the atmosphere is horizontally stratified, attenuation a_φ in a direction defined by an angle of elevation φ may be written:

$$a_\varphi = \frac{a_v}{\sin\varphi}, \tag{11.12}$$

where, a_v is the attenuation in the vertical direction. This expression no longer applies if angle φ is less than 10° (Fig. 11.5).

Noise is particularly intense under certain meteorological conditions, for example if there are clouds or precipitation along the antenna beam. It is possible to calculate the noise if we know the attenuation due to the clouds or precipitation, using a formula similar to Equation (11.10A). In this case, the antenna temperature is given by:

$$T_A = T_p\left(1 - \frac{1}{a}\right) + \frac{T_c}{a}, \tag{11.13}$$

Where T_c is the temperature of clear sky, a is the attenuation due to the rain in the direction of the main lobe (it is assumed that its influence is negligible in other directions) and T_p is the rain temperature. In practice, to take account of the fact that part of the energy going through the rain is scattered and not absorbed, we have to take T_p slightly lower than the rain temperature: 275K is often used.

11.5.2.2. Terrestrial thermal noise

As the antenna temperature corresponds to the integral of the noise energy over a whole sphere, we have to take account of the lower half which 'sees' the Earth's temperature. That temperature is generally taken to be equal to 290K, and the Earth is assumed to be a black body at that temperature.

In fact, this is a pessimistic approximation, as the temperature of the Earth depends on the reflection coefficient and frequency considered, i.e. on the roughness of the Earth's surface and the angle of incidence of the rays. In particular, if the antenna is surrounded by water or if the Earth's surface in its immediate neighbourhood is metallic, the value of 290K is certainly too high, though it is not possible to work out a precise value to replace it.

The noise temperature resulting from the influence of the Earth's surface depends on the mean gain in the remote side lobes. With an ordinary parabolic antenna the mean gain in the remote side lobes is of the order of –5dB with respect to an isotropic source. The corresponding noise temperature is therefore of the order of 100K. For space communications antennas have mean isotropic gain in the remote lobes of the order of –15dB, and the noise to be added to T_a is then 10K.

11.5.2.3. Atmospherics

The term atmospherics is used for damped waves caused by storm discharges. Their frequency spectrum ranges from a few Hertz to twenty or more Megahertz, reducing more or less regularly as a function of frequency. Atmospherics cause disturbance above all to communications using kilometric or hectometric waves.

Atmospherics have been studied with the aim of attempting to reduce the damaging effects that they produce for reception and to determine the minimal field necessary to prevent their having a detrimental influence. Furthermore, it has become apparent that their analysis, particularly in the lower region of the spectrum (a few MHz) where they are most intense, can give interesting information on the composition of the upper atmosphere and on the Earth's magnetosphere, through a study of 'whistlers' (Chapter 9).

Numerous statistical studies have made it possible to determine a mean level of atmospherics for a given region. As the spectrum is broad, the power received is proportional to the bandwidth of the receiver, and consequently the field strength is proportional to the square root of that bandwidth. We can therefore express the mean level of atmospherics in W/kHz or in μV/m for 1kHz of bandwith. It is also possible to express this level in the form of an equivalent noise factor in the receiver, taking as a reference standard for the power, the thermal noise at the ambient temperature (290K) for the same bandwidth.

The curves in Fig. 11.6, drawn from documents provided by the CCIR, give the mean value of the equivalent noise factor of a receiver due to atmospherics. It is immediately clear that the noise factor takes on an enormous value at low frequencies, which makes it obviously futile to use very low noise receivers at these frequencies.

The shape of these curves are a result both of the spectrum of the storm discharges themselves and of propagation effects. It is known that during the night the level decreases regularly as frequency increases, while during the day (around midday) there is a minimum at about 2 to 3MHz, due to the ionospheric absorption which signals from very remote storms undergo. Such absorption is, on the other hand, very reduced at night (see Chapter 9). Beyond 20MHz, the level decreases rapidly, as for all these frequencies the ionosphere becomes transparent, and only local storms are detected.

The curves in Fig. 11.6 correspond roughly to the European region. The levels indicated have to be modified when considering regions at other latitudes. The difference is most marked in the central portion of the curves. For example, for 1MHz the level increases by around 20 to 30dB as we approach the continental equatorial regions and reduces by 10dB as we reach the polar region. For 0.01KHz

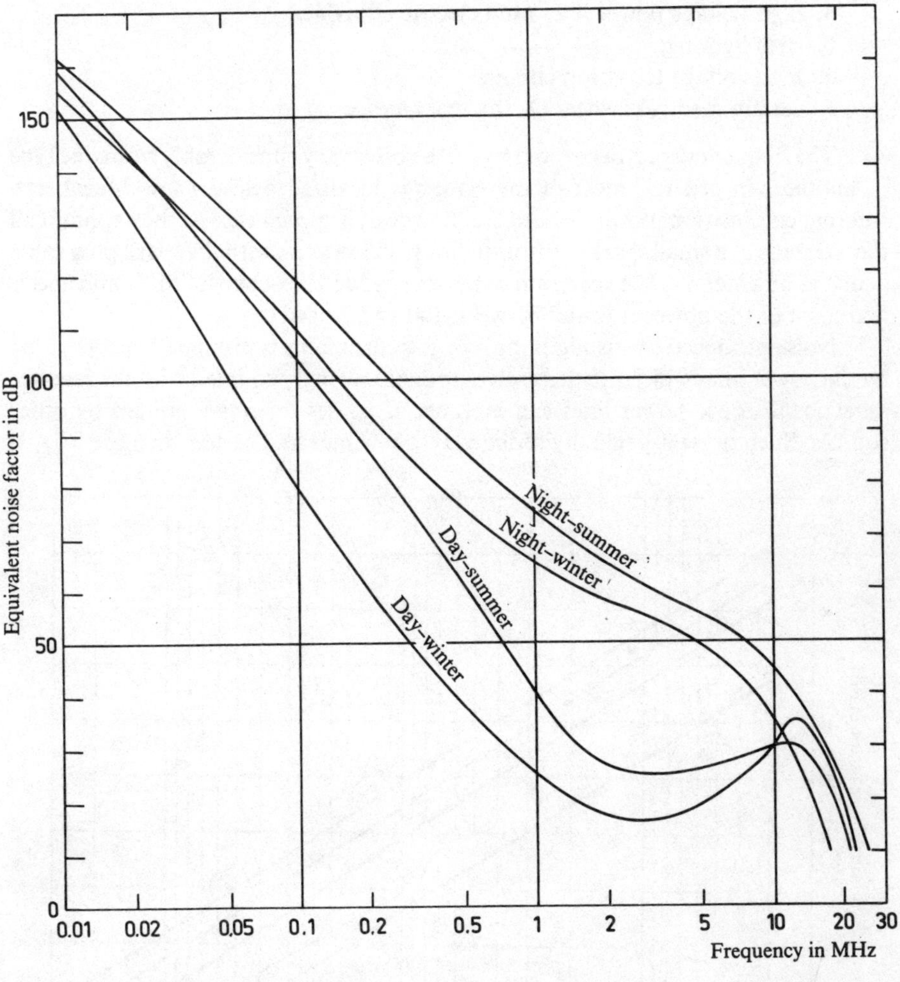

Fig. 11.6. Mean noise level due to atmospherics (storm discharges). European zone.

the amplitude of the variation will not be greater than 5 to 10dB. Noise levels are generally considerably lower in sea areas.

11.5.2.4. Industrial generated noise

Such radio noise has its source in free or damped electric oscillations in all circuits or parts of circuits in which a rapid variation of current takes place. The following are some of the devices producing such noise:

1. motors with collectors;
2. circuit breakers, switches, commutators, relays, etc;
3. electric sparking device for internal combustion engines;

4. high voltage power lines (and electric effluvia);
5. strip lighting;
6. high voltage television circuits;
7. certain medical devices (X-ray machines, etc.).

The frequency spectra covered by these noises vary considerably from one type to another. In general, most of the power is localized below a few Megahertz. Internal combustion motor devices are, however, a special case as they contain all the elements of a small spark transmitter for metric waves, with the spark plug cable acting as an antenna. The spectrum of noises produced reaches 400MHz and above and most of the power is found between 100 and 200MHz.

Noise produced by installations linked to the mains is principally propagated by the power lines which are themselves roughly strongly coupled with the receiver antennas. Electric power lines can even act as guides for noise emitted by other sources. Such noise is generally radiated to a few hundred metres or more.

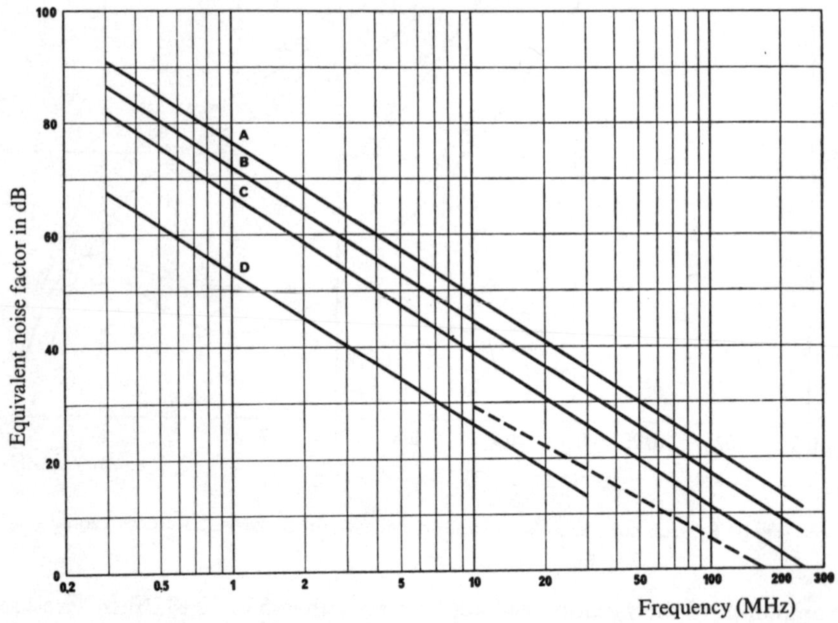

Fig. 11.7. Industrial noise (CCIR).
Median values measured using a short vertical monopole antenna.

A: business areas
B: residential areas
C: rural areas
D: quiet rural areas
– – – Galactic noise.

The level of industrial noise, measured at a particular point, depends essentially on the number and level of activity of industrial or domestic devices near that point. Figure 11.7, drawn from CCIR texts, gives a median value of the industrial noise power in different inhabited areas. These curves should be treated as being only of general value. By way of comparison, the mean value of galactic noise is shown as a dotted line.

11.5.2.4.1. Protection against industrial noise

The development of industrial and domestic usage of electricity has led to such an increase in the level of noise that regulations have had to be adopted to limit it. These regulations are based in part on work carried out by international organizations such as the Union International des Telecommunications (UIT) and the International Electrotechnical Commission (IEC), the second of which has set up a specialist organization in this area, the Special International Committee for Radio Disturbance (CISPR).

From the technical point of view the most efficient method of protection against industrial noise is to prevent it. This is known as suppression of devices. This operation involves sufficient damping of the circuit in which the noise could be produced (resistance in series on the cable to a spark plug), or to prevent the noise leaving the device which has produced it (special casings and low-pass filters on power lines to devices).

If noise is produced and radiated, the only recourse is to use methods to reduce their detrimental effect on a receiver. These include decoupling power lines, the use of directional balanced antennas, maximum selectivity compatible with the quality of the signal, addition of limiting circuits, the use of types of modulation which are of low sensitivity to disturbances, etc.

Appendix 1

Statistical distributions used in radio wave propagation studies

1. Introduction

The propagation of radio waves in a terrestrial environment is mainly associated with a random medium, which makes it necessary to analyse propagation phenomena by means of statistical methods. Such a study is of value only if it is possible to define a statistical model that represents these phenomena sufficiently accurately. Fortunately, it is possible in most cases to give a satisfactory description of signal variations in time and space using known statistical distributions. It is therefore important to know the fundamental properties of the most common probability distributions used in the statistical study of radio wave propagation phenomena.

Probability distributions are described either by a density function or by a cumulative distribution function. The density function, here denoted by $p(x)$ for the random variable X, is such that the probability of X taking a value in the infinitesimal interval x to $x + \mathrm{d}x$ is $p(x)\,\mathrm{d}x$. The cumulative distribution function, denoted by $F(x)$, gives the probability that variable X takes a value less than x, i.e. the functions are related as follows:

$$p(x) = \frac{\mathrm{d}}{\mathrm{d}x}\left[F(x)\right] \tag{1}$$

or

$$F(x) = \int_c^x p(t)\mathrm{d}(t) = \mathrm{Prob}\,(X \leqslant x) \tag{2}$$

where c is the lower limit of the values which X can take.

2. Gaussian or normal distribution

This distribution is applied to a continuous variable of any sign. The probability density is of the type:

$$p(x) = \mathrm{e}^{-T(x)}, \tag{3}$$

$T(x)$ being a non-negative second degree polynomial. If as parameters we use the mean m and the standard deviation σ, $p(x)$ is written in the usual way:

$$p(x) = \frac{1}{\sigma\sqrt{2\pi}} \exp\left[-\frac{1}{2}\left(\frac{x-m}{\sigma}\right)^2\right], \tag{4}$$

hence:

$$F(x) = \frac{1}{\sigma\sqrt{2\pi}} \int_{\infty}^{x} \exp\left[-\frac{1}{2}\left(\frac{t-m}{\sigma}\right)^2\right] dt \tag{5}$$

$$= \frac{1}{2}\left[1 + \operatorname{erf}\left(\frac{x-m}{\sigma\sqrt{2}}\right)\right]$$

with:

$$\operatorname{erf} z = \frac{2}{\sqrt{\pi}} \int_0^z e^{-t^2} dt. \tag{6}$$

The cumulative normal distribution $F(x)$ is generally tabulated in a short form, i.e. with m taken to be zero and σ equal to 1. Table 1 gives the relation between x and $F(x)$ for a number of rounded values of x or $F(x)$.

Table 1.

x	$1 - F(x)$	x	$1 - F(x)$
0	5.000 00 10^{-1}	1.281 55	10^{-1}
1	1.586 55 10^{-1}	2.326 35	10^{-2}
2	2.275 01 10^{-2}	3.090 23	10^{-3}
3	1.349 90 10^{-3}	3.719 01	10^{-4}
4	3.167 12 10^{-5}	4.264 89	10^{-5}
5	2.866 52 10^{-7}	4.753 42	10^{-6}
6	9.865 87 10^{-10}	5.199 34	10^{-7}
		5.612 00	10^{-8}

For the purpose of practical calculations, $F(x)$ can be represented by approximate functions, for example the following which is valid for positive x with a relative error less than

$$1 - F(x) \simeq \frac{\exp(-x^2/2)}{\sqrt{2\pi}(0.661x + 0.339\sqrt{x^2 + 5.51})} \tag{7}$$

The reverse problem, i.e. the calculation of x as a function of $F(x)$ may be solved by the following approximate equation:

$$x \simeq \sqrt{-5.21 \log_{10} 1.53y} - 0.744 - 0.42y + 0.7y^2 \tag{8}$$

with:

$$y = 1 - F(x). \tag{9}$$

A gaussian distribution is most often encountered when values of the quantity considered result from the additive effect of numerous random causes, each of them of relatively slight importance.

If we consider n independent gaussian variables such that, for each of them, $m = 0$ and $\sigma = 1$, the sum of the squares of these variables has a chi-squared distribution defined by

$$p(\chi^2) = \frac{1}{2^{n/2}\,\Gamma\,(n/2)}\, e^{-(\chi/2)^2}\,(\chi^2)^{(n/2)-1} \tag{10}$$

where n is the number of degrees of freedom of the distribution.

In propagation most of the physical quantities involved (power, voltage, fading duration, etc.) are essentially positive quantities and cannot therefore be represented directly by a gaussian distribution. On the other hand, this distribution is used in two important cases:

1. to represent the fluctuations of a quantity about its mean value (scintillation);
2. to represent the logarithm of a quantity. We then obtain the log-normal distribution which is studied later.

Diagrams in which one of the coordinates is a so-called gaussian distribution are available commercially, i.e. the graduation is such that a gaussian distribution is represented by a straight line. These diagrams are frequently used in propagation studies, to represent non-gaussian distributions, even.

3. Log-normal distribution

This is the distribution of a positive variable whose logarithm has a gaussian distribution. It is possible therefore to write directly the probability and the cumulative density functions:

$$p(x) = \frac{1}{\sigma\sqrt{2x}}\,\frac{1}{x}\,\exp\left[-\frac{1}{2}\left(\frac{\ln x - m}{\sigma}\right)^2\right] \tag{11}$$

$$F(x) = \frac{1}{\sigma\sqrt{\pi}}\int_c^x \frac{1}{t}\,\exp\left[-\frac{1}{2}\left(\frac{\ln t - m}{\sigma}\right)^2\right]\,dt \tag{12}$$

$$= \frac{1}{2}\left[1 + \operatorname{erf}\left(\frac{\ln x - m}{\sigma\sqrt{2}}\right)\right].$$

However, in these relations m and ρ are the mean and the standard deviation not of the variable x but of the logarithm of this variable. The characteristic quantities of the variable x can be derived without difficulty. We find:

1. most probable value: $\exp(m - \sigma^2)$;
2. median value: $\exp(m)$;
3. mean value: $\exp(m + \sigma^2/2)$;
4. root mean square value: $\exp(m + \sigma^2)$;
5. standard deviation: $\exp(m + \sigma^2/2)\sqrt{\exp(\sigma^2) - 1}$.

Unlike the gaussian distribution, a log-normal distribution is asymmetrical. In particular, the mean value, the median value and the most probable value (often called the mode) are not identical (Fig. 1).

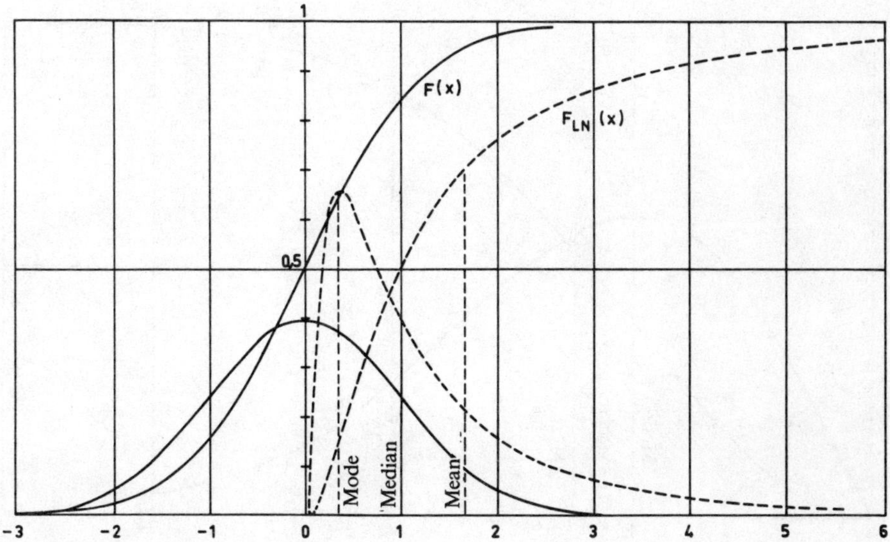

Fig. 1. Normal (—) and log-normal (---) distributions.

The log-normal distribution is often met in connection with propagation, mainly for quantities associated either with a power or field strength level or a time. Power or field strength levels are generally only expressed in decibels so that it is more usual to refer to a normal distribution of levels. In the case of time (e.g. duration of fades), the log-mormal distribution is used explicitly because the natural variable is the second or the minute and not their logarithm.

Since the reciprocal of a variable with a log-normal distribution also has a log-normal distribution, this distribution is sometimes used for rates (reciprocals of time). For example, it is used to represent rainfall rate distribution at least for low and medium rainfall rates.

In comparison with a gaussian distribution, a log-normal distribution can be taken to imply that the numerical values of the variable are the result of the action of numerous causes of slight individual importance which are multiplicative.

4. Rayleigh distribution

The Rayleigh distribution applies to a non-limited positive continuous variable. The probability density and the cumulative distribution are given by:

$$p(x) = \frac{x}{q^2} \exp\left(-\frac{x^2}{2q^2}\right) . \tag{13}$$

$$F(x) = 1 - \exp\left(-\frac{x^2}{2q^2}\right). \tag{14}$$

Figure 2 shows these functions $p(x)$ and $F(x)$.

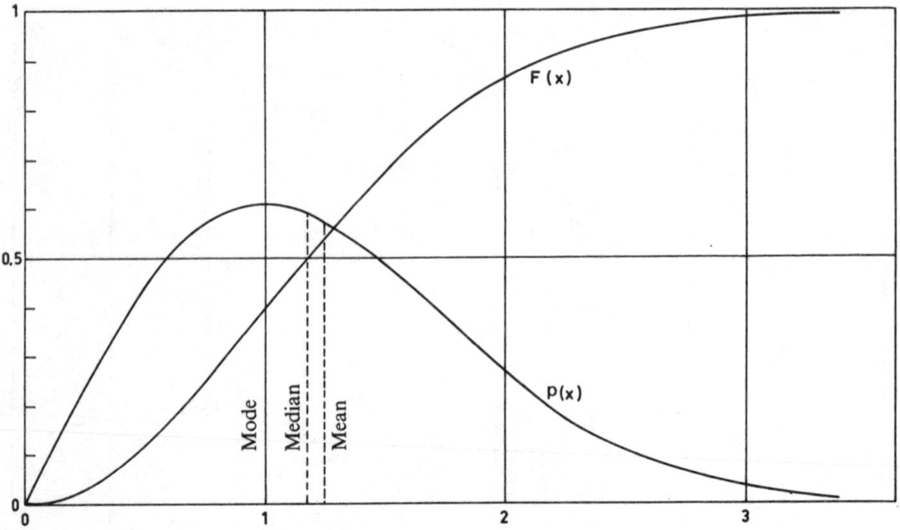

Fig. 2. Rayleigh distribution.

The characteristic values of the variable are as follows:

1. most probable value: q;
2. median value: $q\sqrt{2 \ln 2} = 1.18q$;
3. mean value: $q\sqrt{\pi/2} = 1.25q$;
4. root mean square value: $q\sqrt{2} = 1.41q$;
5. standard deviation: $q\sqrt{2 - \pi/2} = 0.655q$;

The Rayleigh distribution is often only used near the origin, i.e. for low values of x. In this case we have:

$$F(x) \simeq \frac{x^2}{2q^2} \tag{15}$$

This can be interpreted as follows: the probability that the random variable X will have a value of less than X is proportional to the square of this value. If the variable in question is a voltage, its square is proportional to the power of the signal. In other words, on a decibel scale the power 'decreases by 10dB for each decade of probability. This property is often used to determine whether a received level has a Rayleigh distribution, at least asymptotically. It should be noted however that other distributions can behave in the same way.

The Rayleigh distribution is linked with the gaussian distribution as follows. Given a two-dimensional gaussian distribution with two independent variables x and y of mean zero and with the same standard deviation σ, the random variable:

$$r = \sqrt{x^2 + y^2} \tag{16}$$

has a Rayleigh distribution and the most probable value of r is equal to σ. As r represents the length of a vector joining a point in a two-dimensional gaussian distribution to the centre of this distribution, it may be deduced that the Rayleigh distribution represents the distribution of the length of a vector which is the sum of a large number of vectors of low amplitudes whose phases have a uniform distribution.

The Rayleigh distribution applies in particular to scattering phenomena.

5. Combined log-normal and Rayleigh distribution

In some cases the distribution of a random variable can be regarded as the resultant of a combination of two distributions, i.e. a log-normal distribution for long term variations and a Rayleigh distribution for short term variations. The distribution of instantaneous values is obtained by considering a Rayleigh variable whose mean (or root mean square) value is itself a random variable having a log-normal distribution.

If m and σ are used to designate the mean and the standard deviation of the gaussian distribution associated with the log-normal distribution, the following distribution is obtained:

$$1 - F(x) = \frac{1}{\sqrt{2x}} \int_{-\infty}^{+\infty} \exp\left[-x^2 e^{-2\,\sigma u} - \frac{u^2}{2} \right] du. \tag{17}$$

In Equation (17) the standard deviation σ is expressed in nepers. If σ' is used to designate its value in decibels, we have:

$$\sigma = 0.115\sigma' \tag{18}$$

Figure 3 shows a graph of this distribution for a number of values of the standard deviation, m is taken as zero.

This distribution applies mainly to propagation by fluctuation of the medium when the characteristics of the latter have non-negligible long term variations, as for example in the case of tropospheric scatter.

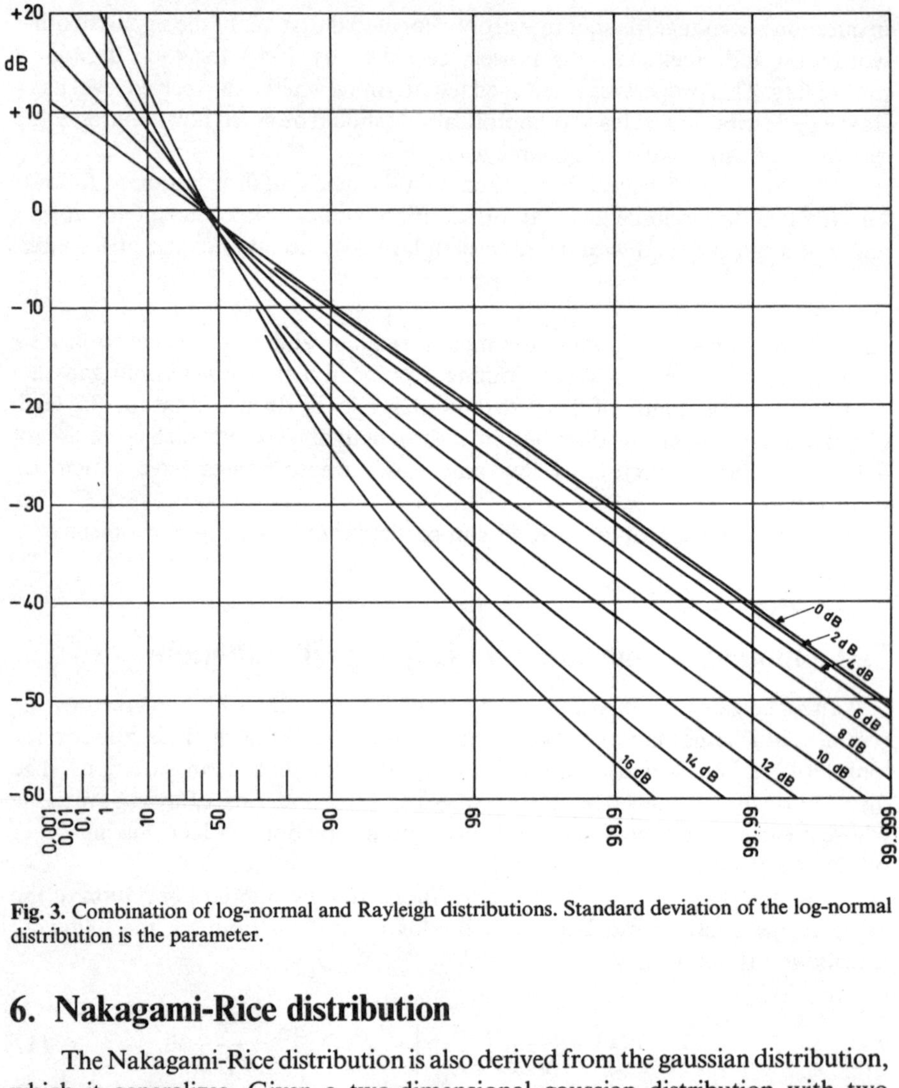

Fig. 3. Combination of log-normal and Rayleigh distributions. Standard deviation of the log-normal distribution is the parameter.

6. Nakagami-Rice distribution

The Nakagami-Rice distribution is also derived from the gaussian distribution, which it generalizes. Given a two-dimensional gaussian distribution with two independent variables x and y and the same standard deviation, the length of a vector joining a point in the distribution to a fixed point different from the centre of the distribution will have a Nakagami-Rice distribution. This distribution can therefore also be considered as the distribution of the length of a vector which is the sum of a fixed vector and a vector whose length has a Rayleigh distribution.

If a is used to designate the length of the fixed vector and σ the most probable length of the Rayleigh vector, the probability density is given by:

$$p(x) = \frac{x}{\sigma^2} \exp\left(-\frac{x^2 + a^2}{2\sigma^2}\right) I_0\left(\frac{ax}{\sigma^2}\right) \tag{19}$$

where I_0 is a modified Bessel function of the first kind and of zero order. I_0 has several possible expressions, e.g.

$$I_0(z) = \frac{1}{\pi}\int_0^{\pi} e^{-z\cos\theta}\, d\theta$$

$$I_0(z) = 1 + \left(\frac{z}{2}\right)^2 + \frac{1}{(2!)^2}\left(\frac{z}{2}\right)^4 + \frac{1}{(3!)^2}\left(\frac{z}{2}\right)^6 +$$

$$I_0(z) = \frac{e^z}{\sqrt{2\pi z}}\left[1 + \frac{1}{8z} + \frac{(1.3)^2}{2!(8z)^3} + \frac{(1.3.5)^2}{3!(8z)^5} + \right]$$

The density function depends on two parameters, but for the purposes of propagation problems it is necessary to choose a relation between amplitude a of the fixed vector and the root mean square amplitude $\sigma\sqrt{2}$ of the random vector. This relation depends on the application envisaged. Two main applications are as follows:

1. Power in the fixed vector is constant, but total power in fixed and random components varies. For studies of the influence of a ray reflected by a rough surface, a represents the field of the direct ray which can be considered constant and is taken as a reference by writing:

$$\rho = \frac{\sigma\sqrt{2}}{a} \text{ and } R = 20 \log \rho \tag{20}$$

   where ρ represents the relative amplitude of the reflected ray. The mean power received is then equal to $(1 + \rho^2)\, a^2$.
2. Total power in fixed and random components is constant, but both components vary. For the purpose of studying multi-path propagation through the atmosphere it can be considered that the sum of the power carried by the fixed vector and the mean power carried by the random vector are constant since the power carried by the random vector originates from that of the fixed vector. If the total power equals 1, then:

$$a^2 + 2\sigma^2 = 1. \tag{21}$$

Let us write:

$$a = \cos\phi \tag{22}$$

$$\sigma\sqrt{2} = \sin\phi \tag{23}$$

The fraction of the total power carried by the random vector is then equal to $\sin^2\varphi$. If X is used to designate the instantaneous amplitude of the resultant vector and x a numerical value of this amplitude, we find that the probability of having an instantaneous level greater than x is given by:

$$\text{Prob}\,(X > x) = 1 - F(x)$$
$$= 2\exp(-\,1/\tan^2\phi)\int_{x/\sin\phi}^{\infty} \nu \exp(-\,\nu^2)\, I_0\left(\frac{2\nu}{\tan\phi}\right) d\nu. \tag{24}$$

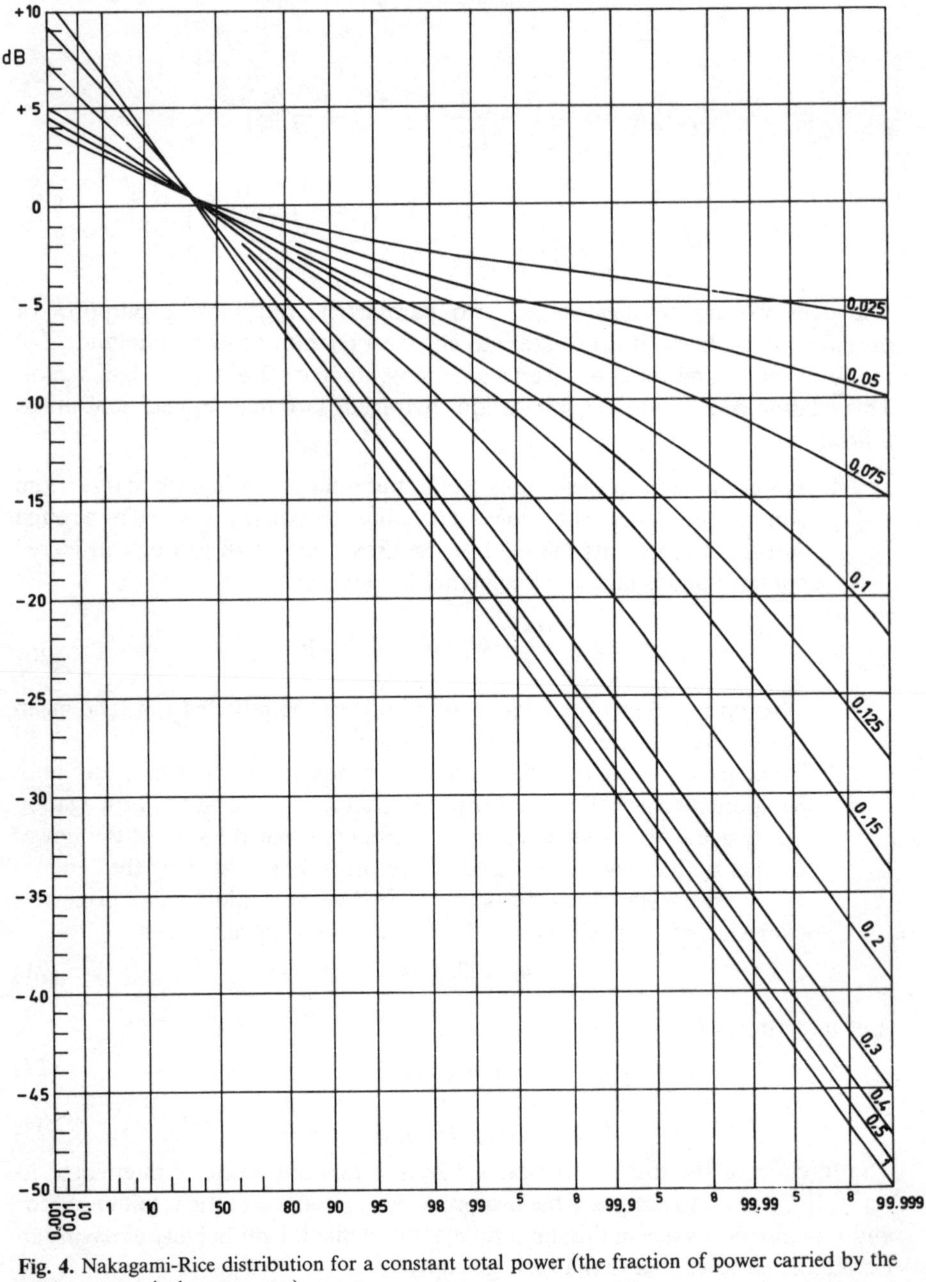

Fig. 4. Nakagami-Rice distribution for a constant total power (the fraction of power carried by the random vector is the parameter).

Figure 4 shows this distribution for different values of the fraction of power carried by the random vector ($\sin^2 \varphi$).

The Nakagami-Rice distribution is shown in Fig. 4 but for the purposes of practical applications use has been made of a decibel scale for the amplitudes, and for the probabilities a scale such that a Rayleigh distribution is represented by a straight line. It will be seen that for values of $\sin^2 \varphi$ above about 0.5, the curves approach a limit corresponding to the Rayleigh distribution. This is because in this case the fixed vector has an amplitude of the same order of magnitude as the random vector and it is practically indistinguishable from it. On the other hand for small values φ, it can be shown that the distribution of the amplitude tends towards a gaussian distribution whose mean and standard deviation are given by:

$$m = \cos \phi \tag{25}$$

$$\sigma = \frac{\sin \phi}{\sqrt{2}} \tag{26}$$

This applies particularly to scintillation.
N.B. Equation (24) and Fig. 4 can be used for case (1) above.

It is sufficient to write:

$$\rho = \tan \phi \tag{27}$$

and to increase the level shown on the y-axis in Fig. 4 by a number of decibels equal to:

$$10 \log (1 + \rho^2).$$

For certain specific applications it may be necessary to draw diagrams in which ρ is shown on the x-axis and the probability is a parameter.

7. Gamma and exponential distributions

Unlike the previous distributions which are derived from the gaussian distribution, the gamma distribution is derived from the exponential distribution of which it is a generalization. It is applied to a positive and non-limited variable. The probability density is:

$$p(x) = \frac{\alpha^\nu}{\Gamma(\nu)} x^{\nu-1} e^{-\alpha x} \tag{28}$$

This distribution depends on two parameters α and ν. However α is only a scale parameter of variable x. Characteristic values of the variable are:

1. mean value: ν/α;
2. root mean square value: $\frac{\sqrt{\nu(1+\nu)}}{\alpha}$;
3. standard deviation: $\frac{\sqrt{\nu}}{\alpha}$.

The integral expressing the cumulative distribution cannot be evaluated in closed form except for integral values of v. On the other hand the following expansions are possible:

Series expansion:

$$F(x) = \frac{1}{\Gamma(v+1)} e^{-\alpha x}(\alpha x)^{v}\left[1 + \frac{\alpha x}{v+1} + \frac{\alpha x^2}{(v+1)(v+2)} + \ldots\right] \quad (29)$$

Asymptotic series expansion:

$$1 - F(x) = \frac{1}{\Gamma(v)} e^{-\alpha x}(\alpha x)^{v-1}\left[1 + \frac{v-1}{\alpha x} + \frac{(v-1)(v-2)}{(\alpha x)^2} + \ldots\right] \quad (30)$$

When $v = 1$ we obtain an exponential distribution. For integer v the asymptotic

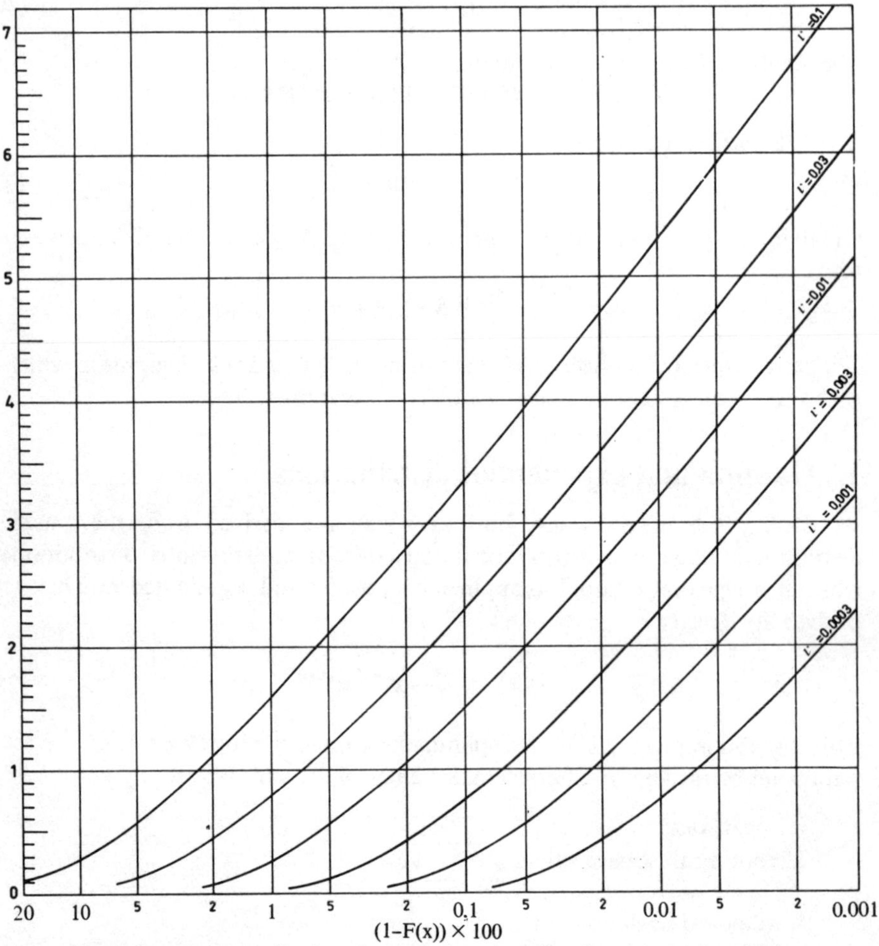

Fig. 5. Cumulative distribution of complementary gamma function $\alpha = 1$, $v < 0.1$.

expansion has a finite number of terms and gives the gamma distribution in an explicit form.

For some propagation applications the useful values of v are very low, of the order of 10^{-2} to 10^{-4}. As v approaches zero:

$$\frac{1}{\Gamma(v)} = \frac{v}{\Gamma(v+1)} \simeq v. \tag{31}$$

It is possible therefore to write for small v and not too small αx:

$$1 - F(x) = v \int_x^{\infty} \frac{e^{-t}}{t} \, dt. \tag{32}$$

For practical calculations it is useful to find an approximation to the above integral, e.g. the following (Boithias, 1984):

$$1 - F(x) = v \frac{e^{-\alpha x}}{0.68 + \alpha x + 0.28 \log \alpha x} \tag{33}$$

which is valid for $v < 0.1$ and $\alpha x > 0.03$.

The complementary gamma distribution for low values of v is shown in Fig. 5. It can be seen that the probability of the variable X being significantly greater than zero is always small. In particular, this explains the use of the gamma distribution to represent rainfall rates, since the total percentage of rainfall time is generally of the order of 2 to 10%.

N.B.: If $2v$ in Equation (28) is assumed to be an integer and if x is given a form such as $\alpha x = X^2/2$ the chi-squared distribution with $2v$ degreees of freedom is obtained.

8. Nagakami m-distribution

This distribution is applied to a non-limited positive variable. The probability density is equal to:

$$p(x) = \frac{2m^m}{\Gamma(m)\Omega^m} x^{2m-1} e^{-(m/\Omega)x^2} \tag{34}$$

is a scale parameter equal to the mean value of x^2:

$$\overline{x^2} = \Omega. \tag{35}$$

The cumulative Nakagami m-distribution is shown in Fig. 6.

This distribution has various relationships with previously mentioned distributions:

1. if a variable has a Nakagami m-distribution, the square of this variable has a gamma distribution;
2. for $m = 1$ we obtain a Rayleigh distribution;
3. for $m = \frac{1}{2}$ we obtain the one-sided normal distribution.

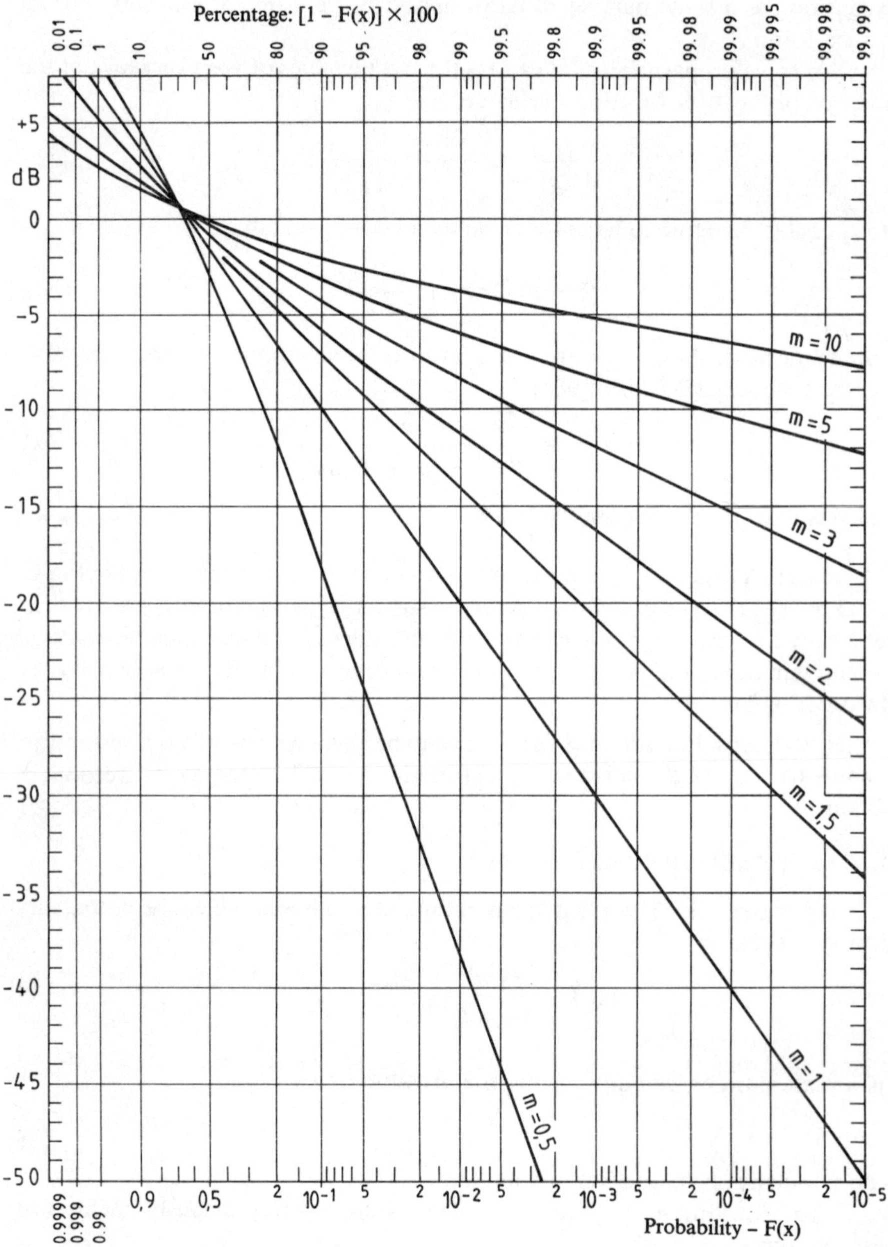

Fig. 6. Nakagami-m distribution: $\overline{x^2} = 1$.

The Nakagami-m and the Nakagami-Rice distribution can thus be regarded as two different generalizations of the Rayleigh distribution. It should be noted that for very low levels the slope of the Nakagami m-distribution tends towards a value which depends on the parameter, unlike the Nakagami-Rice distribution for which the limit slope is always the same (10dB per decade of probability).

In this section m denotes a parameter of the Nakagami m-distribution; it is not a mean value as in previous sections of this appendix.

9. Extreme value distributions: Gumbel distributions

In certain cases we are only interested in the greatest or smallest values a random variable can take, and not in its whole distribution. For example, when we consider the influence of rainfall on wave attenuation, we are chiefly interested in heavy rain, as it is this type of rain that causes attenuation likely to disturb transmissions.

The determination of extreme values is a difficult task in statistics, which may, for example, make it necessary to extrapolate very considerably from a distribution only known in its central portion. Another approach to this problem, which becomes possible as soon as we have a large number of samples, for example from measurements of heavy rainfall over several years in succession, is to study the distribution of extreme values over those years.

It can be shown that such a distribution depends almost entirely on the behaviour of the initial distribution at its extremes and that it tends towards a single asymptotic law, the only qualification being that the initial distribution must meet certain conditions which are in fact often satisfied.

If the decrease in the probability density of variable x is fast enough, as we move away from the mean value, we may obtain a linear transformation

$$y = ax + b \tag{36}$$

such that the asymptotic distribution of the extremes of variable y is given by:

$$F m\,(y) = \text{Prob}\,(Y < y) = \mathrm{e}^{-\mathrm{e}^{-y}} \sum_{f=0}^{m-1} \frac{\mathrm{e}^{-py}}{p!} \tag{37}$$

where m is the rank of the extreme value considered. If we choose $m = 1$, this implies we are considering the most extreme value in each sample, if we choose $m = 2$, we choose the second value in each sample, etc. Figure 7 shows the cumulative distributions for the first six extremes using a diagram where the first extreme is represented by a straight line. It can be shown that a Rayleigh distribution can be represented by a straight line on the same graph, on condition that its log is plotted rather actual values. It is clear, in particular, that when we go from the first to the second extreme, the reduction in the variable value is great for high time percentages.

For applications, and to determine the values of a and b in Equation (36), it is useful to know the mean value and standard deviation of the first few extremes. They are shown in Table 2.

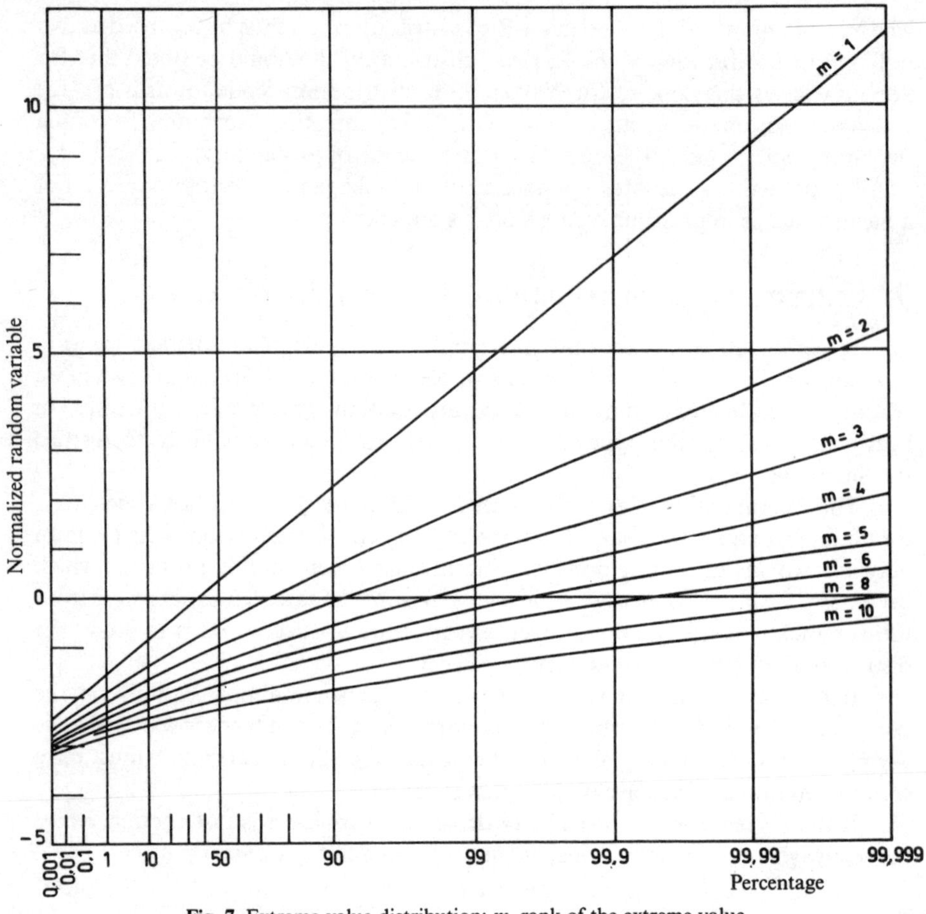

Fig. 7. Extreme-value distribution; m, rank of the extreme value.

Table 2.

m	y	σ
1	+0.577 2	1.282 5
2	−0.422 8	0.803 1
3	−0.922 8	0.628 4
4	−1.256 1	0.532 7
5	−1.506 1	0.470 0
6	−1.706 1	0.425 8

Appendix 2

Summary of the effects of propagation on the use of different frequency ranges

We briefly consider the effects of propagation, from the lowest to the highest radio frequencies. We use the metric terms for the various wavelength ranges. Waves referred to as kilometric are those whose wavelength is between 1 and 10km, and so forth. A correspondence is established between this terminology and that generally used in English, as follows:

3 to 30kHz	myriametric waves	VLF
30 to 300kHz	kilometric waves	LF
0.3 to 3MHz	hectometric waves	MF
3 to 30 MHz	decametric waves	HF
30 to 300MHz	metric waves	VHF
0.3 to 3GHz	decimetric waves	UHF
3 to 30GHz	centimetric waves	SHF
30 to 300 GHz	millimetric waves	EHF

Terms based on codes drawn up during the Second World War for radar are sometimes used: L band, S band, X band, etc. These terms, which are based on no logical rule and do not correspond to clearly defined and generally accepted band limits, should be avoided, except in their application to radar equipment.

1. Myriametric and kilometric waves (VLF and LF)

These waves are characterized by good propagation by diffraction around the Earth's surface, almost independently of the nature of the surface and obstacles on it. At great distances, there is the additional effect of a wave reflected from layer *D*, which is very stable, and possibly a wave reflected by layer *E*. Kilometric waves, previously heavily used for inter-continental telegraph traffic, are currently used for time signals and radio navigation systems, because of their stability, and for radio broadcasting. These can, however, suffer significant absorption during the day. Furthermore, as these waves can penetrate several metres below the surface of the sea, they are used for submarines links. The main problem with these waves and especially with the myriametric wave is to build antennas of sufficiently high efficiency.

Industry generated atmospherics have very high levels, leading to the use of high power transmitters.

2. Hectometric waves (MF)

The range of the diffracted wave falls, and far more quickly over Earth than over sea. Its contribution is dominant only during the summer at around midday and for distances that are not too great. In all other cases, it is the wave reflected by the E layer that is most important, but it is not very stable, particularly for the higher frequencies. It is above all at night that it produces a relatively intense field. Atmospherics (e.g. lightning) is still a source of difficulty, as well as industry generated atmospherics in cities.

The low end of this range is used for sound radio broadcasts with high fields to compensate for the level of atmospherics and the low directivity of receiving antennas. The high end of the range is used for links with ships, as propagation is good over sea and there is no need for large antennas.

3. Decametric waves (HF)

These are the old short waves used especially for transmissions over long distances by reflection from the ionosphere (above all from the F_2 layer). The lower D and E layers play a role that is rather detrimental, as they absorb power.

The use of these waves depends crucially on choosing a frequency and a specific time of day for a link.

Atmospherics have a fairly small effect at these frequencies. The major difficulty is due to deep fading from fluctuations in the ionosphere and mutual interference, as low power can sometimes give serious interference at long range.

4. Metric waves (VHF)

Total reflection from the ionosphere exceeds 30MHz except at certain periods of high solar activity. We are therefore limited to either line of sight propagation or propagation by diffraction around the Earth's surface or around various obstacles. We saw, in particular, the positive effects created by obstacles in the form of sharp edges. Antenna gains are generally insufficient to penetrate deeply into the tropospheric scatter zone. This wave range is chiefly used for television transmission, links with mobiles and point to point links. At the bottom of the range, 30 to 60MHz, propagation by ionospheric scatter is sometimes used, or propagation by reflection from meteor trails.

In this range, atmospherics are almost negligible, but, on the other hand, certain forms of industrial generated atmospherics (from sparking of internal combustion engines) can be troublesome.

5. Decimetric waves (UHF)

Scattering phenomena become more and more significant. These waves are used on line of sight links (television, radio links, radar, radio navigation), or links beyond the horizon by tropospheric scatter. It is, in fact, the most suitable range of waves for this use as the scattered power is relatively high and it is easy to produce transmitting powers of several kilowatts and fairly large antennas (parabolic antennas 10 to 20m in diameter) are used.

Atmospherics and industry generated atmospherics are practically non-existent. The sensitivity of receivers is only limited by their own noise (thermal noise). Consequently, everything is done to reduce this as far as possible by using low noise amplifiers.

6. Centrimetric waves (SHF)

This range of waves is almost exclusively used on line of sight systems (radar, radio relay links), even though the scattering phenomenon is used a little at the low end of the range, from 3 to 4GHz. Since cosmic atmospherics is particularly low, especially between 4 and 7GHz, these frequencies are also used for satellite communications, using very low noise amplifiers. Note, moreover, that attenuation due to rain increases rapidly above 10GHz.

7. Millimetric and shorter waves (EHF)

These waves are currently used in research. There are, however, 8mm radars. Only line of sight propagation is possible, and it too is limited by atmospheric absorption by oxygen and water vapour, as well as by rainfall.

Appendix 3

Main uses of radio waves for telecommunication

1. General

Radiocommunication, i.e. telecommunication using radio waves, not only require agreement on transmission characteristics as for cable telecommunications systems, but also extremely precise international regulations to avoid interference between different services. The determination of technical characteristics and the technical basis for the establishment of regulations is carried out by the Consultancy Committee for Radiocommunications (CCIR) which is an International Union of Telecommunications (IUT) body. The regulations are contained in a book called the Radio Regulations, drawn up at world administrative conferences, based on the technical guidelines drawn up by the CCIR. Wave propogation considerations play an essential role in these technical guidelines.

2. Radiocommunication definitions

The radio frequency spectrum is divided between a certain number of radiocommunication services. The following list, drawn from the definitions contained in the radiocommunication regulations, gives a few uses of radio waves and the services which are allowed access to those parts of the frequency spectrum. Since the introduction of the use of satellites for telecommunication, each service has been split up into a terrestrial service and a satellite service, e.g. the aeronautical mobile service and the aeronautical mobile satellite service. However, in order to simplify the list below, we have not in general included the 'satellite' service.

It also needs to be said that as well as assigning frequencies to telecommunication services, the use of certain limited parts of the spectrum can be permitted for industrial, scientific and medical (ISM) applications, in particular around the following frequencies:

27MHz
40MHz
433MHz
910MHz
2450MHz
5870MHz

2.1. Functions provided by radio waves and defined in the Radio Regulations

Radio waves or Hertzian waves: electromagnetic waves of frequencies arbitrarily lower than 3000GHz, propagated in space without artificial guide.

Radiocommunication: telecommunication by means of radio waves (CONV.).

Terrestrial radiocommunication: any radiocommunication other than space radiocommunication or radio astronomy.

Space radiocommunication: any radiocommunication involving the use of one or more space stations or the use of one or more reflecting satellites or other objects in space.

Radiodetermination: the determination of the position, velocity and/or other characteristics of an object, or the obtaining of information relating to these parameters, by means of the propagation properties of radio waves.

Radionavigation: radiodetermination used for the purposes of navigation, including obstruction warning.

Radiolocation: radiodetermination used for purposes other than those of radionavigation.

Radio direction-finding: radiodetermination using the reception of radio waves for the purpose of determining the direction of a station or object.

Radio astronomy: astronomy based on the reception of radio waves of cosmic origin.

2.2. Radio Services

Fixed service: a radiocommunication service between specified fixed points.

Fixed-satellite service: a radiocommunication service between Earth stations at specified fixed points when one or more satellites are used; in some cases this service includes satellite-to-satellite links, which may also be effected in the inter-satellite service; the fixed satellite service may also include feeder links for other space radiocommunication services.

Aeronautical fixed service: a radiocommunication service between specified fixed points provided primarily for the safety of air navigation and for the regular, efficient and economical operation of air transport.

Space operation service: a radiocommunication service concerned exclusively with the operation of spacecraft, in particular space tracking, space telemetry and space telecommand.

Mobile service: a radiocommunication service between mobile and land stations, or between mobile stations (CONV.).

Land mobile service: a mobile service between base stations and land mobile stations, or between land mobile stations.

Maritime mobile service: a mobile service between coast stations and ship stations, or between ship stations, or between associated on-board communication stations; survival craft stations and emergency position-indicating radiobeacon stations may also participate in this service.

Aeronautical mobile service: a mobile service between aeronautical stations and aircraft stations, or between aircraft stations, stations in which survival craft stations may participate; emergency position-indicating radiobeacon stations may also participate in this service on designated distress and emergency frequencies.

Broadcasting service: a radiocommunication service in which the transmissions are intended for direct reception by the general public. This service may include sound transmissions, television transmissions or other types of transmission (CONV.).

Broadcasting-satellite service: a radiocommunication service in which signals transmitted or retransmitted by space stations are intended for direct reception by the general public.

Radiodetermination service: a radiocommunication service for the purpose of radiodetermination.

Radionavigation service: a radiodetermination service for the purpose of radionavigation.

Maritime radionavigation service: a radionavigation service intended for the benefit and for the safe operation of ships.

Aeronautical radionavigation service: a radionavigation service intended for the benefit and for the safe operation of aircraft.

Radiolocation service: a radiodetermination service for the purpose of radiolocation.

Meteorological aids service: a radiocommunication service used for meteorological, including hydrological, observations and exploration.

Earth exploration-satellite service: a radiocommunication service between Earth stations and one or more space stations, which may include links between space stations, in which:

1. information relating to the characteristics of the Earth and its natural phenomena is obtained from active sensors or passive sensors on Earth satellites;
2. similar information is collected from airborne or Earth-based platforms;
3. such information may be distributed to Earth stations within the system concerned;
4. platform interrogation may be included.

This service may also include feeder links necessary for its operation.

Meteorological-satellite service: an Earth exploration-satellite service for meteorological purposes.

Standard frequency and time signal service: a radiocommunication service for scientific, technical and other purposes, providing the transmission of specified frequencies, time signals, or both, of stated high precision, intended for general reception.

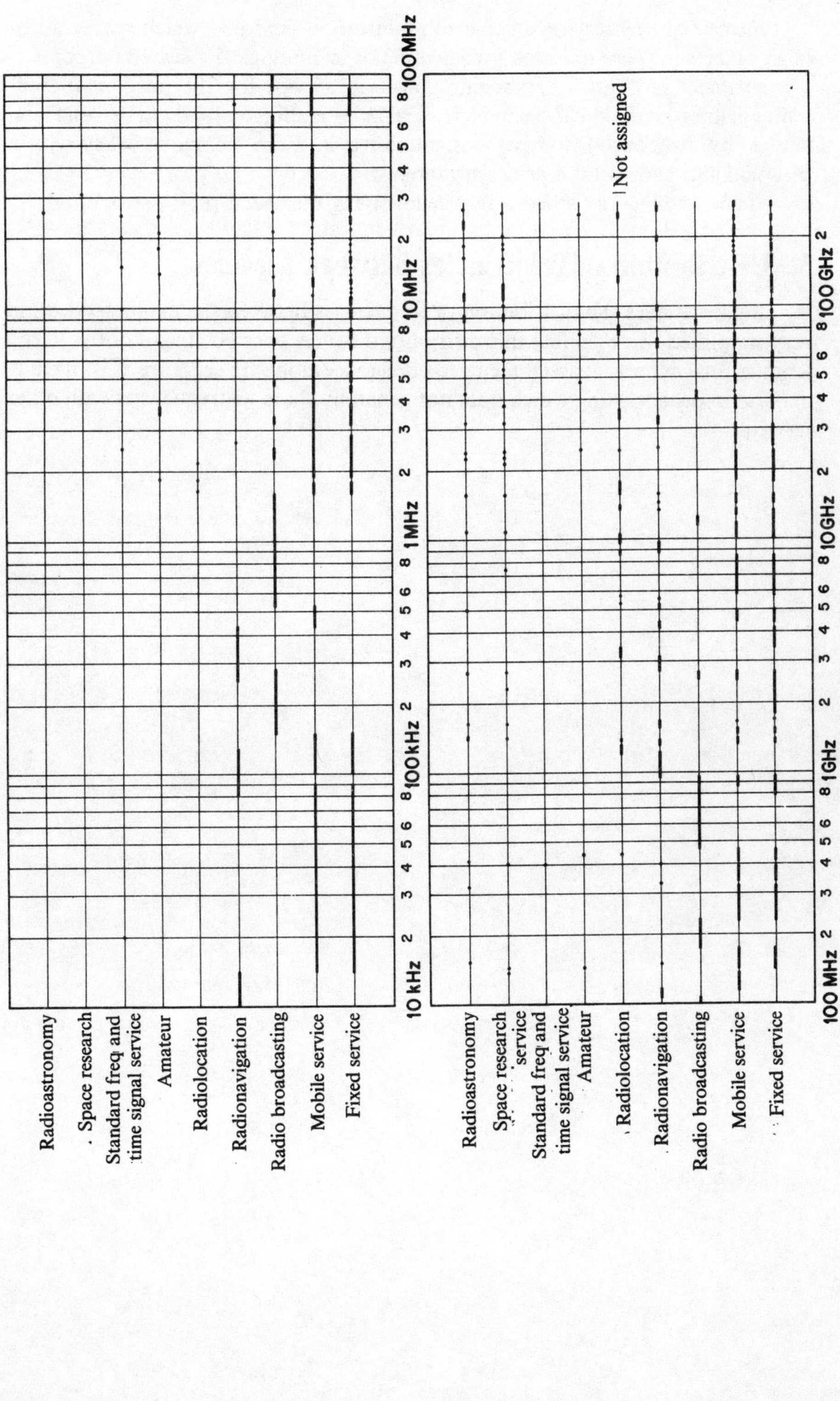

Fig. 1. Frequency band assignments.

Space research service: a radiocommunication service in which spacecraft or other objects in space are used for scientific or technological research purposes.

Amateur services: a radiocommunication service for the purpose of self-training, intercommunication and technical investigations carried out by amateurs, that is, by duly authorized persons interested in radio technique solely with a personal aim and without pecuniary interest.

Radio astronomy service: a service involving the use of radio astronomy.

3. Distribution of frequencies between services

Figure 1 gives the distribution of frequencies between the major services, in very simplified form (in fact this distribution covers over 200 pages of the Radio Regulations). Note that assignments to one service alone are very rare. It is virtually only radio broadcasting which does not generally share its frequencies with other services.

Bibliography

There are many papers that describe propagation but they are variable in quality. For example, some simply give the results of measurements carried out on a given link at a particular frequency and for a limited period of time, whereas others bring together the results obtained by various observers in different countries and attempt to deduce general laws and prediction methods from them.

As far as possible, the references below fall into the second category. However, the first few references for each chapter give information about the relevant CCIR texts, i.e. Recommendations (Rec) or Reports (Rep) as these texts, and the Reports in particular, review the state of the art of the subject in question and generally contain a long list of references. Moreover, since these texts are revised every four years, their reference list is current.

Chapter 2

Beckmann, P; Spizzichino, A (1963) *The scattering of electromagnetic waves from rough surfaces* Pergamon.
Booker, H G (1982) *Energy in electromagnetism* Peter Peregrinus Ltd.
Born, M; Wolf, E (1975) *Principles of optics* Pergamon.
Boudouris, G *Propagation tropospherique* Centre de documentation universitaire.
Bouix, M (1966) *Les discontinuités du rayonnement éléctromagnetique* Dunod.
De Santo, S A (1981) Scattering from a rough surface. *Radio Sci.* **16**, 6.
Jouguet, M (1973) *Ondes éléctromagnetiques, Tome 1* Dunod.
Kraus, J D; Carver, C R (1984) *Electromagnetics* McGraw Hill.
Miller, A R, *et al.* (1984) New derivation for the rough surface reflection coefficient and for the distribution of sea wave elevations. *IEE Proc.* **131**, part H, April.

Chapter 3

ITU, Electrical characteristics of the Earth. *CCIR Vol V* Rep 229.
ITU, Influence of terrain irregularities and vegetation on tropospheric propagation. *CCIR Vol V* Rep 236.
ITU, World atlas of ground conductivities. *CCIR Vol V* Rep 717.
ITU, Propagation in tunnels and mines. *CCIR Vol V* Rep 880.
Bothias, L (1972) *Calcul par nomogrammes de la propagation des ondes* Eyrolles.
Boithias, L; Battesti, J; Fimbel, J (1976) Estimation du niveau de l'onde réflechie sur le sol dans la transmission par faisceau Hertzien. *Ann. Télécommun.* **31** No 5–6.
Burrows, C R; Attvood, S S (1949) *Radiowave propagation* Academic Press.
Kerr, D E (1965) *Propagation of short-radio waves* Dover.
Saxton, A; Lane, J A (1952) Electrical properties of sea water. *Wireless Eng.* **29**, Oct.

Wait, J R (1970) *Electromagnetic waves in stratified media* Pergamon Press.
Wentworth, F L; Cohn, M (1964) Electrical properties of sea ice at 0.1 to 30Mc/s. *Radio Sci.* **68D**.

Chapter 4

ITU, Definitions of terms relating to progagation in the troposphere. *CCIR Vol V* Rec 310.
ITU, The formula for the radio refractive index. *CCIR Vol V* Rec 453.
ITU, Radiometeorological data. *CCIR Vol V* Rep 563.
ITU, Effects of tropospheric refraction on radiowave propagation. *CCIR Vol V* Rep 718.
ITU, Cross polarization due to the atmosphere. *CCIR Vol V* Rep 722.
Barnett, N T (1972) Multipath progagation at 4, 6, 11GHz. *Bell Syst. Tech. J.* **51**, Feb.
Bean, B R; Dutton, E J (1966) Radiometeorology. *NBS Monograph No 92* US Govt Printing Office, Washington.
Bean, B R; Cahoon, B A; Samson, C A; Thayer, G D (1966) A world atlas of atmospheric radiorefractivity. *ESSA Monograph No 1* NTIS Springfield, USA.
Boithias, L (1979) Multipath propagation in line of sight links: *Electron. Le.* **17**, 7, March.
Boithias, L (1981) Distribution statistique des niveaux réçus en propagation par trajets multiples troposphériques. *Ann. Telecommun.* May-June.
Booker, H G; Walkinshaw, W (1946) The mode theory of tropospheric refraction and its relation to waveguide and diffraction. *Meteorological factors in radiowave propagation* Physical Society, London.
Gossard, E (1977) Refractive index variance and its height distribution in different air masses. *Radio Sci.* **12**, Jan.
Hall, M P M (1979) Effects of the troposphere on radiocommunication. *IEE Electromagnetic Waves Series* Peter Peregrinus Ltd.
Kerr, D E (1951) *Propagation of short radiowaves* McGraw Hill.
Morita, K (1970) Prediction of Rayleigh fading occurrence probability on line of sight microwave links. *Review of Electrical Communication Laboratory* **18**, Nov-Dec.
Tatarski, V I (1961) *Wave propagation in a turbulent medium* McGraw Hill, New York.

Chapter 5

ITU, Attenuation by atmospheric gases. *CCIR Vol V* Rep 719.
ITU, Attenuation by hydrometeors. *CCIR Vol V* Rep 721.
ITU, Scattering by precipitations. *CCIR Vol V* Rep 882.

ITU, Cross-polarization due to the atmosphere. *CCIR Vol V* Rep 722.

Barbaliscia, F; Fedi, F (1980) Rainfall intensity statistical properties and protection techniques. *Proceeding of URSI Com. F Symposium* Lennoxville, Canada.

Bean, B R; Dutton, E J (1966) Radiometeorology. *NBS Monograph No g2* US Govt Printing Office, Washington.

Boithias, L (1980) Au sujet de la distribution statistique des intensités de pluie. *Ann. Telecommun.* **35**, No 9–10, Sept–Oct.

Brussard, G (1979) Attenuation due to rain on a slant path, *Alta Freq.* **48**, p 104–109.

Cox, D (1981) Depolarization of radio waves by atmospheric hydrometeors on Earth space paths: review. *Radio Sci.* **16**, 5, Sept-Oct.

Crane, R K (1977) Prediction of the effects of rain on satellite communication systems. *PIEEE* **63**, 3, March.

Crane, RK (1980) Prediction of the attenuation by rain: *IEEE Trans. Commun.* **28**, 9 part II, Sept.

Drufuca, G (1974) Rain attenuation statistics for frequencies above 10GHz from rain gauge observations. *J. Rech. Atmos.* **1–2**.

Fedi, F (1979) Attenuation due to rain on terrestrial paths. *Alta Freq.* **49**, 4.

Gunn, G D; Kinzer, R (1949) The terminal velocity of fall for water droplets in stagnant air. *J. Meteorol.* **6**.

Gunn, K; East, T (1954/80) The microwave properties of precipitation particles. *Q. J. R. Meteorol. Soc.*

Hall, M P M (1979) Effects of the troposphere on radio-communication. *IEE Electromagnetic Waves Series* Peter Peregrinus

Kerr, O E (1946) The dielectric properties of water vapour. *Meterological Factors in Radiowave Propagation.* Physical Society, London.

Laws, J O; Parsons, D A (1945) The relation of drop size to intensity. *Trans. Am. Geophys Union* **24**, p 452–460.

Marshall, J S; Palmer, W (1948) The distribution of raindrops with size *J. Meteor.* **5**, Aug.

Mie, G (1949) *Ann Physik* **25**, 377.

Oguchi, T (1977) Scattering properties of Pruppacher and Pitter form raindrops and cross-polarization due to rain. *Radio Sci.* 12, 1.

Oguchi, T; Hosoya, Y (1974) Scattering properties of oblate raindrops and cross-polarization of radio wave due to rain. *J. Radio Res. Labs. Japan* **21**, 191–259.

Olsen, R L (1981) Cross-polarization during precipitations on terrestrial links: a review. *Radio Sci.* **16**, 5, Sept–Oct.

Olsen, R L; Rogers, D V; Hodge, D B (1978) The aR^b relation in the calculation of rain attenuation. *IEEE Trans. A. P.* **26**, 2.

Rice, P L; Homberg (1973) Cumulative time statistics of surface point rainfall rate. *IEEE Trans. Commun.* **21**, 1131–1136.

Saxton, J A (1946) The dielectric properties of water vapour. *Meteorological factors in radiowave propagation* Physical Society, London.

Van De Hulst, H C (1957) *Light scattering by small particles* J Wiley and Sons.

Chapter 6

ITU, Propagation data and prediction methods required for line of sight radio relay systems. *CCIR Vol V* Rep 338.

ITU, Progagation data and prediction methods required for Earth-space telecommunication systems. *CCIR Vol V* Rep 564.

Barnett, W T (1970) Microwave line of sight propagation with and without frequency diversity. *Bell Syst. Tech. J.* **45**, Oct.

Battesti, J and Boithias, L (1964) Etude des sections longues dans les faisceaux Hertziens en visibilité. *Ann. Telecommun.* **19**, July-Aug.

Boithias, L (1979) Protection contre les evanouissements dus aux conduits troposphériques sur les liasisons en visibilité. *Ann. Telecommun.* **34**, Sept-Oct.

Boithias, L; Battesti, J (1967) Protection contre les evanouissements sur les faisceaux Hertziens en visibilité. *Ann. Telecommun.* **22**, Sept-Oct.

Boithias, L; Battesti, J (1970) Calcul de la qualité de transmission sur les faisceaux Hertziens de téléphonie utilisant la modulation de frequences. *Ann. Telecommun.* **25**, April-May.

Fedi, F (1981) A simple method for predicting rain attenuation statistics on terrestrial and Earth-space paths. *Fundazione Ubo Bordoni* June.

Greenstein, L J; Czekaj, B A (1980) A polynomial model for multipath fading channel responses: BSTJ, Vol 59, 7.

Hosoya, Y; Satch, K; Nagatsu, T (1980) Progagation characteristics on slant. *Rev. Elec. Comm. Labs*, **28**, 7-8.

Kalinin, A I (1979) Propagation on terrestrial and space links *Sviaz, Moscow* (in Russian).

Martin, L (1980) Study of the selectivity of multiple trajectories. *URSI Symposium* Lennoxville, Canada.

Martin, L (1983) Rate of change of propagation medium transfer frequencies during selective fadings. *URSI Symposium* Louvain, Belgium.

Rooryck, M (1979) Validity of a two-path model for calculating quality on a digital radio link. *Electron. Lett.* **15**, 24.

Rooryck, M (1983) Modelisation de la propagation avec trajets multiples, Application d'un modèle a deux rayons. *Onde Electri.* **63**, 8–9.

Rummler, W D (1979) A new selective fading model, application to progagation data. *BSTJ* **58**, 5, May-June.

Rummler, W D (1982) A statistical model of multi-path fading on a space diversity radio channel. *BSTJ* **612**, Nov.

Sasakio, O; Tadashi, A (1979) Multi-path delay characteristics on line of sight microwave radio systems. *IEEE Trans. Commun.* **27**, Dec.

Sylvain, M; Lavergnat, J (1985) Modelling the transfer function in medium bandwidth radio channel during multipath progagation. *Ann. Telecommun.* **40**, 11–12, Nov–Dec.

Vigants, A (1975) Space-diversity engineering. *Bell Syst. Tech. J.* **54**, Jan.

Chapter 7

ITU, Ground wave progagation curves for frequencies between 10KHz and 30GHz. *CCIR, Vol V* Rec 368.

ITU, Progagation by diffraction. *CCIR, VOL V* Rep 715.

ITU, The phase of the ground wave. *CCIR, Vol V* Rep 716.

Assis, M S (1971) A simplified solution to the problem of multipath diffraction over rounded obstacles. *IEEE Trans. A. P.* **19**, March.

Blomquist, A (1975) Seasonal effects on ground wave propagation in cold regions. *J. Glaciol.* **15**.

Bremmer, H (1949) *Terrestrial radio waves* Elsevier.

Deygout, J (1966) Multiple knife-edge diffraction of microwaves. *IEEE Trans. A. P.* Vol AP14, p 480–489.

Dougherty, H T; Wilkerson, R E (1967) Determination of antenna height against microwave diffraction fading. *Radio Sci.*

Epstein, J; Peterson, D N (1953) An experimental study of wave propagation at 850 MC/s *PIRE* 595–611, May.

Furutsu, K (1957) (1959) Wave propagation over an irregular terrain. *J Radio Res. Lab. Japan* **4**, 16–1957; **4**, 18–1957; **6**, 23–1959.

Hacking, K (1970) UHF propagation over rounded hills. *PIEE* **117**.

Hill, D A; Wait, J R (1980) Ground wave attenuation function for a spherical Earth with arbitrary surface impedance. *Radio Sci.* **15**, 3.

Millington, G (1949) Ground wave progagation over an unhomogeneous smooth Earth. *PIEE* **96**, Part III, Jan.

Millington, G; Hewitt, R; Immirzi, F S (1962) Double knife-edge diffraction in field strength prediction. *Proc. Instn Elect. Engin.* March.

Rice, S O (1954) Diffraction of plane radio waves by a parabolic cylinder. *Bell Sys. Tech. J.* March.

Stokke, K N (1975) Some graphical considerations on Millington's method for calculating field strength over unhomogeneous Earth. *Telecomun. J.* **42**, No III.

Vigants, A (1981) Microwave radio obstruction failing. *BSTJ* **60**, 6, July–Aug.

Volger, L E (1982) An attenuation function for multiple knife-edge diffraction. *Radio Sci.* **17**, 6.

Wait, J R; Conda, A M (1959) Diffraction of electromagnetic waves by smooth obstacles for grazing angles. *J. Res. NBS* **63D**, Sept–Oct.

Chapter 8

ITU, Effects of tropospheric refraction on radiowave propagation. *CCIR, Vol V* Rep 718.

ITU, Propagation data and prediction methods required for transhorizon radio relay systems. *CCIR, Vol V* Rep 238.

ITU, Propagation by ionospheric scatter. *CCIR, Vol VI* Rep 260.

Battesti, J, Boithias, L; Misme P (1968) Calcul des affaiblissements en propagation

transhorique à partir de paramètres radiométéorogiques. *Ann. Telecommun.* **23**, 5–6, May–June.
Battesti, J Boithias, L (1971) Nouveaux éléments sur la propagation par les hétéogénéités de l'atmosphère. *Ann. Telecommun.* **26**, 1–2, Jan–Feb.
Bello, P A (1969) A troposcatter channel model. *IEEE Trans.* A *P.* **17**.
Boithias, L; Battesti, J (1964) Etude expérimentale de la baisse de gain d'antennes dans les liaisons transhorizon. *Ann. Telecommun.* **19**, Sept–Oct.
Boithias. L; Battesti, J (1964) Les faisceaux Hertziens transhorizon de haute qualité. *Ann. Telecommun.* **20**, Sept–Oct.
Bothias, L and Battesti, J (1983) Propagation due to tropospheric inhomogeneities. *IEE Proceedings* **130**, F No 7, Dec.
Booker, H G; Gordon, W E (1950) A theory of radio scattering in the troposphere. *PIRE* **38**.
Cairns, J B (1979) A review of digital troposcatter links. *Proceedings of the conference Télécom 79*, UIT Geneva.
Hall, M; Misme, P (1975) Gain degradation on a 25m paraboloidal aerial on 2GHz transhorizon radio paths. *PIEE* **122**.
NBS, Technical Note 101 (revised), AD 682820 and 682821, NTIS, Springfield VA, USA.
Sunde, E D (1968) Intermodulation distortion in analog FM troposcatter systems. *Bell Syst. Tech. J*, **43-1**, Part 2.
Tsuzuki, T; Kajihara, H; Sand, Y; Kurobe, A (1979) Characteristics of intermodulation noise due to multipath propagation of a troposcatter link. *Trans IEEE Japan* **62-B**, 8.

Chapter 9

ITU, CCIR Atlas of ionospheric characteristics. *CCIR, Vol VI*, Rep 340.
ITU, Long-term ionospheric propagation prediction. *CCIR, Vol VI*, Rep 255.
ITU, Long-distance ionospheric propagation without intermediate ground reflection. *CCIR, Vol VI*, Rep 250.
ITU, Short-term forecasting of critical frequencies, operational maximum usable frequencies and total electron content. *CCIR, Vol VI*, Rep 888.
ITU, Ionospheric effects upon Earth-space propagation. *CCIR, Vol* Rep 263.
ITU, Ionospheric cross polarization. *CCIR*, *Vol VI* Rep 574.
ITU, Ionospheric modification by ground-based, high power radio transmissions. *CCIR, Vol VI* Rep 728.
ITU, Prediction of sky-wave field strength between 150 and 1600KHz. *CCIR*, *Vol VI* Rec 435.
ITU, Sky-wave propagation and circuit performace at frequencies between about 30KHz and 500KHz. *CCIR*, *Vol VI* Rep 265.
ITU, Sky-wave propagation and circuit performance of frequencies below about 30KHz. *CCIR*, *Vol VI* Rep 895.
Blackband, W T (Ed) Propagation factors in space communications. *Conference*

AGARD 1965 Technivision, Maidenhead.
Budden, K G (1961) *Radiowaves in the ionosphere* Cambridge University Press.
Budden, K G (1985) *The propagation of radio waves, the theory of radiowaves of low power in the ionosphere and magnetosphere* Cambridge University Press.
Davies, K (1965) Ionosphere radiopropagation. *NBS Monograph 80.*
Donnely, R F (1979) (editor) *Solar terrestrial prediction proceedings* Boulder (4 Volumes), NOAA Boulder: USA, April.
Giraud, A; Petit, M (1975) *Physique de l'ionosphère* PUF.
Quemada, D (1968) *Ondes dans les plasmas* Herman.
Rishbeth, Garriott (1969) *Introduction to ionosphere physics* Acedemic Press.

Chapter 10

ITU, VHF and UHF propagation curves for the frequency range from 30 to 1000MHZ, broadcasting services. *CCIR*, *Vol V* Rep 370.
ITU, Propagation statistics required for broadcasting services using the frequency range 30 to 1000MHz. *CCIR*, *Vol V* Rep 239.
ITU, Propagation curves for aeronautical services using the VHF, UHF and SHF bands. *CCIR*, *Vol V* Rep 528.
ITU, Methods and statistics for estimating field strength values in the land mobile services using the frequency range 30MHz to 1GHz. *CCIR, Vol V* Rep 567.
ITU, Propagation data required for terrestrial broadcasting and point to multipoint communications systems in the frequency bands above 10GHz. *CCIR*, *Vol V* Rep 562.
ITU, Propagation data for broadcasting from satellites. *CCIR*, *Vol V* Rep 565.
ITU, Propagation data for maritime mobile-satellite systems for frequencies above 100MHz. *CCIR*, *Vol V* Rep 884.
ITU, Propagation data for land mobile-satellite systems for frequencies above 100MHz. *CCIR*, *Vol V* Rep 1009.
ITU, The evaluation of propagation factors in interference problems between stations on the surface of the Earth at frequencies above about 0.5GHz. *CCIR*, *Vol V* Rep 569.
ITU, Propagation data required for the evaluation of coordination distance in the frequency range 1 to 40GHz. *CCIR*, *Vol V* Rep 724.
ITU, Method for calculating sporadic E field strength. *CCIR*, *Vol VI* Rep 534.
Crane, R K (1974) Bistatic scatter from rain. *IEEE Trans. A. P.* **22**, March.
Crane, R K (1981) A review of transhorizon propagation phenomena. *Radio Sci.* **16**, No 5.
Doble, J (1981) Interference on digital radio links due to transhorizon propagation. *Electron. Lett.* **17**, 12, June.
Dougherty, H T; Hart, B A (1979) Recent progress in duct propagation predictions. *IEEE Trans. A.P.* **27**, No 4.
Neessen, J de Haas, J (1981) Measured statistics of transmission loss due to ducting on transhorizon links at 6.4 and 7.4GHz *Ann. Telecommun.* **36**, Jan–Feb.

Vogel, W S; Smith, E K (1985) Propagation considerations in land mobile satellite transmission. *Microwave. J.* Oct.

Chapter 11

ITU, Characteristics and applications of atmosphere radio noise data. *CCIR*, *Vol VI* Rep 322.
ITU, Radio noise within and above the ionosphere. *CCIR Vol VI* Rep 342.
ITU, Measurement of atmospheric radio noise from lightning *CCIR*, *Vol VI* Rep 254.
ITU, Radio emission from natural sources above about 50MHz. *CCIR*, *Vol V* Rep 720.
ITU, Man-made noise. *CCIR*, *Vol VI* Rep 258.
Kraus, J D (1966) *Radio astronomy* McGraw Hill.
Tsang, L *et al.* (1979) Theory of microwave thermal emission from a layer of cloud or rain. *IEEE Trans. A. P.* **25**, 5.

Appendix 1

ITU, Statistical distribution in radio wave propagation. *CCIR*, *Vol V*, Rep
Johnson, N L; Kotz, S (1970) *Distributions in statistics* John Wiley and Sons.
Nakagami, N (1960) The m-distribution, general formula of intensity distribution of rapid fadings. *Statistical methods in radio wave propagation* by W C Hoffman, Pergamon Press.
Norton, K A *et al.* (1955) The probability distribution of the amplitude of a constant vector plus a Rayleigh distributed vector. *PIRE* **43**, No 10, Oct.
Patel, J J (1976) *Handbook of statistical distribution* M Dekker.

Appendixes 2 and 3

ITU, Radio Regulations: Chapters 1 and 3.

INDEX